TELECOMMUNICATIONS MEASUREMENTS, ANALYSIS, AND INSTRUMENTATION

Other Prentice-Hall Books by Dr. KAMILO FEHER

ADVANCED DIGITAL COMMUNICATIONS:
Systems and Signal Processing Techniques
1987

DIGITAL COMMUNICATIONS:
Satellite/Earth Station Engineering
1983

DIGITAL COMMUNICATIONS:
Microwave Applications
1981

TELECOMMUNICATIONS MEASUREMENTS, ANALYSIS, AND INSTRUMENTATION

Dr. KAMILO FEHER

Professor, Electrical and Computer Engineering
University of California, Davis
Davis, California 95616
Director, Consulting Group, DIGCOM, Inc.

and Engineers of Hewlett-Packard Ltd.

South Queensferry, West Lothian, Scotland

PRENTICE-HALL, INC.
Englewood Cliffs, New Jersey 07632

Library of Congress Cataloging-in-Publication Data

Telecommunications measurements, analysis, and
    instrumentation.

    Includes bibliographies and index.
    1. Telecommunication systems—Measurement.
I. Feher, Kamilo. II. Hewlett-Packard Limited.
TK5102.5.T395 1987   621.38   86-15110
ISBN 0-13-902404-2

Editorial/production supervision and
    interior design: Reynold Rieger
Cover design: Lundgren Graphics, Ltd.
Manufacturing buyer: Gordon Osbourne

Printed in the United States of America

10 9 8 7 6 5 4 3 2 1

ISBN 0-13-902404-2 025

Prentice-Hall International (UK) Limited, *London*
Prentice-Hall of Australia Pty. Limited, *Sydney*
Prentice-Hall Canada Inc., *Toronto*
Prentice-Hall Hispanoamericana, S.A., *Mexico*
Prentice-Hall of India Private Limited, *New Delhi*
Prentice-Hall of Japan, Inc., *Tokyo*
Prentice-Hall of Southeast Asia Pte. Ltd., *Singapore*
Editora Prentice-Hall do Brasil, Ltda., *Rio de Janeiro*

CONTENTS

2 DIGITAL SIGNAL PROCESSING (DSP) TECHNIQUES 65

Dr. Kamilo Feher

3 DIGITAL SIGNAL PROCESSING IN TELEPHONE CHANNEL MEASUREMENTS AND INSTRUMENTATION   90

David Dack and Bob Coackley

4 OVERCOMING INTRINSIC UNCERTAINTIES IN PCM CHANNEL MEASUREMENTS   108

M. B. Dykes

5   ERROR PERFORMANCE ANALYSIS OF DIGITAL
    TRANSMISSION SYSTEMS   136

Peter Huckett and Geoff Thow

8 FDM SYSTEM TESTING AND NETWORK SURVEILLANCE   **239**

Boyd Williamson and Robin Myles

9 ANALOG MICROWAVE DIAGNOSTIC MEASUREMENTS   314

Guy Douglas

11 PHASE-NOISE MEASUREMENTS   361
Catharine M. Merigold

FOREWORD

Modern telecommunications measurements, systems analysis, and related instrumentation techniques are covered in this valuable book. It is the first book to offer comprehensive coverage of modern telecommunications measurements with balanced presentation of digital and analog transmission and processing techniques. Dr. Feher's original contributions and understanding of practical digital communication systems combined with the telecommunications knowledge and experience of his coauthors, who are outstanding engineers of Hewlett-Packard, contributed to this text.

This book will provide invaluable help to telecommunications engineers and advanced communications students who wish to know the principles and also understand the problems related to measurements and systems analysis. This book can contribute to the successful development, performance analysis, and maintenance of a new generation of modern telecommunications systems.

Introductory material covers the principles and applications of digital transmission systems, of digital signal processing principles and applications. This is followed by chapters on specific measurements and instrumentation techniques.

I believe this book will also be welcomed by practicing professionals and academics in the telecommunications and instrumentation fields as a major contribution to their rapidly changing technologically based industry.

T. L. LEMING
Senior Vice President
MCI Telecommunications Corp.
Washington, D.C.

PREFACE

The purpose of this book is to present the engineering considerations necessary for the comprehension of modern telecommunication measurement and related instrumentation and analysis techniques. A book covering these subjects is needed and, to the best of my knowledge, does not exist.

Experienced telecommunications engineers as well as new graduates are frequently given experimental systems analysis, laboratory and factory measurement, and field performance monitoring-maintenance tasks. An efficient execution of these assignments requires a solid comprehension of digital and/or analog communication systems and measurement techniques. Interpretation of the measured results requires a comprehension of the principles, capabilities, and limitations of the available instrumentation.

By the 1980s most major telecommunication systems developments used digital transmission and processing techniques. However, there are many high-capacity analog systems in operation. Digital communication systems frequently have to be operated in a mixed digital and analog systems environment. For these reasons the main emphasis of this book is on digital communications measurements analysis and instrumentation; however, analog transmission and related instrumentation problems are also covered.

Currently, there are only a few North American universities and colleges teaching courses in this important field. As a consequence, a number of otherwise well-trained engineers are lacking knowledge of measurement techniques and related instrumentation. Therefore, they tend to shy away from experimental analysis and, for this reason, sometimes do not get involved in some of the more challenging tasks of modern telecommunications systems measurements and design.

Engineers and managers employed by operators of telecommunications networks, system and equipment designers employed by manufacturers of telecommunications equipment and instrumentation, manufacturing engineers and managers, consulting engineers, engineers engaged in research and development of telecommunications systems, and the technical staff of government agencies concerned with the regulation of telecommunications will find this book to be valuable in their work. This book is also suitable for use as a text or as a reference at universities and other technical institutions.

Analog transmission system designs are based on principles discovered before the 1970s. These systems have been adequately covered in many telecommunications books, and universities have been offering comprehensive analog communications courses for many years. For these reasons most practicing engineers have extensive experience and knowledge of analog systems. Thus we do not see the justification for including a review chapter presenting the principles and applications of analog systems.

During the 1980s many new digital communication and signal processing systems concepts and techniques have been developed at an ever increasing rate, and only during this time have courses in these areas been introduced at many universities. Thus many senior practicing engineers have not had the benefits of formal and systematic instruction in digital communications fundamentals.

I wish to emphasize that this is not an academic book in the sense of analytical communications or measurement theory. Many books cover communications theory; however, they do not stress the measurements, experimental analysis, and instrumentation problems related to communications systems.

In terms of coverage of the total field of telecommunications measurements, analysis, and instrumentation, the book is rather specialized. However, the material does cover the most important techniques required in the measurement of modern telecommunications systems.

I feel it is appropriate to state how this book was conceived. Engineers of Hewlett-Packard's South Queensferry Division (Scotland) have been presenting Telecommunications Symposiums in the United States and Canada on recent developments in telecommunications measurements. As a participant at one of their symposiums, I learned valuable information for my telecommunications consulting activities as well as for my research and teaching at the University of Ottawa and Stanford University. Following my stay with Hewlett-Packard in Scotland and after numerous meetings and extensive correspondence, we defined the outline of this book, which is based on the notes prepared for the symposium. Throughout the preparation of individual chapters and the overall manuscript, we found that we needed a significant effort to improve our manuscript to have a more logical, uniform, and understandable presentation of the material covered.

To facilitate the comprehension of later chapters, which requires a background in digital communications and signal processing, I included Chapters 1 and 2, presenting a review of this rapidly evolving field. For these chapters I extracted material from recent technical papers as well as from various sections of my books on digital communications.

Digital transmission system fundamentals and applications are reviewed in Chapter 1. In our review we highlight the most important concepts required for the comprehension of more advanced digital communication system measurements. Baseband transmission systems, digital modulation-demodulation techniques and integrated systems digital network (ISDN) performance objectives are highlighted. Error-free-seconds (EFS), bit-error-rate (BER), and availability objectives of ISDN and Radio Systems, as recommended by international organizations, are also presented.

Digital signal processing (DSP) techniques used for Analog-to-Digital (A/D) conversion of audio (speech), television and frequency division multiplex (FDM) signals are reviewed in Chapter 2. Related echo-suppression and cancellation techniques as well as digital-speech interpolation systems are also covered.

In Chapter 3 you are exposed to digital signal-processing (DSP) techniques in telephone channel measurements and instrumentation. Requirements and advantages of DSP (filtering) techniques for analog and digital telephone channel instrumentation applications are highlighted. Signal-to-noise, digital filter distortion requirements, and filtering implementation techniques are described.

You learn the techniques of overcoming intrinsic uncertainties in PCM channel measurements in Chapter 4. Problems related to the PCM quantizing mechanism, gain, and noise generation in measurements are presented. Averaging techniques, related errors, and measurement methods are described.

Error performance analysis of digital transmission systems is presented in Chapter 5. Out-of-service and in-service testing, measurement equipment and related error analysis of the BER, EFS, errored second, and error distribution are described.

You learn about the limits and the measurement of jitter tolerance in digital transmission systems in Chapter 6. The fundamental jitter tolerance in regenerative digital transmission systems is defined and analyzed. Jitter in asynchronous multiplex-demultiplex (MULDEX) systems is described. A description of intrinsic jitter, measurement techniques for jitter tolerance, and highlights of modern instruments are presented.

Digital radio measurement techniques are described in Chapter 7. This chapter examines the application of the constellation display to the diagnosis of error margins and the generation of BER versus C/N and C/I plots using additive noise and interference. The effectiveness of countermeasures against multipath fading is measured by radio signatures. This chapter contains a wealth of information, presented for the first time, for comprehensive digital radio diagnostic/measurement techniques.

Analog frequency division multiplex (FDM) system testing and network surveillance are described in Chapter 8. A brief review of FDM system principles and architecture is followed by an in-depth study of FDM system measurement and experimental analysis techniques. The newest generation of related instruments (hardware and software) is also highlighted.

In Chapter 9 you are exposed to analog microwave diagnostic measurements. White noise measurements as a diagnostic aid to microwave radio trouble-shooting, differential phase and differential gain as well as the relationships between the shapes of measured curves, obtained by microwave link analysis and noise-power-ratio test sets, are highlighted.

Enhanced microwave radio measurement techniques, which employ a tracking downconverter and extend the measurement bandwidth, are described in Chapter 10. Residual distortion components are also analyzed.

In the final chapter, Chapter 11, phase-noise measurements for analog and digital communication systems are described. As the performance of modern communication systems improves, the phase-noise requirements of the signal sources and of the transmission systems often become the limiting factor for the overall system performance. Frequently used phase-noise definitions, the most common measurement techniques, and their applicability are described.

I would be pleased to hear from you. If you have any questions, comments, or suggestions related to the content of this book, or problems related to the topics covered, please feel free to write to me.

DR. KAMILO FEHER
Davis, California

1

DIGITAL TRANSMISSION SYSTEMS

DR. KAMILO FEHER

*University of California, Davis,
Davis, California 95616
and
Consulting Group, DIGCOM, Inc.*

1.1 INTRODUCTION

Digital transmission techniques are reviewed in this chapter. In our review of the principles of operation and applications of these techniques, we highlight some of the most important concepts required for the comprehension of the more advanced digital signal and system measurement, analysis, and instrumentation problems, which are described in Chapters 3 through 7. Illustrative digital signal processing and transmission system configurations are also presented. The references provide a detailed description of these systems.

Since the early 1960s, architectures and standards for digital signal processing, multiplexing, and transmission of voice, data, television, facsimile, and other signals have been established worldwide. **Integrated digital networks** (IDNs), where the term *integrated* refers to the commonality of digital techniques used in the transmission of voice and data signals, are already in use. The basic building block of the telephony IDNs is the 64-kb/s-rate **pulse code modulation** (PCM) processor of speech signals. The 64-kb/s digitized voice signal has been used as a standard in circuit switching, digital multiplexing, and transmission systems. Service-dedicated networks have been conceived and implemented to provide, in a cost-effective manner, a limited range of services centered around the main service application to which they are devoted. The proliferation of different types of digital networks (e.g., a circuit switched IDN for telephony, a circuit switched IDN for bulk data, a packet switched IDN for bursty data) impose a burden on both network users and providers. This is particularly true when taking into account the multiplicity of access interfaces and access facilities.

These circumstances, together with the rapid growth of communication system capabilities and technologies, have stimulated the emergence of the **integrated services digital network** (ISDN) concept [Decina, 1.9]. *An ISDN can be characterized by its three main features:*

1. End-to-end digital connectivity
2. Multiservice capability (voice, data, video)
3. Standard terminal interfaces

In general, the ISDN should be based on and evolve from the 64-kb/s telephony IDN, including digital subscriber loop facilities. The telephony IDN provides interconnection with current service-dedicated facilities (such as data packet switching and wideband switching). It will progressively incorporate, according to technology evolution and economic considerations, additional network functions and features, including those of any other service-dedicated facilities. The transition from the existing network to comprehensive ISDN may require a long period of time. In the initial stages of evolution, new network and system architectures will arise to improve the efficiency of integrated voice, data, and wideband communications [Decina, 1.9].

In Fig. 1.1, a typical digital transmission system, including the multiplexing equipment, is illustrated. In standard PCM equipment, 8-bit time-slot interleaving of 64-kb/s digitized (analog-to-digital, or A/D) converted channels is used. Primary

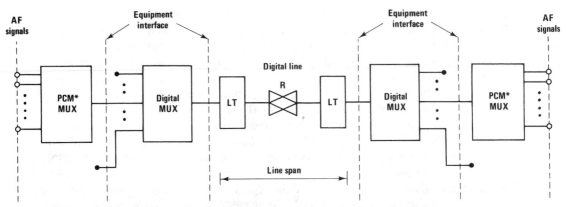

*Digital multiplex equipment based on adaptive–reduced bit rate signal–processing techniques has also been developed.

AF   Audio frequency signals
R   Repeater
LT   Line terminal
MUX Multiplexer/demultiplexer

Figure 1.1 Typical digital transmission system including PCM digital MUX and line termination equipment [Decina and Roveri, 1.9].

multiplexed signals based on the North American standard have a capacity of 24 PCM channels, whereas PCM multiplex systems, which meet the CCITT recommendations (in Europe and many other parts of the world), have a 30-channel capacity. Time division digital multiplexing equipment is used to transform lower-rate channels into higher-rate channels. Standard digital transmission rates and their equivalent PCM voice-channel capacities are listed in Table 1.1. Note that in addition to the standard bit rates, other bit rates are also in frequent use. For example, the DS-1C rate is for the transmission of two multiplexed DS-1 rate signals. The multiplexed 3.152-Mb/s rate signal has the capacity of 48 PCM voice channels. For the transmission of two time division multiplexed DS-3 rate

TABLE 1.1   Standard Digital Transmission Rates and PCM Voice-Channel Capacities

Hierarchy level no.	Standard transmission rates		
	U.S./Canada (Mb/s)	Japan (Mb/s)	Europe (CCITT) (Mb/s)
1	1.544	1.544	2.048
2	6.312	6.312	8.448
3	44.736	32.064	34.368
4	274.176	97.728	139.264
5	—	396.200	560–840
Number of PCM voice channels (capacity)			
Hierarchy level no.	U.S./Canada	Japan	Europe (CCITT)
1	24	24	30
2	96	96	120
3	672	480	480
4	4032	1440	1920
5	—	5760	7680–11520
Industry terms/abbreviations			
1.544 Mb/s	DS-1 or T-1		
6.312 Mb/s	DS-2 or T-2		
44.736 Mb/s	DS-3 or T-3		
274.176 Mb/s	DS-4 or T-4		
2.048 Mb/s	CEPT-32		
Other frequently used rates			
3.152 Mb/s	DS-1C		
90 Mb/s(approx.)	(For 2 × DS-3)		
120 Mb/s(approx.)	(For 59 × CEPT-32)		

signals, a 90-Mb/s rate system is required. Currently, most North American high-capacity digital radio systems are operated at the 90-Mb/s rate. In international high-capacity satellite systems, such as INTELSAT's **time division multiple access** (TDMA), a transmission rate of 120 Mb/s is used [Feher, 1.4]. Adaptive delta-modulation, adaptive-PCM, linear-predictive-coding (LPC), and other advanced digital signal-processing techniques enable lower bit-rate A/D and digital-to-analog (D/A) conversion of speech signals. These techniques, combined with digital speech interpolation (DSI), lead to improved overall system utilization and more cost-efficient applications. Signal-processing techniques are reviewed in Chapter 2.

In Fig. 1.1, the complete transmission system is shown simply as *digital line* or *line span*. This line span can contain regenerative sections of coaxial cable or optical fiber systems, terrestrial **line-of-sight** (LOS) microwave sections, and one or more satellite system sections. In Fig. 1.2, a simplified diagram of a transmission system is illustrated. Each repeater contains a receiver-transmitter pair. Repeaters may be analog or digital (regenerative). An illustrative digital transmitter contains a bipolar to non-return-to-zero (NRZ) converter, a scrambler (randomizer), and a forward-error-correction (FEC) encoder. For baseband system applications (e.g., coaxial wideband cable systems), a line coder and a low-pass filter (LPF) are also required. For modulated system applications such as terrestrial microwave systems, satellite systems, and frequency translated cable system applications, a modulator and a band-pass filter (BPF) are required. The receiver performs the inverse signal processing functions of the transmitter. The **regenerator,** also known as *threshold-comparator* or *clocked-slicer,* regenerates the noise-free baseband source signal. In Section 1.2, we describe the principles of baseband and modulated data-transmission systems.

1.2 BASEBAND TRANSMISSION TECHNIQUES

1.2.1 Concepts and System Configurations

A typical digital baseband transmitter and receiver are illustrated in Fig. 1.3. The binary source or customer data source may be encoded in an FEC encoder. FEC is particularly suitable for applications in which the performance of the uncoded transmission system is not satisfactory. In the next block, the channel encoder converts the binary source signal to the desired multilevel (M-level) signal. The transmit low-pass filter (LPF_T) band limits the channel coded (also known as line coded) signal, which in turn is fed to the transmit amplifier. The transmission link, not shown in Fig. 1.3, may contain a number of baseband amplifiers and filters. These amplifiers, combined with the transmission medium (e.g., the low-pass effect of a long-haul cable system), distort the transmitted signal. The thermal noise generated in the amplifiers and interference from other systems is illustrated by the summing network. The receive low-pass filter (LPF_R) contains an amplitude and phase equalizer. The equalized signal [Qureshi, 1.36]

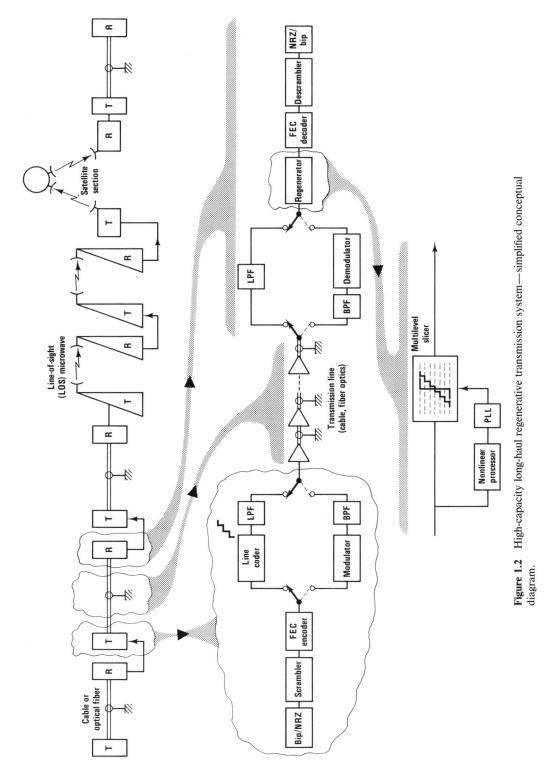

Figure 1.2 High-capacity long-haul regenerative transmission system—simplified conceptual diagram.

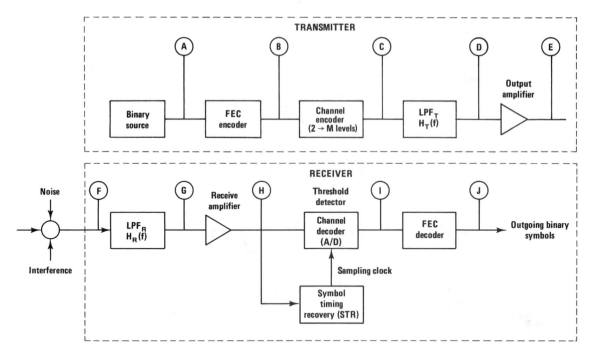

Figure 1.3   Regenerative baseband transmission section.

is fed into the regenerator. The **regenerator,** also known as *channel decoder* and *threshold detector,* is a sampled A/D converter. The symbol-timing-recovery (STR) circuit regenerates the symbol-timing (sampling) clock [Feher, 1.5]. The FEC decoder corrects certain types of transmission errors and removes the redundant bits introduced by the FEC encoder.

Baseband waveforms in a binary regenerative transmission system are illustrated in Fig. 1.4. A sample of a synchronous binary NRZ data stream (waveform), Ⓐ in Fig. 1.4—corresponding to Ⓐ in Fig. 1.3—illustrates the binary source. Here we assume uncoded, binary transmission; that is, the FEC encoder and channel encoders are bypassed. An ideal bandlimited, **intersymbol-interference** (ISI) and noise-free signal sample is assumed at Ⓓ. Furthermore, we assume that the receive LPF_R contains an ideal adaptive-equalizer [Qureshi, 1.36]. With these assumptions at Ⓖ, an ISI-free signal is present; however, the signal is corrupted by noise. The channel decoder, for the case of binary transmission, is a 1-bit sampled A/D converter. The illustrated A/D converted (regenerated) data stream (see signal Ⓘ in Fig. 1.4) contains 2 inverted (errored) bits. These inverted bits (shaded in the figure) are caused by noise samples that exceed in the sampling instants the magnitude of the received signal. In a well-designed and well-maintained regenerative system, the probability of occurrence of these errored bits is small—typically in the 10^{-6} to 10^{-10} **bit error rate** (BER) range.

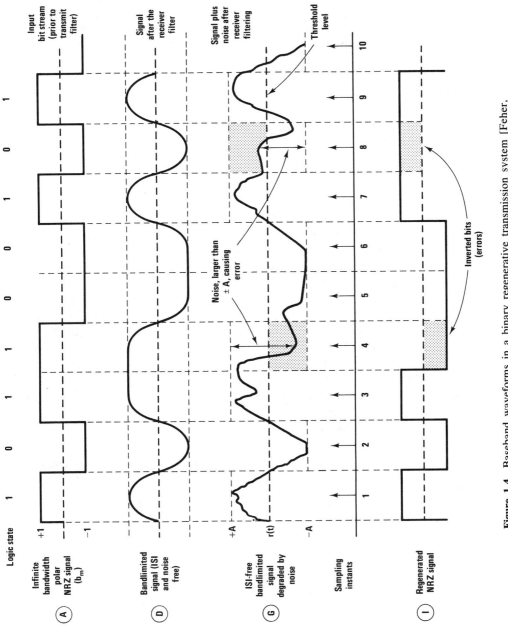

Figure 1.4 Baseband waveforms in a binary regenerative transmission system [Feher, 1.4].

1.2.2 Binary Systems

In binary systems, one transmitted symbol contains 1 bit of information. The theoretical limit, or *Nyquist rate,* is 2 *symbols/s/Hz* [Feher, 1.5]. For binary signals, this means that we can transmit only 2 b/s/Hz. In other words, an f_b = 10-Mb/s rate binary transmission system requires a theoretical minimal bandwidth, or *Nyquist bandwidth,* of 5 MHz and a practical bandwidth of about 6 MHz.

1.2.3 PAM and PRS-Multilevel Baseband Systems

Many system applications require a considerably higher spectral efficiency than is possible with binary signaling. An increased spectral efficiency can be achieved by channel encoding (conversion) of the binary signal into a *M*-level signal, also known as a **pulse amplitude modulated** (PAM) or PAM-baseband processed signal (see Fig. 1.5). To generate an M = 4-level PAM signal, two consecutive bits are paired to form a symbol. Thus the output symbol rate f_s of the M = 4-level PAM (for short: 4-PAM) converter is one-half of the input bit rate $f_b = 1/T_b$; that is, $f_s = 1/T_s = \frac{1}{2}T_b$. Transmission at the Nyquist rate (2 symbols/s/Hz) leads to a spectral efficiency of 4 b/s/Hz. Similarly, in an 8-PAM system, 3 bits form a symbol; that is, $T_s = 3T_b$. For this reason, the theoretical spectral efficiency of an 8-level PAM system is 6 b/s/Hz. In the measured example, we used 16 PAM. In this case, $T_s = 4T_b$; see Fig. 1.5(b). In Table 1.2, the theoretical and practical spectral efficiency of PAM and also of **partial-response systems** (PRS) is listed [Feher, 1.5]. Note that in the case of PAM systems, the practical spectral efficiency, defined at the 35- to 50-dB out-of-band attenuation point, is about 20% lower than the theoretical Nyquist rate. The required signal-to-noise ratio, S/N, in the Nyquist bandwidth for a BER of 10^{-8} is also listed. It is assumed that the probability density of the noise is Gaussian.

Example 1.1

> If we wish to transmit a bit rate of f_b = 90 Mb/s with an 8-PAM system, then the minimal Nyquist channel bandwidth is f_N = 15 MHz. However, it is impossible to design **brick-wall** filters (later we will call these filters α = 0 *steepness* or *roll-off* factor filters). With current state-of-the-art filter design, a roll-off factor of α = 0.2 is achievable. This means that for the f_b = 90 Mb/s 8-PAM baseband example, the filter guard-band attenuation reaches about 50 dB at $f_c = f_n (1 + \alpha) = (15$ MHz)$(1 + 0.2)$ = 18 MHz. The practical spectral efficiency of this system (defined at the 50-dB spectral attenuation point) is $f_b \div f_G$ = 90 Mb/s $\div$ 18 MHz = 5 b/s/Hz. This spectral efficiency is 20% below the spectral efficiency of the ideal brick-wall Nyquist channel. ∎

In baseband PRS and also in **quadrature partial-response systems** (QPRS), a controlled amount of ISI is intentionally introduced in order to simplify the filter design and particularly the phase-equalization problem. These systems, discovered by Dr. A. Lender in the early 1960s, found numerous applications

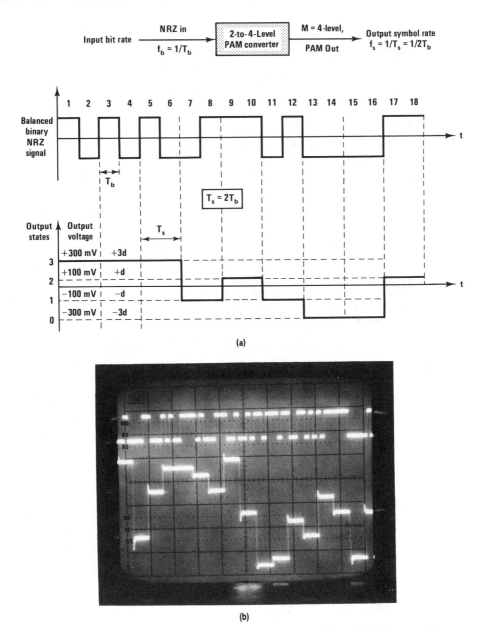

(a)

(b)

Figure 1.5 Binary NRZ source and corresponding multilevel PAM signals. (a) Binary NRZ input and corresponding M = 4-level PAM output. (b) Binary input and corresponding M = 16-level PAM output. $T_s = 4T_b$, horizontal scale: 5 ms/division; vertical scale: 50 mV/division. In the measured 16-level output signal, in addition to the multilevel discrete signal, "spikes" (impulses) are also noticed. These high-frequency imperfections can be easily removed by low-pass filtering.

TABLE 1.2 PAM and PRS* Baseband Spectral Efficiency and S/N Requirement [based on Feher, 1.4]

Number of levels	Theoretical spectral efficiency [b/s/Hz][1]	Practical spectral efficiency [35 dB atten][2]	Required S/N[3],[4] for $P_{(e)}$ = BER = 10^{-8}
2-PAM	2	1.66 [−20%]	15
3-PRS	2	2.6 [+30%]	17.5
4-PAM	4	3.33 [−20%]	21.5
7-PRS	4	4.4 [+10%]	24
8-PAM	6	5 [−20%]	28
15-PRS	6	6.25 [+ 5%]	31
16-PAM	8	6.66 [−20%]	34
31-PRS	8	8 [0%]	37
32-PAM	10	8.33 [−20%]	40

*Partial response systems, invented by Dr. A. Lender, are also known as **generalized duobinary** systems. For a comprehensive reference, see [Lender, 1.10].

[1]Assuming ideal phase equalized brick-wall filters. Transmission at the Nyquist rate of 2 symbols/s/Hz for PAM and for PRS.

[2]Assuming α = 0.2 raised-cosine-filters; attenuation at 20% above Nyquist frequency is 35 to 50 dB with state-of-the-art filter designs. Note PRS transmission is practically possible above the Nyquist rate. Terms in brackets indicate practical achievable rate (without complex equalization) in reference to the Nyquist rate as a percent.

[3]Defined in the Nyquist bandwidth (within 1 dB accurate).

[4]Entries in this table apply also to modulated systems. See Table 1.3.

in terrestrial microwave, cable, and other data-transmission systems. Due to the intentionally introduced, controlled ISI, these systems are also known as **correlative** and **duobinary** systems. An advantage of these systems is that correlative coding enables transmission at the Nyquist rate and even at higher-than-Nyquist binary signaling rates. A detailed description of PR and QPR transmission systems, principles, and applications may be found in [Lender, 1.10].

1.2.4 Bandlimited Baseband Systems and Spectral Efficiency

A digital transmission system is more spectrally efficient if it has the capability of transmitting a greater number of bits per second in a given bandwidth. The bandwidth is frequently normalized to 1 Hz, and the **spectral efficiency** is expressed in bits per second per hertz. Table 1.2 indicates that higher-state systems having a larger number of output levels have an increased spectral efficiency. However, higher-level systems require a higher S/N. For example, Table 1.2 indicates that a simple binary (2-level) system has a practical spectral efficiency of 1.66 b/s/Hz, whereas the 8-level PAM system has a 5 b/s/Hz practical spectral efficiency. To achieve this spectral efficiency improvement, an increased S/N from 15 dB to 28 dB is required, assuming that BER $= 10^{-8}$ has to be maintained.

In his landmark patents on channel characteristics in the 1920s, Nyquist defined the minimum channel bandwidth requirements and also the general channel characteristics required for ISI-free reception. An in-depth practical study of the Nyquist channel transmission theorems combined with the study of matched filter (correlation) receivers is given in [Feher, 1.4]. For data-transmission systems, the following are the most frequently used Nyquist theorems.

1.2.5 Nyquist's First Theorem (Minimum Bandwidth–Brick-Wall Channel)

THEOREM.   If synchronous impulses having a transmission rate of f_s symbols per second are applied to an ideal, linear-phase brick-wall low-pass channel having a cutoff frequency of $f_N = f_s/2$, then the responses to these impulses can be observed independently, that is, without ISI.

Note. For ISI-free transmission of f_s rate rectangular pulses, an ($x/\sin x$)-shaped amplitude equalizer has to be added to the ideal brick-wall channel. Thus it is possible to have an ISI-free transmission rate of 2 symbols/s/Hz.

1.2.6 Nyquist's Second Theorem (Vestigial Symmetry Theorem)

THEOREM.   The addition of a skew-symmetrical, real-valued transmittance function $Y(w)$ to the transmittance of the ideal low-pass filter maintains the zero-axis crossings of the impulse response. These zero-axis crossings provide the necessary condition for ISI-free transmission. The symmetry of $Y(w)$ about the cutoff frequency w_N (Nyquist radian frequency $w_n = 2\pi f_N$) of the linear-phase brick-wall filter is defined by

$$Y(w_n - x) = -Y(w_n + x) \qquad 0 < x < w_n$$

where $w_n = 2\pi f_N$.

In order to visualize the practical importance of these theorems, the ensuing description is presented for the binary case; however, the *Nyquist transmission theorems apply to the multilevel case as well.*

Consider the ideal brick-wall channel model of Fig. 1.6. The cutoff frequency, also known as the Nyquist frequency, is defined to be $f_N = 1/2T_s = f_s/2$, where T_s is the unit symbol duration. (In binary systems, the symbol rate equals the

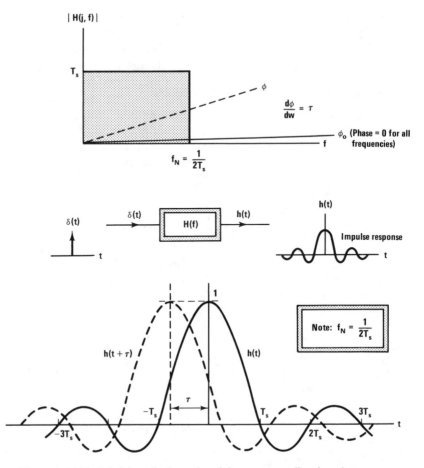

Figure 1.6 Ideal brick-wall channel and its corresponding impulse response [Feher, 1.4].

bit rate; thus $T_s = T_b$, where T_b is the unit bit duration. In multilevel systems $T_s = T_b \log_2 M$, where M is the number of signaling levels.) The impulse response of the channel, $h(t)$, is given by the inverse Fourier transform of the transfer function $H(f)$:

$$H(f) = \begin{cases} T_s & |f| \leq \dfrac{1}{2T_s} \\ 0 & |f| > \dfrac{1}{2T_s} \end{cases} \tag{1.1}$$

$$h(t) = F^{-1}[H(f)] = \int_{-\infty}^{\infty} H(f)e^{j2\pi ft}\, df \tag{1.2}$$

$$h(t) = \frac{\sin(2\pi f_n t)}{2\pi f_n t} = \frac{\sin(\pi t/T_s)}{\pi t/T_s} \tag{1.3}$$

(Note that the phase of $H(f)$ is 0 for all frequencies.) From this impulse response,

$$h(nT_s) = \begin{cases} 1 & \text{for } n = 0 \\ 0 & \text{for } n = \pm 1, \pm 2, \pm 3, \ldots \end{cases} \tag{1.4}$$

Thus the impulse response attains its full value for $t = nT_s = 0$ and has 0 crossings for all other integer multiples of the symbol duration. If the ideal brick-wall channel has a nonzero but linear phase (dashed line in Fig. 1.6), the impulse response is shifted by an amount that equals the channel delay. This delay is $\tau = d\phi/dw$; and for linear-phase filters, is constant over all frequencies. As the shape of the impulse response is the same as with $\tau = 0$, there will be no further distortion introduced. The sampling instants in the receiver are shifted by τ seconds.

Note that the conceptual channel input (also known as filter excitation), $\delta(t)$, has an infinitesimally short duration, whereas the output (i.e., the impulse response) has an infinite duration. The bandlimited channel stretches the impulse beyond the T_s interval and deforms the input signal. A desirable property of the described impulse response is that it has zero values for integer multiples of T_s. Now we should be in a position to see that in an $f_n = 1/2T_s$-wide brick-wall channel, it is possible to transmit and detect synchronous, random impulses at a rate $f_s = 1/T_s = 2f_N$. Theoretically, the detection of any of these symbols can be performed without any interference from the previously sent or the subsequent impulse patterns. This situation is known as ISI-free transmission.

Unfortunately, the described minimum-bandwidth Nyquist channels are not realizable. An infinite number of filter sections would be required to synthesize the infinite attenuation slope of the brick-wall channel. Additionally, the decay in the lobes of the time domain response is very slow. This, in turn, would cause a prohibitively large ISI degradation for smallest filtering or symbol timing imperfections.

To alleviate these problems and to define more practical channel characteristics, Nyquist introduced a theorem on vestigial symmetry (Nyquist's second theorem).

The **raised-cosine function** satisfies Nyquist's vestigial symmetry requirement. Filter designers frequently approximate raised-cosine channel characteristics for ISI-free impulse transmission. The amplitude response of this channel is given by

$$|H(j\omega)| = \begin{cases} 1 & 0 \leq \omega \leq \dfrac{\pi}{T_s}(1 - \alpha) \\[2mm] \cos^2\left\{\dfrac{T_s}{4\alpha}\left[\omega - \dfrac{\pi(1 - \alpha)}{T_s}\right]\right\} & \dfrac{\pi}{T_s}(1 - \alpha) \leq \omega \leq \dfrac{\pi}{T_s}(1 + \alpha) \\[2mm] 0 & \omega > \dfrac{\pi}{T_s}(1 + \alpha) \end{cases} \tag{1.5}$$

where $\omega = 2\pi f$ and α is the channel roll-off factor. For practical systems that are employed for the transmission of $f_s = 1/T_s = 2f_N$-rate synchronous rectangular pulses, an $(x/\sin x)$-shaped amplitude equalizer has to be added. Thus the desired channel transfer function for ISI-free *pulse* transmission (such as that of NRZ and of multilevel PAM signals) is given by

$$H(j\omega) = \begin{cases} \dfrac{\omega T_s/2}{\sin(\omega T_s/2)} & 0 \leq \omega \leq \dfrac{\pi}{T_s}(1-\alpha) \\[3ex] \dfrac{\omega T_s/2}{\sin(\omega T_s/2)} \cos^2\left\{\dfrac{T_s}{4\alpha}\left[\dfrac{\omega-\pi(1-\alpha)}{T_s}\right]\right\} & \dfrac{\pi}{T_s}(1-\alpha) \leq \omega \leq \dfrac{\pi}{T_s}(1+\alpha) \\[3ex] 0 & \omega > \dfrac{\pi}{T_s}(1+\alpha) \end{cases}$$

$$(1.6)$$

where α is the roll-off factor in the raised-cosine channel equation. For $\alpha = 0$, an unrealizable minimum-bandwidth filter having a bandwidth equal to $f_N = \frac{1}{2}T_s$ is obtained. For $\alpha = 0.5$, a 50% excess bandwidth is used, whereas for $\alpha = 1$, the transmission bandwidth is double the theoretical minimum bandwidth. The amplitude characteristics for various values of the bandwidth parameter α are shown in Fig. 1.7. Theoretically, at the frequency $f = (1 + \alpha)f_N$, the attenuation has an infinite value. For practical realizations, an attenuation of 20 to 60 dB is specified depending upon the adjacent channel interference allowance.

Example 1.2: $\alpha = 0.2$ Filtered 16-PAM Baseband Spectral Measurement

In Fig. 1.8, the measured spectrum of a raised-cosine filtered NRZ random data stream is illustrated. In this measurement, an $\alpha = 0.2$ roll-off raised-cosine filter characteristic is used. In the experimental setup, an $f_b = 800$-kb/s rate binary NRZ data stream is converted into a 16-level PAM baseband signal. As each 16-PAM symbol contains 4 bits of information, the corresponding 16-PAM symbol rate is $f_s = 200$k Baud (or 200k symbols/second). The Nyquist frequency is $f_N = 0.5f_s = 100$ kHz. A theoretical $\alpha = 0.0$ filter would filter out all spectral components above 100 kHz. From the power spectral measurement result, we note that all spectral components beyond 120 kHz are attenuated by 55 dB. In other words, the $\alpha = 0.2$ filter has a practical attenuation of 55 dB at 20% above the 100 kHz Nyquist frequency. ■

Note. The filter amplitude response and the measured output power spectral density are practically identical. This is so, as the raised-cosine filter is preceded by an $x/\sin x$ amplitude equalizer. The power spectral density of the 16-PAM signal at the output of this amplitude equalizer (known also as aperture equalizer) is flat.

1.2.7 Spectra of Random and of PRBS Data

Random equiprobable data streams are transmitted in most digital communications systems. A/D converted voice and video signals, the output of time division multiplex (TDM) equipment and other data subsystems are scrambled. These signals are random or closely resemble truly random data streams. Probability

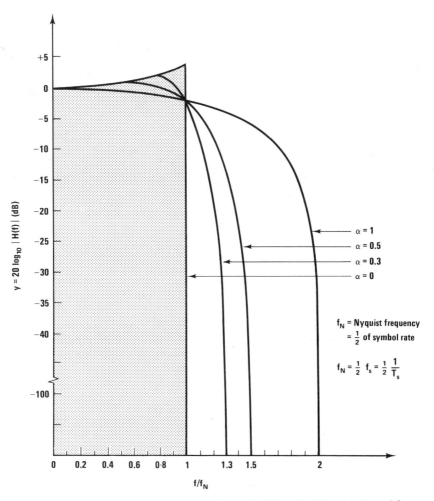

Figure 1.7   Amplitude characteristics (plotted in dB) of the Nyquist channel for rectangular pulse transmission. The cascaded amplitude response of the transmit and receive filters is illustrated [Feher, 1.5].

of error, $P(e)$, error-free-second (EFS), and error-containing-second (ECS) measurement techniques have been developed [Feher, 1.5] to evaluate the **in-service** performance of truly random data streams. Here the term *in-service* implies that the customer data transmission must not be interrupted in order to perform field performance measurements.

For laboratory, production, and field tests, more accurate measurements can be performed by means of **off-line,** or **out-of-service,** measurements. These out-of-service $P(e)$, EFS, and ECS measurements utilize a **pseudorandom binary-sequence** (PRBS) generator and a corresponding error-rate detector. Sufficiently long PRBS sequences approach the characteristics of truly random data sequences.

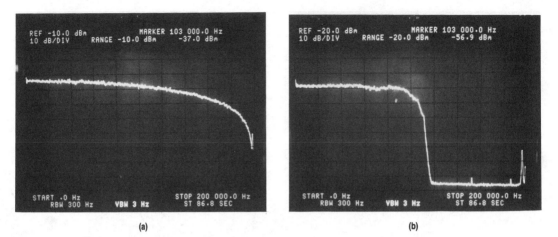

Figure 1.8 Power spectral density of random 16-level PAM baseband signals having a transmission rate of 200,000 Baud (800 kb/s). (a) Main lobe of the unfiltered signal. As expected, the main lobe "zero" is at 200 kHz. (b) Filtered output spectrum. In the measurement, Karkar Electronics, Inc. $\alpha = 0.2$ raised-cosine channel filters preceded by $(x/\sin x)$-shaped equalizers are used.

In this section, frequently used equations of PRBS and truly random sequences are presented. We assume *synchronous* data transmission, that is, a system in which data transitions may occur only at integer multiples of the symbol duration, T_s. Note that in binary systems, the bit duration T_b equals the symbol duration T_s; that is, $T_b = T_s$.

The power spectral density of a PRBS, $w_{\text{PR}}(f)$, consists of discrete spectral lines separated by $1/LT_s$ Hz. It is given by

$$w_{\text{PR}}(f) = \frac{L+1}{L^2}\left(\frac{\sin \pi f T_b}{\pi f T_b}\right)^2 \sum_{\substack{n=-\infty \\ n \neq 0}}^{\infty} \delta\left(f - \frac{n}{LT_b}\right) + \frac{1}{L^2}\delta(f) \qquad (1.7)$$

where $L = 2^N - 1$ is the period of the pseudorandom sequence and N is the number of shift-register stages in a maximal-length pseudorandom code generator. A pseudorandom code generator (scrambler) and corresponding descrambler (BER detector) are illustrated in Fig. 1.9 [Feher, 1.4; Proakis, 1.7].

Example 1.3: Measured Periodic and PRBS Data Spectra

Measured periodic and PRBS data patterns and their corresponding power spectra are illustrated in this example. For all cases, an $f_b = 400$-kb/s rate binary source is used. The following four data patterns are displayed in Fig. 1.10.

1. 1:1 Periodic stream with 1s and 0 states alternating.
2. 1:3 Periodic stream with 1s and three 0 states alternating.
3. 1:7 Periodic stream with 1s and seven 0 states alternating.
4. $2^6 - 1$ Segment of a PRBS data pattern generated by a 6-bit maximal length PRBS shift register having a sequence length of $2^6 - 1 = 63$ bits.

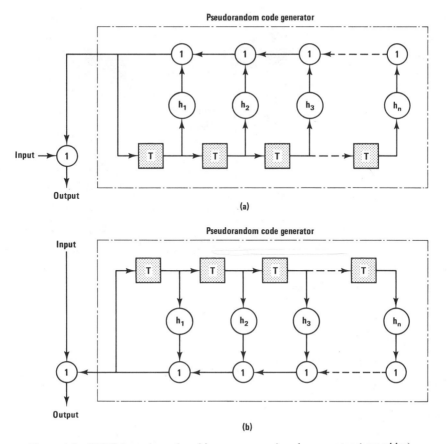

Figure 1.9   PRBS (pseudorandom binary sequence) code generator (scrambler) and corresponding error detector (descrambler). (a) Scrambler. (b) Descrambler. $h_1, h_2 \ldots h_N$ represent the shift register internal connections that determine the code sequence.

The corresponding power spectrum of the periodic 1:1 data pattern is displayed in Fig. 1.11(a). The fundamental frequency (first) harmonic component in the Fourier series of the $f_b = 400$-kb/s data stream is at $f_1 = 200$ kHz. Note that $f_1 = f_b/2$; that is, *the first harmonic is at one-half of the bit rate* and not at the bit rate frequency. This is so because the first harmonic in a Fourier series is the reciprocal of the signal period (T_p); that is, $f_1 = 1/T_p$. In this case the period has a 2-bit duration; i.e., $T_p = 2T_b$. The signal in this and in the following figures does not contain a dc (0 frequency) component. The spike at 0 Hz indicates the 0-Hz start sweep frequency of the spectrum analyzer. The theoretical harmonic components of an ideal symmetrical 400-kb/s rate 1:1 pattern are at 200 kHz, 600 kHz, 1000 kHz . . . ; that is, in the Fourier series only the odd harmonics are present. However, the measurement result, displayed in Fig. 1.11(a), indicates that the second, fourth, sixth, . . . , harmonics (400, 800, 1200 kHz . . .) are also present. In this case these harmonics are attenuated by about 40 dB. (Why?) Why do we measure these even-

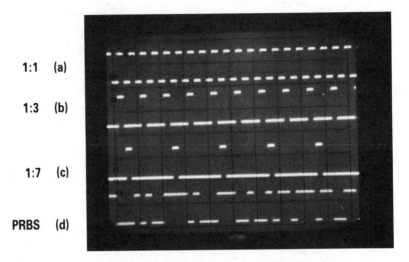

1:1   (a)

1:3   (b)

1:7   (c)

PRBS   (d)

Figure 1.10 Oscilloscope display of illustrative period data patterns and of a PRBS data segment.

order harmonics? Even-order harmonics are present due to imperfections of the data source or of the spectrum analyzer. A small lack of symmetry in the data source—that is, if the duration of a transmitted 1 state is not exactly the same as that of a transmitted 0 state or if the pulse rise time is not exactly the same as the fall time—is among frequent causes of the even-order harmonics. High-power test patterns may cause the spectrum analyzer to operate in a nonlinear (overload) mode, thus causing the display of even-order harmonics that are not present in the test pattern.

In Fig. 1.11(b) and (c), $f_b = 400$-kb/s rate periodic signals with 1:3 and 1:7 patterns are displayed. Note that the fundamental frequency and the separation between the discrete components is decreased with the increase in the period of a "word" ($3 + 1 = 4$-bit word and $7 + 1 = 8$-bit word). The spacing between spectral lines is further reduced in the spectral display of a longer PRBS.

In Fig. 1.12, the spectra of an $f_b = 400$-kb/s rate PRBS is displayed. In this case, the sequence (word) length is $2^6 - 1 = 63$ bits. In the measured results spectrum analyzer *stop* frequencies of (a) 600 kHz and (c) 2 MHz are used. The lowest stop frequency, Fig. 1.12(a), leads to a 60-kHz/division display. In this case, the discrete spectral components are visible. Due to the increased measurement bandwidth, the discrete power spectrum appears like a continuous (sin x)/x-shaped spectrum in Fig. 1.12(b) and 1.12(c). Caution has to be exercised in the interpretation of these measurement displays. The "envelope" of the power spectral density has a sin x/x shape; however, the power spectral density consists of discrete spectral lines. For many applications, the distance between these lines (frequency separation) must be small. The expanded spectral display of Fig. 1.13 indicates that the distance between adjacent spectral lines is approximately 6.4 kHz. This measured spacing corresponds to the theoretical spacing computed by equation (1.7)—that is, 400 kb/s $\div$ 63 = 6.34 kHz. In frequency division multiplexed (FDM) telephony systems, adjacent voice channels are separated by only 4 kHz. A PRBS word having a

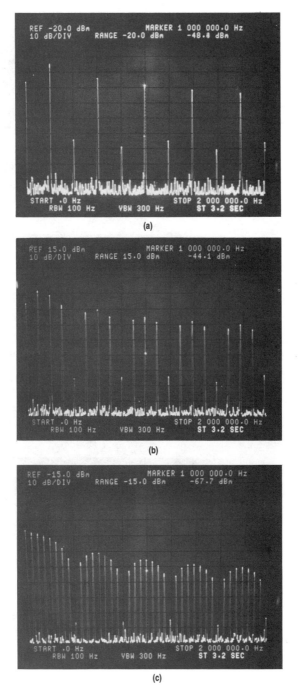

Figure 1.11   Spectra of periodic f_b = 400-kb/s rate periodic signals. (a) 1 : 1. (b) 1 : 3. (c) 1 : 7 pattern.

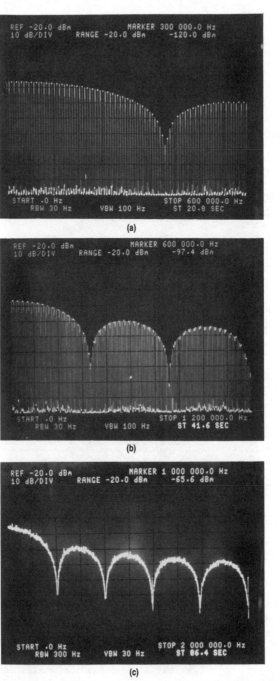

Figure 1.12   Spectra of PRBS (sequence length 63 bits)-f_b = 400-kb/s rate data streams. (a) 0 to 600 kHz display. (b) 0 to 1.2 MHz display. (c) 0 to 2 MHz display.

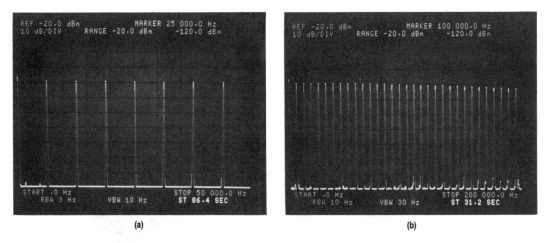

(a)	**(b)**

Figure 1.13 Expanded spectral display of PRBS (sequence length 63 bits) f_b = 400-kb/s data streams corresponding to Fig. 1.12. (a) 0 to 50 kHz. (b) 0 to 200 kHz.

length of 63 bits would not be sufficient to measure the interference from an f_b = 400-kb/s rate source into the FDM system, because the interleaved discrete spectral components would not be present in all telephone channels. Particularly for higher-data-rate systems, longer PRBS words have to be used. It is common to use a PRBS length of $2^{20} - 1$. ■

1.2.8 Eye Diagrams: ISI, Data Transition, and Clock Jitter

The quality of digital transmission links is frequently evaluated by means of **eye diagrams,** or **eye patterns.** Eye diagrams are useful in systems design and also in field measurements. They provide an excellent qualitative means to display system or hardware imperfections and to provide a first-order approximation of system performance. An eye-diagram measurement setup and an oscilloscope display are given in Fig. 1.14. For the display of 2-level eye diagrams, the bit rate clock f_b is connected to the external trigger of the oscilloscope and the 2-to M-level converter is bypassed. For multilevel (M-level) PAM and PRS systems, the multilevel signal is fed into the y-input of the oscilloscope, and the oscilloscope is triggered by the symbol rate clock. The front trigger delay adjustment, conveniently available on most oscilloscopes, assures that the displayed eye pattern is centered on the screen. The horizontal time base is set approximately equal to the symbol duration. The inherent persistence of the cathode-ray tube displays the superimposed segments of the signal. This display is known as the *eye diagram,* or *eye pattern.*

A computer-generated eye diagram of an ideal raised-cosine channel having a roll-off factor $\alpha = 0.3$ is illustrated in Fig. 1.15. A closer examination of this eye diagram indicates that in the sampling instants, all traces cross the normalized

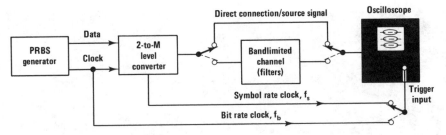

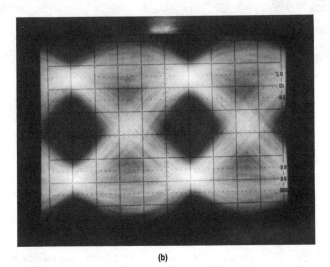

(b)

Figure 1.14 Eye-diagram measurement setup and measured binary eye pattern (diagram). Horizontal scale: 1 μs/division; vertical scale: 100 mV/division. In the eye diagram measurement, an $\alpha = 0.2$ roll-off factor raised-cosine filter manufactured by Karkar Electronics, Inc. is used. The transmitted bit rate is $f_b = 200$ kb/s.

+1 or -1 levels; in other words, there is no ISI. Furthermore, note that the 0-threshold-level crossings occur at different time instants; that is, the 0 crossings are pattern dependent. The peak-to-peak time difference of the 0 crossings is known as peak-to-peak **data-transition jitter**, D_{jpp}. Data-transition jitter frequently is measured and specified in seconds or as a ratio, D_{jpp}/T_s, specified in percent, where T_s is the symbol duration. The peak-to-peak jitter in the computer-generated eye diagram (Fig. 1.15) corresponds to 35% of the symbol duration. Thus the peak-to-peak jitter is 35.5%. In the measured example of an $\alpha = 0.2$ raised-cosine channel (see Fig. 1.14), $D_{jpp}/T_s = 2.4\ \mu$s: 5 μs = 48%. In general, steeper filters (smaller roll-off parameter α) lead to a larger data-transition jitter.

Data-transition jitter by itself does not necessarily lead to performance degradation. However, data-transition jitter combined with **clock jitter** or clock static offset or both may lead to significant performance degradation. The impact

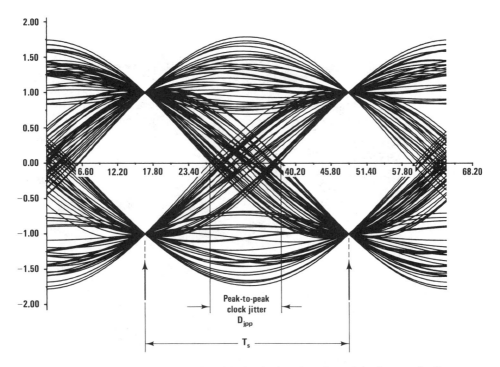

Figure 1.15 Eye diagram of an ideal raised-cosine channel having a roll-off factor $\alpha = 0.3$. In this computer simulated eye-diagram, the peak-to-peak data transition jitter D_{jpp} corresponds to 35% of the symbol duration T_s—that is, $D_{jpp}/T_s = 0.35$.

of clock jitter on the probability of error performance of a multilevel system is illustrated in Fig. 1.16. In many products, the symbol-timing clock jitter is practically independent of the data-transition jitter exhibited in the eye diagram.

Example 1.4

Let us assume that the ratio of peak-to-peak clock jitter, C_{jpp}, to symbol duration, T_s, in Fig. 1.16 is 6%. In other words, the peak excursion of the sampling instant from the ideal sampling instant is 3% of the symbol duration, T_s. The decision distance for an ideal sampling instant is 1000 mV, or 100%. For the worst-case sampling instant, as seen from our illustrative example, Fig. 1.16, the reduced decision distance is 700 mV. Thus we obtain an upper bound on the *performance degradation* of 20 log 1000/700 = 3 dB. From this example we note that a narrower raised-cosine channel, having a smaller roll-off parameter, α, is more sensitive to clock jitter than a wider-band channel. Multilevel systems are also more sensitive than binary systems. ■

Nyquist's $\alpha = 1$ roll-off raised-cosine channel characteristics lead to **ISI and data-transition jitter-free** (IJF) signals. Such signals are fairly robust in regard to *clock jitter*. Another class of IJF signals, generated by means of nonlinear

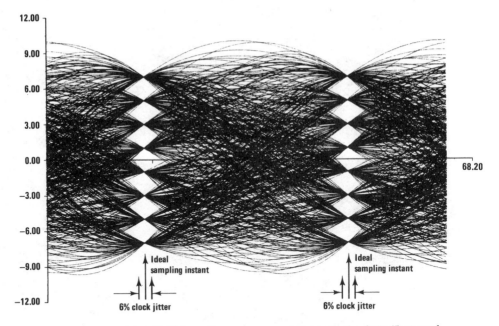

Figure 1.16   8-level PAM eye diagram corresponding to the in-phase (*I*) or quadrature (*Q*) channel of a 64-state (8 × 8) QAM (64-QAM) modem. Computer-generated eye diagram of an ideal $\alpha = 0.2$ filtered raised-cosine channel. A 6% peak-to-peak *clock jitter* is assumed.

switching filters, by pulse-overlapping techniques, or both, has been introduced by Feher and others [Feher, 1.4]. These IJF signals found numerous applications in power-efficient nonlinearly amplified satellite communication systems.

Clock jitter, C_j, may be generated in the STR circuit (see Fig. 1.3) or by the transmission system. The received bandlimited signal does not contain discrete spectral components. For this reason, a typical STR circuit contains a nonlinear element, which generates a discrete tone at the symbol rate, followed by a narrow-band circuit (passive or phase-locked loop (PLL)), which filters out the desired clock frequency. Figure 1.17 illustrates a simplified STR block diagram and the measured power spectra of demodulated 256-QAM signals. In the measurement, an $f_b = 1.6$-Mb/s transmission rate is used. The 256-state (16 × 16) quadrature amplitude modulated (QAM) signal, 256-QAM for short, contains 16 baseband levels in the in-phase, *I*, and 16 levels in the quadrature, *Q*, channels, respectively. Each level is formed by 4 bits. For our example the symbol rate of the *I*-channel (or *Q*-channel) demodulator is

$$f_s = 1600 \text{ kb/s} \div 2 \text{ (number of channels)} \div 4 \text{ (number of bits/symbol)}$$
$$= 200,000 \text{ symbols/s}$$
$$= 200,000 \text{ Baud}$$

The measured photographs indicate that the $f_s = 200,000$-Baud received signal has no discrete spectral components (Ⓐ in Fig. 1.17) and that all spectral

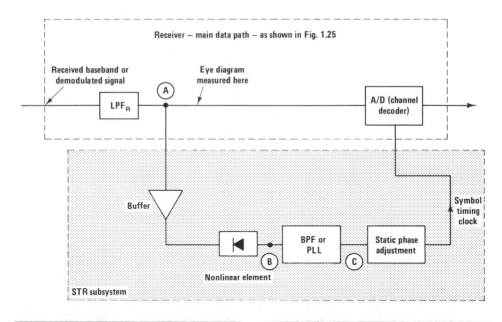

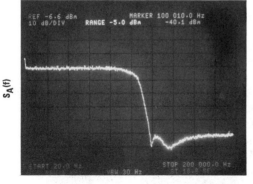

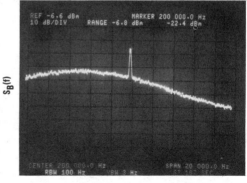

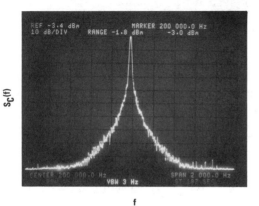

Figure 1.17 STR subsystem and corresponding power spectral density
measurement results. In the measurements, an $\alpha = 0.2$ filtered raised-cosine channel used for the transmission
of an $f_b = 1.6$ Mb/s rate 256-QAM
modem is used. A 1.6 Mb/s bit rate
corresponds in this case to $f_s = 200$
ksymbols/seconds; I-channel of the 256-
QAM demodulator = i.e., 1600 kb/s:
2 (number of channels): 4 (bits/symbols)
= 200 kBaud. Courtesy of Karkar
Electronics, Inc.

components above 120 kHz are attenuated by at least 55 dB. After nonlinear processing, ⑧, a discrete component at $f_s = 200$ kHz is obtained. In addition to this desired STR clock frequency, a continuous spectrum and other undesired discrete components are also generated by most nonlinear circuits. A narrow band-pass filter or PLL removes most undesired spectral components and provides a relatively clean symbol-timing clock frequency to the main data path A/D converter, that is, channel decoder.

Unfortunately, clock jitter, also known as clock **phase noise,** is generated in the STR process. We note from the measured spectrum at © that in addition to the desired discrete spectral component, an undesired continuous power spectrum is also present. The total integrated power of this continuous spectrum (©) is the cause of clock jitter and is also known as STR *phase noise, or phase jitter.*

Transmission systems may contain a number of frequency-translating subsystems, such as upconverters and downconverters. All oscillators generate some phase noise. The system-introduced phase noise, particularly in low-bit-rate mobile system and satellite system applications, may also lead to increased STR clock jitter [Feher, 1.32]. A detailed description of phase-noise measurements is given in Chapter 11.

1.2.9 Sample and Hold (S & H) Eye-Diagram Measurements and Display

Performance degradation, caused by the *combined effect* of ISI and clock jitter, can be estimated by the display of a **sample and hold** (S & H) eye diagram. An S & H eye-diagram setup and corresponding eye diagrams are illustrated in Fig. 1.18. An S & H 16-level ISI-free eye diagram has 16 equidistant horizontal levels, assuming that the sampling clock has no jitter and no static phase error. However, if a clock jitter is introduced, then the ISI-free S & H eye display contains 16 *smeared* horizontal lines. The spread around the nominal, equidistant positions of these lines is proportional to STR jitter. A similar smeared display is also due to ISI with and without clock jitter. S & H eye diagrams are useful in the design of transmission systems and for qualitative and also quantitative monitoring and trouble-shooting of system performance.

1.2.10 $P_e = f(S/N)$ *Performance and b/s/Hz Efficiency of Ideal Baseband Systems*

In an M-level baseband transmission system, one or more bit errors are measured if, at the sampling instant, the noise η exceeds in amplitude the distance from the nominal received level to the nearest decision threshold level d; that is, $|\eta| > d$. A simplified model of an ideal M-level PAM system is illustrated in Fig. 1.19. The ideal received signal level is the value of the filtered and sampled received signal when there is no ISI or noise in the transmission system. The raised-cosine Nyquist filtered system has an *optimal performance* in an additive white Gaussian noise (AWGN) transmission environment *if the receiver is matched*

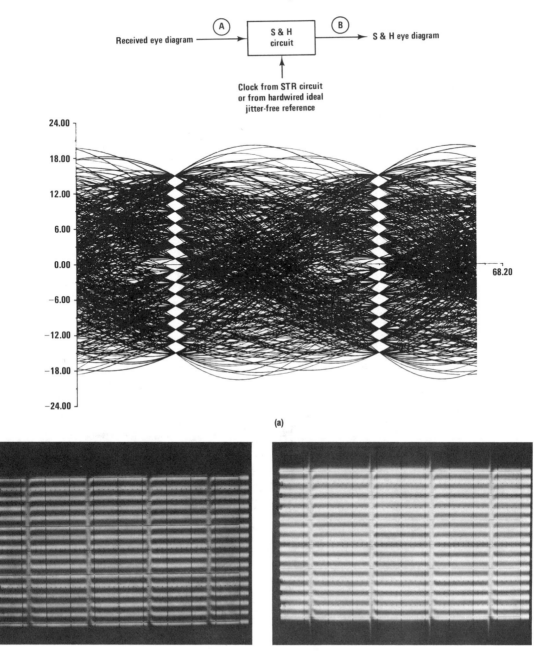

(a)

(b)   (c)

Figure 1.18   S&H and conventional eye-diagram measurement setup and corresponding measured displays. (a) Ideal 16-level eye diagram. (b) S&H 16-level eye diagram (no jitter, ideal STR clock simulated by hardwired clock). (c) S&H 16-level eye diagram with some clock jitter.

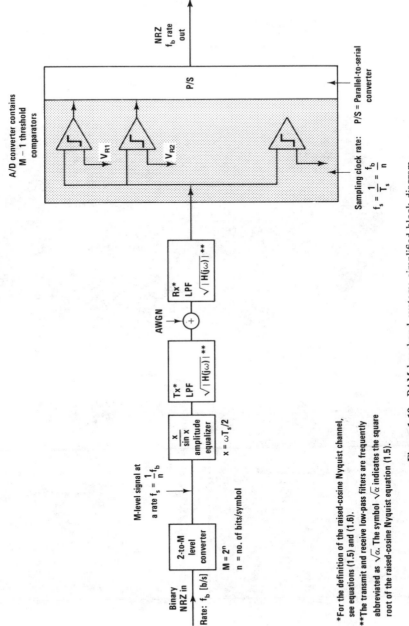

Figure 1.19 PAM baseband system: simplified block diagram.

Binary
NRZ in

Rate: f_b [b/s]

2-to-M
level
converter

$M = 2^n$
n = no. of bits/symbol

M-level signal at
a rate $f_s = \frac{1}{n}f_b$

$\frac{x}{\sin x}$
amplitude
equalizer

$x = \omega T_s/2$

Tx*
LPF

$\sqrt{|H(j\omega)|}$ **

AWGN

Rx*
LPF

$\sqrt{|H(j\omega)|}$ **

A/D converter contains
M − 1 threshold
comparators

V_{R1}

V_{R2}

P/S

NRZ
f_b rate
out

Sampling clock rate:

$f_s = \frac{1}{T_s} = \frac{f_b}{n}$

P/S = Parallel-to-serial
converter

*For the definition of the raised-cosine Nyquist channel,
see equations (1.5) and (1.6).

**The transmit and receive low-pass filters are frequently
abbreviated as $\sqrt{\alpha}$. The symbol $\sqrt{\alpha}$ indicates the square
root of the raised-cosine Nyquist equation (1.5).

28

to the transmitter. In bandlimited systems, the matched filter has optimal performance; that is, it leads to ISI-free transmission *if it satisfies* the Nyquist transmission criteria. An in-depth discussion of filter partitioning, matched filtering, and the possible equivalence of matched filtered and Nyquist filtered systems is presented in [Feher, 1.4]. A simple $P_e = f$ (S/N) performance derivation of optimal PAM systems is given in [Feher, 1.5] and most modern communications textbooks. The end result is as follows.

$$P_e = 2\left[1 - \frac{1}{M}\right] Q\left(\sqrt{\frac{3}{M^2 - 1} \frac{S}{N}}\right) \tag{1.8}$$

where $M = 2, 4, 8 \ldots$

$$Q(x) = \frac{1}{\sqrt{2\pi}} \int_x^\infty e^{-t^2/2} \, dt \tag{1.9}$$

The numerical values of the function Q can easily be obtained from tables of complementary error functions, denoted by erfc, by the following equations:*

$$\text{erfc}(y) \triangleq \frac{2}{\sqrt{\pi}} \int_y^\infty e^{-z^2} \, dz = 2Q(\sqrt{2}y), \qquad y > 0 \tag{1.10}$$

and

$$\text{erfc}(y) = 1 - \text{erf}(y) \tag{1.11}$$

In equation (1.8), the S/N term represents the average signal-to-noise power ratio in the Nyquist band—that is, in one-half of the symbol rate band. The $P_e = f(S/N)$ performance of M-level PAM systems is illustrated in Fig. 1.20.

Example 1.5

Determine the Nyquist bandwidth and the required S/N for an $f_b = 15$-Mb/s rate transmission system if the available baseband bandwidth is 3 MHz and the specified $P_e = 10^{-8}$. How many transmission levels are required?

Solution: We require a practical spectral efficiency of 15 Mb/s: 3 MHz = 5 b/s/Hz. In order to have the best possible performance, that is, lowest P_e, we have to select a transmission system that satisfies this spectral efficiency requirement and also has the largest decision distance to the nearest threshold level. From Table 1.2 and Fig. 1.20, we note that the 4-level PAM system has a theoretical spectral efficiency of only 4 b/s/Hz and an S/N requirement of 22.5 dB, whereas the 8-level PAM has a theoretical efficiency of 6 b/s/Hz and a requirement of $S/N = 28.5$ dB. Thus we have no choice but to choose an 8-level PAM system, which requires an increased S/N. In 8-level PAM, each symbol level is formed by 3 bits;

---

* Also see Chapter 10 for a numerical listing of the erfc function. In some references, the error function is somewhat differently defined.

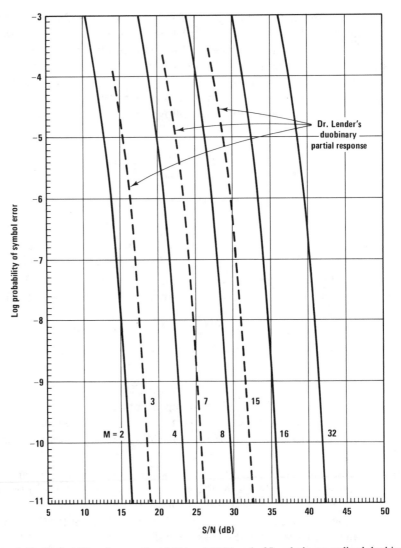

Figure 1.20 Probability of error, P_e, of M-level PAM and of Lender's generalized duobinary (partial response) baseband systems (3, 7, and 15 levels). The rms S/N is measured in the Nyquist bandwidth, after the receive LPF at the regenerator input. Accuracy of these curves is within 1 dB (± 0.5).

thus $f_s = f_b/3$ (because $T_s = 3\,T_b$). Therefore, the corresponding Nyquist bandwidth is

$$f_N = \frac{f_s}{2} = \frac{f_b/3}{2} = \frac{15\ \text{Mb/s} \div 3}{2} = 2.5\ \text{MHz} \tag{1.12}$$

With ideal brick-wall filters ($\alpha = 0$ in equations 1.5 and 1.6), a 2.5-MHz baseband bandwidth would be sufficient. With practical $\alpha = 0.2$ (20% roll-off factor) channel filters, the f_b = 15-Mb/s rate 8-level signal is transmitted in a practical bandwidth of 2.5 × (1 + 0.2) = 3 MHz. This 3-MHz bandwidth includes the out-of-band attenuation of the α = 0.2 roll-off channel. ■

We note that the measured **noise bandwidth** of the receiver (see Fig. 1.19) *is identical to the Nyquist bandwidth* of the receive LPF ($f_s/2$), assuming that the overall Nyquist channel is equally partitioned (matched) between the transmitter and receiver [Feher, 1.4]. In other words,

$$
\boxed{
\begin{array}{l}
\text{Measured} \qquad\qquad\quad \text{computed} \\
\text{PAM receiver} \;\; = \text{Nyquist bandwidth} \; (f_s/2) \\
\text{noise bandwidth} \qquad \textit{for any } \alpha \; (0 \leqslant \alpha \leqslant 1)
\end{array}
}
\qquad (1.13)
$$

Thus in Example 1.5, the receiver noise bandwidth is 2.5 MHz, whereas the total channel bandwidth is 3 MHz. (Why?) From Fig. 1.20, we note that at P_e = 10^{-6}, the M = 4-level PAM system requires an S/N ratio of approximately 8 dB higher than the M = 2-level NRZ system, whereas the M = 8-level system requires an S/N ratio 14 dB higher than the M = 2-level system. Thus increased spectral efficient systems require a considerably higher S/N ratio in order to attain the same P_e performance as binary systems.

In Fig. 1.21, the theoretical spectral efficiency in bits/second/hertz as a function of the available S/N ratio at P_e = 10^{-4} and P_e = 10^{-8} is illustrated. Note that ideal α = 0 roll-off minimum bandwidth Nyquist channel model is assumed. Practical PAM systems have a lower spectral efficiency than illustrated in Fig. 1.21, whereas practical generalized duobinary (PRS) may attain a higher spectral efficiency than shown in the figure. Due to residual ISI, a practical system requires a higher S/N than illustrated in Fig. 1.21. During measurements, particular attention has to be paid to the noise bandwidth of the receiver and also to the probability density function and the crest factor (peak factor) of the noise source. A detailed description of problems related to measurement of the noise crest factor is given in Chapter 7.

1.2.11 P_e Performance in Amplitude and Group Delay Distorted Channels

Most practical transmission systems require a higher S/N than illustrated in the theoretical curves of Fig. 1.20. Nonideal amplitude response and imperfect group delay (GD, also known as envelope delay) of practical channel filters may be a severe cause of performance degradation. Illustrative P_e = $f(S/N)$ measurement results are compared to a theoretical curve in Fig. 1.22. The theoretical curve, (1), of a 1.544-Mb/s 8-level system is the same as the one in Fig. 1.20. In the measurement of curve (2), we use well-equalized channel filters, which closely resemble the α = 0.2 filtered matched raised-cosine channel. In the measurements of curves (3) and (4), we have more amplitude and GD distortion in the channel.

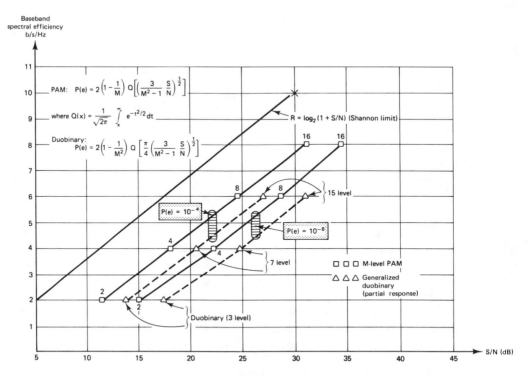

Figure 1.21 Spectrum efficiency (in b/s/Hz) of PAM and generalized duobinary (partial-response) systems as a function of the available S/N at $P_e = 10^{-4}$ and $P_e = 10^{-8}$. The average S/N is specified in the Nyquist bandwidth which equals one-half of the symbol rate. Ideal = 0 roll-off filtering has been assumed. (With permission of Prentice-Hall, Inc. [Feher, 1.5].)

Increased amplitude and GD distortions may lead to an increased performance degradation, particularly for small values of P_e in the frequently critical 10^{-7} to 10^{-10} range.

For badly distorted channels, an **error floor** may be measured (also known as **residual-error rate**). For example, in the measurement of curve (4), Fig. 1.22, we could not obtain a lower P_e than 10^{-5} even with dramatically increased S/N.

The physical reason for the performance degradation is that in the sampling instants, there is a time-variable ISI component in addition to the desired signal component. Deviations from the amplitude or ideal phase of a raised-cosine Nyquist channel lead to ISI. Certain "worst-case" binary signal patterns lead to the worst-case ISI patterns, also known as *peak ISI*. The probability of occurrence of the worst-case patterns is typically small, and for this reason, the error floor is frequently very low—i.e., nonmeasurable. Error floors may also occur due to excessive phase noise, clock jitter, or system nonlinearities.

The **group delay** is defined as the first derivative of the channel phase

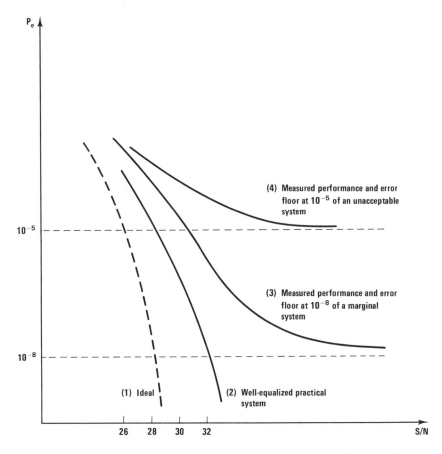

Figure 1.22 Measured $P_e = f(S/N)$ performance of a well-equalized practical system, of a marginal and of an unacceptable M = 8-level PAM system. An "error floor" at 10^{-8} is considered marginal, whereas a 10^{-5} error floor (residual error rate) is unacceptable for most system applications.

characteristics, $\beta(f)$. Thus the group delay, $d(f)$, is given by

$$d(f) = -\frac{1}{2\pi}\frac{d\beta(f - f_c)}{df} \tag{1.14}$$

where f_c is the center frequency of a modulated bandpass system; it equals zero in baseband systems.

   We define the parameter designations for quadratic (parabolic), linear, and sinusoidal group delay distortions in Fig. 1.23. Note that d_{max} represents the maximum *GD at the edge of the band*—that is, at frequencies $f_c \pm f_{max}$. At these limits, the channel filter attenuation is infinity; that is, the raised-cosine pulse spectrum at and beyond these frequencies is zero. For *practical systems,*

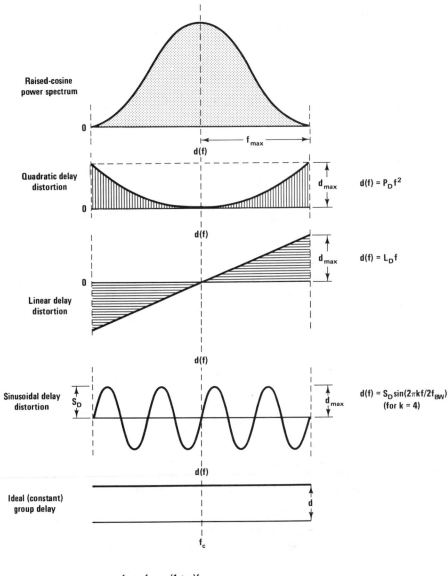

$f_{BW} = f_{max} = (1 + \alpha)f_N$
where f_N = Nyquist frequency
f_c = carrier frequency (in baseband systems $f_c = 0$)

Figure 1.23 Parameter designations for quadratic, linear, and sinusoidal group delay distortion with raised cosine (Nyquist) pulse spectrum at the input of the threshold detectors. The double-sided radio-frequency bandwidth is $2f_{BW} = 2F_{max}$; the baseband bandwidth is $F_{BW} = (1 + \alpha)F_N$, where F_N is the Nyquist frequency (bandwidth) and α is the roll-off parameter of the Nyquist channel.

those frequencies at which the channel filters reach 10- to 15-dB attenuation are considered to be the band limits. To study the impact of group delay distortions on the $P_e = f(S/N)$ curves, we assume that there is *no amplitude distortion* in the channel; that is, the absolute value (amplitude transfer function) of the system satisfies the ideal raised-cosine channel requirements of equations (1.5) and (1.6). With this assumption, an ideal raised-cosine power spectrum is *measured* at the threshold detector inputs. The GD parameters, defined in Fig. 1.23, could be viewed as *residual GDs* due to imperfect channel equalization. The **GD distortions** are defined as follows:

$$d(f) = \begin{cases} L_D \cdot f & \text{for linear GD} \\ P_D \cdot f^2 & \text{for parabolic GD} \\ S_D \cdot \sin(2Kf/2f_{\text{BW}}) & \text{for sinusoidal GD} \end{cases} \qquad (1.15)$$

where $f_{\text{BW}} = f_{\max} = (1 + \alpha)f_N$ and f_N is the Nyquist bandwidth in baseband. This bandwidth definition of $(1 + \alpha)f_N$ (for GD distortion) is more appropriate than f_N, as it includes the critical effect of the GD at the edge of the filter attenuation band. We define the maximum group delay, τ_m, in the filter bandwidth ($2f_{\text{BW}}$) as follows:

$$d_{\max} \triangleq \begin{cases} L_D(2f_{\text{BW}}) & \text{ns for linear GD} \\ P_D(f_{\text{BW}})^2 & \text{ns for parabolic GD} \\ S_D & \text{ns for sinusoidal GD} \end{cases} \qquad (1.16)$$

Illustrative **amplitude distortions** include linear, parabolic, and sinusoidal amplitude distortions. These parameters are defined by:

$$A(f) = \begin{cases} L_A \cdot f & \text{for linear amplitude distortion} \\ P_A \cdot f^2 & \text{for parabolic amplitude distortion} \\ S_A \cdot \sin\left(\dfrac{2\pi Kf}{2f_{\text{BW}}}\right) & \text{for sinusoidal amplitude distortion} \end{cases} \qquad (1.17)$$

The corresponding maximal amplitude distortion is

$$A_m \triangleq \begin{cases} L_A(2f_{\text{BW}}) & \text{for linear amplitude distortion} \\ P_A(f_{\text{BW}})^2 & \text{for parabolic distortion} \\ S_A & \text{for sinusoidal distortion} \end{cases} \qquad (1.18)$$

In the investigation of the impact of amplitude distortions on the $P_e = f(S/N)$ curves, we assume that there is no GD distortion in the channel; that is, $d(f) = $ constant. Amplitude and/or GD distortions degrade the system performance. The performance degradation of coherent quadrature phase-shift keying (QPSK) modulated and QAM systems is the same as that of their corresponding baseband PAM systems, assuming that the channel filters are symmetrical; that is, the band-pass channel model has an equivalent low-pass model [Feher, 1.4]. Computer-generated performance degradation curves of binary NRZ baseband systems and of corresponding coherent QPSK systems, for illustrative GD degradations, are shown in Fig. 1.24. Note that at $P_e = 10^{-9}$, the degradation is more severe than at 10^{-6}.

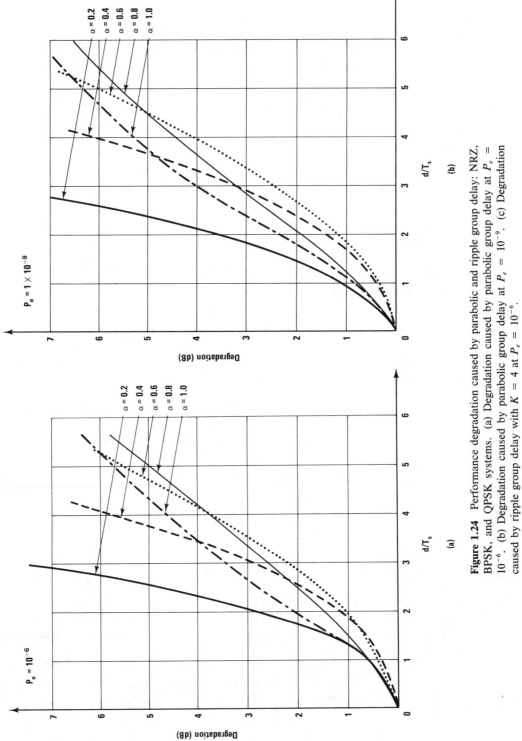

Figure 1.24 Performance degradation caused by parabolic and ripple group delay: NRZ, BPSK, and QPSK systems. (a) Degradation caused by parabolic group delay at $P_e = 10^{-6}$. (b) Degradation caused by parabolic group delay at $P_e = 10^{-9}$. (c) Degradation caused by ripple group delay with $K = 4$ at $P_e = 10^{-6}$.

36

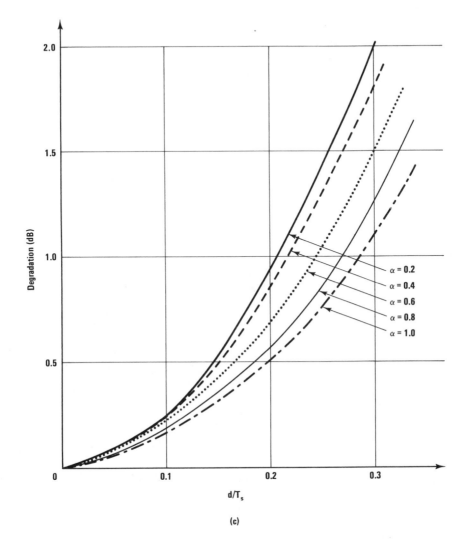

(c)

The performance degradation of higher-state baseband and modulated systems is more severe, particularly if narrow fillers are used. Illustrative results are given in Section 1.3.

1.3 DIGITAL MODULATION-DEMODULATION (MODEM) TECHNIQUES

In this section, we review the principles of the most frequently used digital **modulation-demodulation (modem)** techniques:

QPSK: quadrature (or quaternary) phase-shift keying

QAM: quadrature amplitude modulation

QPRS: quadrature partial-response signaling (modulation)

APK: amplitude phase keying; this technique is practically the same as QAM. The term QAM is more frequently in use in the current literature.

Having mastered the fundamentals of baseband transmission techniques reviewed in Section 1.2.1, you are now well equipped to read the following review of digital modem techniques. A detailed description of modem design techniques and applications can be found in [Feher, 1.5 and 1.4].

1.3.1 QPSK, QAM, and QPRS Principles

A block diagram of a QAM suppressed-carrier modulator is illustrated in Fig. 1.25. The same generic block diagram may be used in the design of QPSK- and QPRS-modulated signals. To simplify our initial discussion, we assume that the premodulation LPFs and also the postmodulation BPF are bypassed; in other words, there are no bandlimiting filters in the modulator.

The f_b-rate binary baseband source is commuted into two binary symbol streams, each having a rate of $f_b/2$. In the design of QAM modulators, the following 2-to-L-level baseband converter converts these $f_b/2$-rate data streams into L-level PAM signals having a symbol rate of

$$f_s = \frac{f_b/2}{(\log L)} \qquad \text{symbols/second} \qquad (1.19)$$

For example, if the source bit rate is $f_b = 6$ Mb/s, then the commuted in-phase and quadrature channel binary baseband streams have a rate of $f_b/2 = 3$ Mb/s. If an $M = 16$-ary QAM modulated signal is desired, these commuted binary streams are converted into $L = 4$-level PAM baseband streams, having a symbol rate of 3 Mb/s: $\log_2 4 = 1.5$ million symbols/s. Four PAM levels of the baseband in-phase (I) and of the baseband quadrature (Q) channels drive the double-sideband-suppressed carrier DSB-SC-AM modulator (mixer) baseband ports. The local oscillator (LO) provides the required quadrature (90° phase shifted) IF or RF carrier signals. As the I- and Q-channels are independent, it is possible to generate any of the $4 \times 4 = 16$ solid dots indicated in Fig. 1.25(b). Let us assume that we terminate the quadrature Q baseband channel. In this case, the 4-level in-phase I-channel PAM baseband signal, after DSB-SC-AM modulation, generates a vector-state space diagram, also known as a **constellation diagram,** as illustrated in Fig. 1.25(c). Thus in this case we have a one-dimensional 4-level 4-DSB-SC-AM system. Now, let us further simplify our modulator. In addition to terminating the Q-channel baseband input, we decide to bypass the 2 to L-level PAM converter in the I-baseband signal path. In this case, we have the simplest digital modulator, namely, a 2-level DSB-SC-AM modulator. The

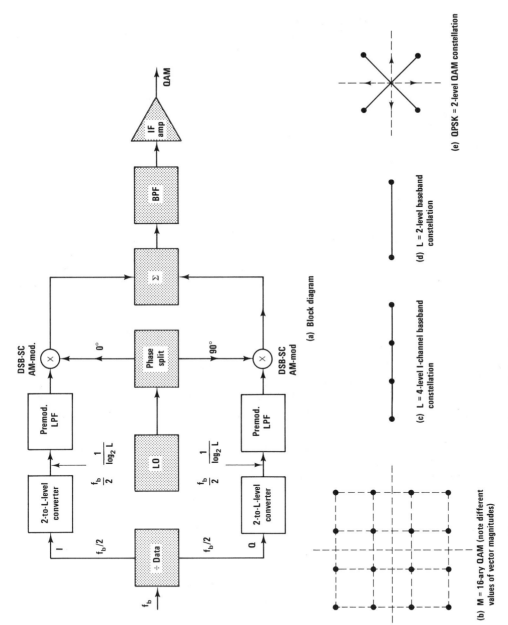

(a) Block diagram

(b) M = 16-ary QAM (note different values of vector magnitudes)

(c) L = 4-level I-channel baseband constellation

(d) L = 2-level baseband constellation

(e) QPSK = 2-level QAM constellation

Figure 1.25 Block and constellation diagram of a QAM modulator. The same generic block diagram may be used in the design of QPSK- and QPRS-modulated signals.

corresponding constellation diagram contains only two dots (see Fig. 1.25(d)), namely, $+A \cos w_c t$ and $-A \cos w_c t$. This can be noticed from the fact that a balanced NRZ baseband signal has only two possible states, $+1$ and -1, and that the DSB-SC-AM modulator is a *product-modulator,* which multiplies (time domain multiplication) the $+1$ or -1 baseband input by the local oscillator signal $A \cos w_c t$. Multiplication in the time domain corresponds to convolution in the frequency domain. Convolution of the baseband data spectrum with the unmodulated carrier wave is equivalent to frequency translation. Thus the NRZ spectrum is translated around the carrier frequency, f_c. Our discussion of the modulator, having a constellation diagram of Fig. 1.25(d), leads to the following statement:

$$\boxed{\text{BPSK} = \text{DSB-SC-AM}} \tag{1.20}$$

namely, a binary phase-shift-keyed modulator is identical (at least in theory) with a double-sideband-suppressed carrier-amplitude modulator *if the baseband drive signal is an unfiltered balanced NRZ data stream.* In other words, multiplication by the $+1$ NRZ state corresponds to a $0°$ shift of the unmodulated carrier, whereas multiplication by the -1 NRZ state corresponds to a $180°$ phase-shifted carrier.

Now, let us connect *both* the I- and Q-channels to the mixer inputs and bypass the 2-to-L-level PAM converters. In this case, the baseband drive signals to both mixers will be binary, balanced NRZ signals. The bit rate of the individual channels is $f_b/2$. The in-phase mixer would lead to a one-dimensional constellation, as illustrated in Fig. 1.25(d), whereas the quadrature mixer would lead to the same type of constellation rotated by $90°$. In all instants of time, both mixers are driven, leading to a resultant two-dimensional constellation, shown in Fig. 1.25(e). We just generated a four-phase state constellation. The phase states may change instantly by integer multiples of $90°$; thus we have a QPSK system. However, a somewhat different view of the same modulator reveals that we have two DSB-SC-AM modulators in quadrature. The resultant modulated signal is a QAM signal. Thus we can state that

$$\boxed{\text{QPSK} = \text{4-QAM}} \tag{1.21}$$

if the baseband drive signal of both of the I- *and* Q-*channel modulators is an unfiltered balanced binary data stream.* Even though the block diagrams of *M*-ary QAM and of *M*-phase PSK systems are similar, the performances of higher-state (more than four-state) QAM systems and PSK systems are not identical. For example, the constellation diagram and the performance of a 16-QAM system are not identical to those of a 16-PSK system [Feher, 1.4].

The block diagram used for the generation of *M*-ary QAM and *M*-phase PSK systems can be used for the generation of *N*-ary **QPRS**. For example, partial response "cosine-shaped" low-pass filters in the *I*- and *Q*-channels of Fig. 1.25 convert 2-level binary baseband signals into 3-level partial-response signals having a controlled amount of ISI. In this case an $N = 3 \times 3 = 9$-ary QPRS-modulated signal is obtained. An advantage of partial response baseband and of QPRS-modulated signals is that they may be transmitted above the Nyquist rate, that is, in a radio-frequency bandwidth that is less than the symbol rate [Lender, 1.10; Wu, 1.14]. This property has been exploited in a number of practical systems. For example, one of the largest operational microwave systems, developed by Bell-Northern Research, employs a QPRS modulation technique. This Canadian coast-to-coast (Atlantic to Pacific) microwave system of approximately 8000 km has a capacity of 90 Mb/s in a 40-MHz radio frequency bandwidth. Thus the spectral efficiency of this radio system is 90 Mb/s : 40 MHz = 2.25 b/s/Hz. This efficiency is 2.25 : 2 = 1.125, or 12.5% higher than the spectral efficiency of theoretical QPSK systems [Lender, 1.10] operated at the Nyquist rate. In general, pulse-shaping bandlimitation in a modem is achieved by low-pass filters and/or band-pass filters (see Fig. 1.25). Note that quadrature double-sideband modulators have the same radio-frequency spectral efficiency as in-phase *or* quadrature baseband systems. Also, the spectral efficiency and the performance of the double-sideband QAM system is identical to that of a single-sideband QAM digital (SSB) systems. This is so because in QAM systems there are two sidebands and also two independent baseband channels. In SSB systems there is only one baseband stream (in the quadrature SSB implementation, the quadrature baseband stream does not contain independent information). The required modulated ratio of carrier power to noise power, C/N, measured in the Nyquist IF or RF bandwidth is equivalent with the baseband S/N requirement. For this reason, the entries in Table 1.2 also apply to modulated systems.

Demodulation is accomplished by multiplication of the received modulated signal by a *coherent* recovered carrier and a 90° shifted (quadrature) carrier. A coherent demodulator block diagram is illustrated in Fig. 1.26. The in-phase and quadrature multipliers, followed by low-pass filters, are essentially amplitude and phase-detectors. At the low-pass filter outputs, the baseband eye diagrams are measured. The carrier recovery circuit generates an unmodulated carrier frequency from the received modulated data signal. As the transmitted spectrum does not contain a discrete tone, nonlinear processing such as quadrupling, remodulation, or Costas loops is required in the implementation of carrier recovery circuits [Feher, 1.4]. Carrier recovery circuits tend to introduce **phase noise.** In general, phase noise may degrade the system performance and is considered in many satellite, line-of-sight microwave, and mobile radio systems a major contributor to system performance degradation. Particularly in relatively low bit-rate systems, operated at rates of less than 64 kb/s, we noticed that phase noise may cause an "error floor" in the measured BER curve, such as illustrated in Fig. 1.22. Phase-noise measurements are described in Chapter 11.

The demodulated baseband signals, corrupted by AWGN, phase noise, ISI, and possible co-channel or adjacent channel interference, are fed to threshold

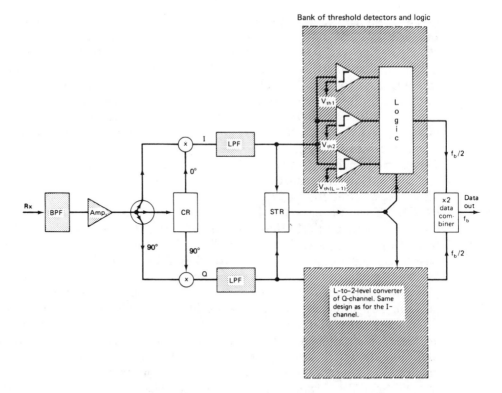

Figure 1.26 *M*-ary QAM or PSK demodulator block diagram. For synchronous demodulation, the carrier recovery (CR) circuit generates an unmodulated carrier, which is coherent (frequency and phase) with the transmitted carrier (Prentice-Hall, Inc. [Feher, 1.5]).

detectors. The binary outputs of the threshold detector outputs are sampled at the symbol rate and fed to logic circuits and to a parallel-to-serial data converter (combiner).

1.3.2 Constellation Diagrams: Measurements and Interpretation

Constellation diagrams or constellation displays provide a useful insight into data-transmission impairments of modulated signals. Constellation displays can be viewed as an extension of eye-diagram measurements, which are described in Section 1.2. A sampling oscilloscope or a conventional oscilloscope provided with an intensity-modulation feature (known also as *Z*-axis input) can be used for constellation measurements. If both *I* and *Q* baseband signals are sampled at the symbol rate and displayed in the horizontal (*X*) and vertical (*Y*) direction on the oscilloscope, then the resulting display represents the constellation of the signal states. For example, 16 states are displayed for an ideal 16-QAM modem in Fig. 1.25(b).

Ideally, a 16-QAM constellation would consist of 16 small bright dots, but because of radio hardware, general system imperfections, noise, and interference, the display actually may consist of 16 small clusters. For this reason such displays are also known as **cluster scope** displays.

A detailed description of the principles of constellation diagrams, measurements, and their applications for system identification is presented in Chapter 7.

1.3.3 $P_e = f(C/N)$ *and* $P_e = f(E_b/N_O)$ *Performance: Ideal Channel*

The probability of error performance, P_e, of *M*-ary QAM, *M*-ary PSK, and *N*-ary QPRS systems is derived in the references [Feher, 1.5; Proakis, 1.7]. The final results are illustrated in the $P_e = f(C/N)$ curves shown in Fig. 1.27. Note that the mean C/N ratio is specified in the **double-sided Nyquist bandwidth,** that is, the symbol rate bandwidth. For a 90-Mb/s 8-PSK radio system, this bandwidth corresponds to 30 MHz, whereas in the case of a 16-QAM system, also having a capacity of 90 Mb/s, this bandwidth corresponds to 22.5 MHz. The measured double-sided Nyquist bandwidth is independent of the α roll-off factor of the raised-cosine filter. In C/N measurements, particular attention should be given to the type of instrumentation used. A *true* power meter (or **rms voltmeter**) must be used to avoid the possibility of making measurement errors. A number of voltmeters, power meters, and spectrum analyzers have built in peak detectors. For sinusoidal waveforms having a 3-dB difference between the peak and rms voltage, these instruments give correct results. However, modulated and bandlimited QPSK, QAM, and QPRS systems may have a peak-to-rms voltage ratio in the range of 3 to 15 dB. Furthermore, the noise *crest factor* of typical sources, described in Chapter 7, is typically more than 15 dB. For such modulated signals and noise sources, conventional peak-detector-equipped instruments could lead to unacceptably high measurement errors.

Relationship Between E_b/N_O and C/N

In many technical papers, textbooks, and customer requirements, the requirement that $P_e = f(E_b/N_O)$ is specified instead of $P_e = f(C/N)$. The supplier has to demonstrate to the customer's satisfaction that a specified $P_e = f(E_b/N_O)$ performance is satisfied, where

$$E_b = C \cdot T_b = \text{average energy of a modulated bit}$$

and

$$N_O = \text{noise power spectral density (noise in 1-Hz bandwidth)}$$

Conventional power meters measure average carrier power, C, and average

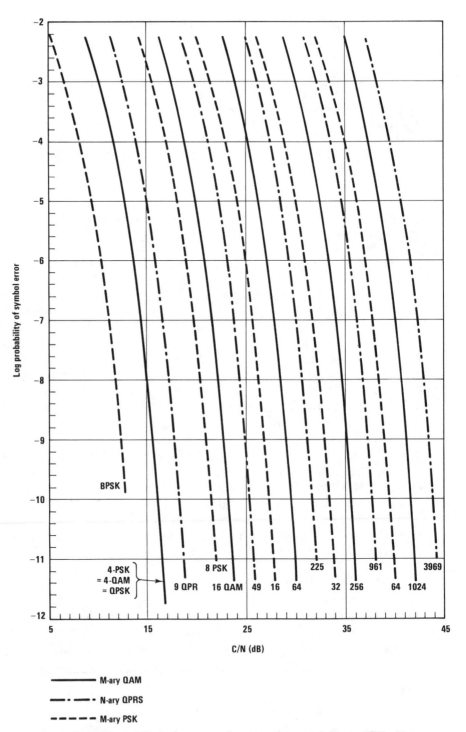

Figure 1.27 Probability of error performance curves of *M*-ary PSK, *M*-ary QAM, and *N*-ary QPR modulation systems vs. the carrier-to-thermal noise ratio. White Gaussian noise in the double-sided Nyquist bandwidth is specified [Kucar and Feher, 1.15].

noise power, N. The following simple relations are useful for the E_b/N_O and C/N transformations:

$$E_b = CT_b = C\left(\frac{1}{f_b}\right) \tag{1.22}$$

$$N_o = \frac{N}{B_w} \tag{1.23}$$

$$\frac{E_b}{N_o} = \frac{CT_b}{N/B_w} = \frac{C/f_b}{N/B_w} = \frac{CB_w}{Nf_b} \tag{1.24}$$

$$\boxed{\frac{E_b}{N_o} = \frac{C}{N} \cdot \frac{B_w}{f_b}} \tag{1.25}$$

The E_b/N_O ratio equals the product of the average C/N ratio and the receiver noise bandwidth-to-bit-rate ratio (B_w/f_b). It should be noted, of course, that any C/N measuring instrument can be recalibrated to read E_b/N_O directly, if required [Feher, 1.4].

> Why is $P_e = f(E_b/N_O)$ more frequently specified instead of the requirement $P_e = f(C/N)$?

The $P_e = f(C/N)$ measurement and C/N specification is *meaningless* unless the noise bandwidth of the receiver is carefully specified. The noise bandwidth of many practical receivers is different from the double-sideband Nyquist bandwidth. Furthermore, postmodulation receive low-pass filters may change the effective receiver-noise bandwidth, particularly if it is measured at IF or RF after a receive BPF. To enable a comparison with theoretical modem performance and also to compare modems manufactured by different suppliers, it has been found that the $P_e = f(E_b/N_O)$ specification leads to more-accurate system measurements. Note that E_b/N_O *is a normalized quantity, independent of the bandwidth* of the receiver. The C/N and E_b/N_O requirements of satellite systems are particularly stringent. For example, in one satellite network a 1-dB increase in the E_b/N_O requirement would lead to an investment of more than $1 million.

In the following simple example adopted from Feher [Feher, 1.5], we illustrate the relationship between E_b/N_O and C/N and the C/N dependence on the receiver-noise bandwidth. An ideal 8-PSK (8-ary) modem is assumed.

Example 1.6

A number of operational microwave systems employ 8-ary PSK modulation techniques. What is the theoretical C/N ratio requirement if a $P(e) \leq 10^{-5}$ is required? (a) Assume that the receive BPF noise bandwidth is 30 MHz. (b) Assume that this bandwidth has been increased to 45 MHz. (c) How much is the C/N requirement if the noise is measured prior to the receive BPF in a bandwidth equaling the bit rate? (d) How much is the E_b/N_O requirement in decibels for the cases in (a) through (c)?

Solution: (a) In this case, the noise bandwidth equals the minimum double-sided Nyquist bandwidth (90 Mb/s ÷ 30 MHz = 3 b/s/Hz, the theoretical limit of 8-ary PSK). The C/N requirement for $P(e) = 10^{-5}$ (reading from Fig. 1.27) equals 18.5 dB.

(b) The E_b/N_O requirement is computed first. From equation (1.25) we obtain:

$$\frac{E_b}{N_O} = \left(\frac{C}{N}\right)_{30\text{ MHz}} + 10 \log \frac{30\text{ MHz}}{90\text{ Mb/s}} = 18.5\text{ dB} - 4.8\text{ dB} = 13.7\text{ dB} \qquad (1.26)$$

That is,

$$\frac{E_b}{N_O} = 13.7\text{ dB} \qquad \text{for } P(e) = 10^{-5} \qquad (1.27)$$

Now, the C/N ratio in a 45-MHz bandwidth is calculated:

$$\left(\frac{C}{N}\right)_{45\text{ MHz}} = \frac{E_b}{N_O}\text{ (dB)} + 10 \log \frac{f_b}{\text{BW}}$$
$$= 13.7\text{ dB} + 3\text{ dB} = 16.7\text{ dB} \qquad (1.28)$$

(c) In this case, BW = f_b; thus,

$$\frac{E_b}{N_O} = \left(\frac{C}{N}\right)_{90\text{ MHz}} = 13.7\text{ dB} \qquad (1.29)$$

(d) The E_b/N_O requirement equals 13.7 dB in all three cases. ■

1.3.4 Spectral Efficiency of PSK, QAM, SSB, and QPRS Modems

The theoretical and practical spectral efficiency of illustrative M-ary, PSK, QAM, SSB, and QPRS systems is summarized in Table 1.3. The following notes are important in the interpretation of the entries in this table.

1.3.5 P_e = f(C/N) Performance in Distorted Channels

Parameter designations for quadratic, linear, and sinusoidal GD distortions are defined in Fig. 1.23. Similar definitions apply to amplitude distortions. Computer-calculated and measured C/N degradations due to amplitude and delay slope on a 45-Mb/s offset QPSK, 50% raised-cosine system are illustrated in Fig. 1.28. Degradations due to amplitude-slope distortion (decibels per hertz) on QPSK, offset-QPSK, 8-PSK, 16-QAM, and offset 16-QAM systems are shown in Fig. 1.29.

For recently developed higher-ary 64-QAM and 256-QAM systems, we present computer-generated data in Figs. 1.30–1.32. The results are presented for 90- and 120-Mb/s rate systems; however, with a simple scaling of the bit rate and GD parameters, these results also apply to other transmission rates.

TABLE 1.3   Spectral Efficiency and C/N Requirement of M-ary PSK, QAM, SSB, and QPRS Systems

Systems (number of states)	Theoretical-RF spectral efficiency [b/s/Hz][1]	Practical-RF spectral efficiency [35 dB atten.][2]	Required C/N[3],[4] for $P_e = \text{BER} = 10^{-8}$
4-QAM; 4-PSK; 2-SSB	2	1.66 [−20%]	15
9-QPRS	2	2.6 [+30%]	17.5
16-QAM; 4-SSB[5]	4	3.33 [−20%]	21.5
16-PSK	4	3.33 [−20%]	26.5
49-QPRS	4	4.4 [+10%]	24
64-QAM; 8-SSB[5]	6	5 [−20%]	28.5
225-QPRS	6	6.25 [+5%]	31
256-QAM; 16-SSB[5]	8	6.66 [−20%]	34
961-QPRS	8	8 [0%]	37
1024-QAM; 32-SSB	10	8.33 [−20%]	40

[1]Ideal-phase equalized brick-wall filters are assumed. Transmission is at the Nyquist rate of 1 symbol/s /Hz for QAM and for QPRS systems.

[2]Assuming $\alpha = 0.2$ raised-cosine filters, attenuation at 20% above Nyquist frequency is 35 to 60 dB with state-of-the-art filter designs. Note that QPRS transmission is practically possible above the Nyquist rate. Terms in brackets indicate practical achievable rate (without complex equalization) in reference to the Nyquist rate in percent.

[3]Defined in the double-sided Nyquist bandwidth (within 1 dB accuracy).

[4]Entries in this table also apply to the corresponding equivalent baseband systems; see Table 1.2.

[5]L-state SSB systems have the same theoretical spectral efficiency and the same C/N requirement as $M = L^2$-state QAM systems.

Figure 1.32 illustrates that the C/N degradation is increasing if a lower P_e performance is required. In operational systems we have been measuring somewhat larger degradations than illustrated in the above computer-simulation results. This is so because in practical modem and data transmission system measurements, other hardware imperfections (in addition to channel group delay and/or amplitude imperfections) are also present.

1.4 EFS, BER, AND AVAILABILITY OBJECTIVES: ISDN AND RADIO SYSTEMS

The *Bit Error Rate* (BER), *Error-Free-Second* (EFS), and availability objectives of digital connections that may form part of an *Integrated Services Digital Network* (ISDN) are highlighted in this section.

Two major committees of the International Telecommunications Union (ITU)—namely, the CCITT and CCIR—have been developing performance rec-

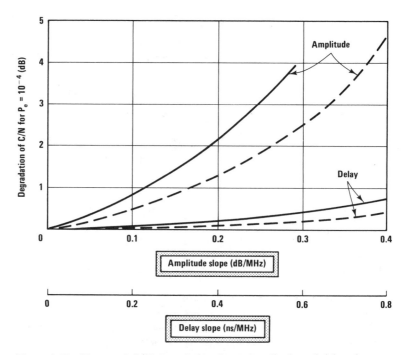

Figure 1.28 Measured C/N degradation due to amplitude and delay slope on a 45-Mb/s offset QPSK, 50% raised-cosine system. Solid line, measured; dashed line, computed [Morais, 1.6].

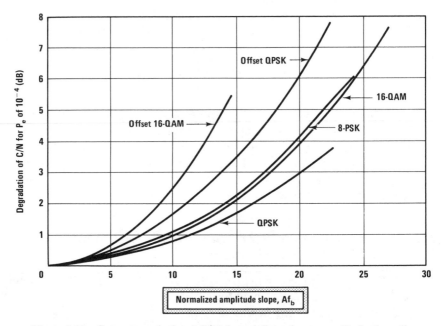

Figure 1.29 Computer-calculated C/N degradations due to amplitude slope distortion; A is amplitude slope (dB/Hz); f_b is bit rate bandwidth (Hz); S is average signal power at receiver filter input; N is average white Gaussian noise (WGN) power in a given bandwidth at receiver filter input [Morais, 1.6].

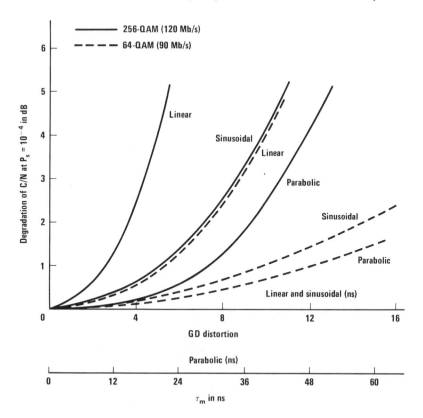

Figure 1.30 Degradation of C/N versus τ_m for linear, parabolic, and sinusoidal GD distortions for 256-QAM and 64-QAM with symbol rate of 15 Mbaud and $\alpha = 0.4$ raised-cosine filters, where

$$\tau_m = L_D \cdot (2f_{\mathrm{BW}_2})\ \text{ns (linear)}$$
$$= P_D \cdot (f_{\mathrm{BW}})\quad \text{ns (parabolic)}$$
$$= S_D \qquad\qquad \text{ns (sinusoidal)}$$
$$f_{\mathrm{BW}} = (1 + \alpha)f_N = 1.4 \cdot f_N = 10.5\ \text{MHz}$$

Computer simulation results performed at University of Ottawa [Wu and Feher, 1.2].

ommendations for digital and analog communication systems. These recommendations have been adopted as mandatory systems specifications by most European and other countries throughout the world. In the United States and Canada, somewhat different performance objectives and system specifications have been developed by AT&T, Bellcore, Bell-Canada, and other organizations. However, with evolution toward ISDN and the rapid growth of intercontinental telecommunications facilities, the International CCITT/CCIR recommendations are also used more and more frequently in North American domestic system specifications.

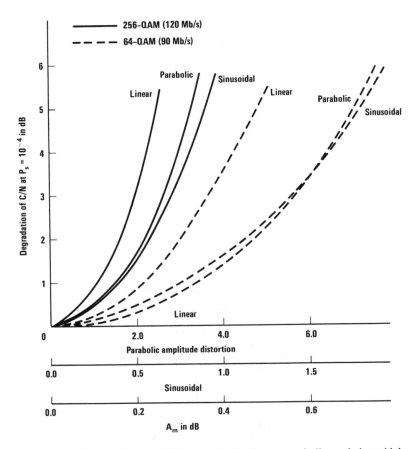

Figure 1.31 Degradation of C/N versus A_m for linear, parabolic, and sinusoidal *amplitude* distortions for 256-QAM and 64-QAM, with symbol rate of 15 Mbaud and $\alpha = 0.4$ raised-cosine filters, where

$$A_m = L_A(2f_{BW_2}) \text{ dB (linear)}$$
$$= P_A(f_{BW}) \text{ db (parabolic)}$$
$$= S_A \text{ dB (sinusoidal)}$$

Computer simulation results performed at University of Ottawa [Wu and Feher, 1.2].

System specifications have a direct impact on the hardware and system design philosophy and on the measurement techniques and maintenance of digital communications facilities.

1.4.1 ISDN Performance Objectives

The BER performance studies of baseband and modulated systems described in previous sections apply to individual regenerative sections, that is, sections between a digital transmitter and a demodulator (regenerator in the case of

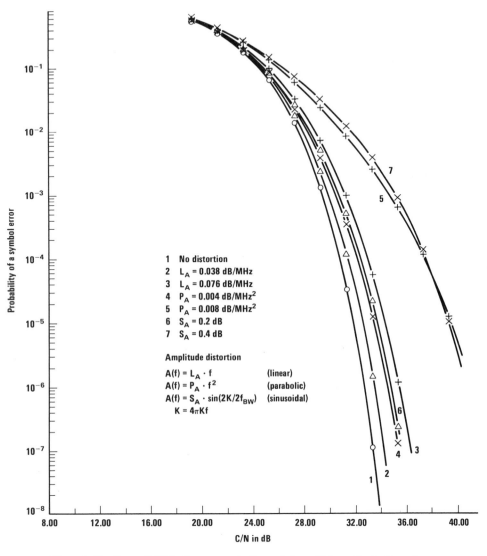

Figure 1.32 P_s vs. C/N for linear, parabolic, and sinusoidal amplitude distortions for 256-QAM with a bit rate of 120 Mb/s, that is, 15 Mbaud and $\alpha = 0.4$. (Noise is defined in the double-sided Nyquist bandwidth; that is, the equivalent noise bandwidth is 15 MHz.) [Wu and Feher, 1.2].

baseband systems). In a complex long-distance national or international data connection, such as an ISDN, many regenerative transmit/receive sections, multiplex, and switching subsystems may be required. For this reason we highlight the overall ISDN performance objectives (EFS and BER) for a 27,500-km hypothetical reference connection of a *64-kb/s data stream* and also the objectives of a 2500-km radio section, which may form part of the connection. A more detailed analysis and description of the measurement principles of BER and EFS is presented in Chapter 5.

Error performance on an international digital connection forming part of an ISDN has been defined by the CCITT [CCITT, 1.21]. The rationale for the development of error performance objectives may be summarized as follows:

1. Services in the future may expect to be based on the concept of an ISDN.
2. Errors are a major source of degradation in that they affect voice services in terms of distortion of voice and data-type services in terms of lost or inaccurate information or reduced throughput.
3. Although voice services are likely to predominate, the ISDN is required to transport a wide range of service types and it is therefore desirable to have a unified specification.

Error performance objectives in international ISDN connections are summarized in Table 1.4. The following definitions and explanations form an integral part of the performance objectives.

TABLE 1.4   CCITT Error Performance Objectives for an International 64-kb/s ISDN Connection*

Part	Performance measurement	Objective
(a)	BER $< 10^{-6}$ for $T_O = 1$ min	$> 90\%$ of 1-min intervals to have 38 or fewer errors
(b)	BER $< 10^{-3}$ for $T_O = 1$ s	$> 99.8\%$ of 1-s intervals to have less than 64 errors
(c)	BER $= 0$ for $T_O = 1$ s (EFS)	$> 92\%$ error-free seconds

TOTAL MEASUREMENT TIME, $T_m = 1$ month

Notes:

(1) The time intervals mentioned should be derived from a fixed time pattern not specifically related to the occurrence of errors.

(2) Total time T_L has not been specified since the period may depend upon the application. A period of the order of any one month is suggested as a reference.

(3) The alternative use of the *error-free decisecond* (EFdS) is currently under study. Should this take place, the objective for classification (c) would need to be based on the EFdS equivalent value of the EFS figure together with an assessment of the customer's perception of circuit quality.

(4) In a practical measurement, a small number of 10-min integration periods may include periods when the connection is judged to be unavailable and/or contain seconds where the number of errors exceeds 63.

(5) Further study needs to be carried out to determine whether the 10-min average period is acceptable (possible network implications). An alternative value of 5 min is also under consideration.

*27,500 km hypothetical reference connection. Based on Table 1/G.821 in the CCITT *Draft Revision*.

1. The performance objectives are stated for each direction of a 64-kb/s circuit-switched connection used for voice traffic or as a *bearer channel* for data-type services.

2. The 64-kb/s circuit-switched connection referred to is an all-digital **hypothetical reference connection** (HRX) and is given in Figs. 1.33 and 1.34. It encompasses a total length of 27,500 km.

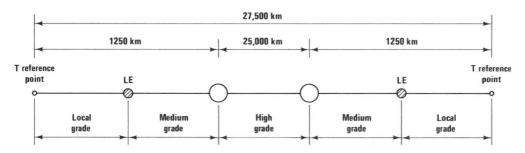

Figure 1.33   Circuit quality demarcation of longest HRX.

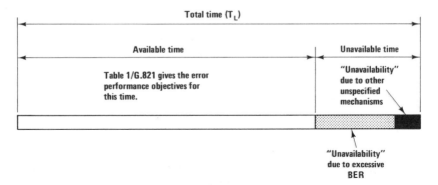

Figure 1.34   Line diagram showing the subdivision of total time T_L. *Note:* The periods of available and unavailable time correspond to *up time* and *down time*, respectively, as defined in Recommendation G.106.

3. The performance objective is stated in terms of an error performance parameter, which is defined as follows:

The percentage of averaging periods T_O during which the number of errors exceeds a threshold value. The percentage is assessed over a much longer time interval T_L (see Note (2) to Table (1.4)).

It should be noted that total time (T_L) is split into two parts, namely, time for which the connection is deemed to be available and that time when it is unavailable; see Fig. 1.34.

Requirements relating to the permissible percentage of *unavailable* time will be the subject of a separate recommendation.

4. The following objectives are specified:
   (a) a BER of less than $1 \cdot 10^{-6}$ for $T_O = 1$ min;
   (b) a BER of less than $1 \cdot 10^{-3}$ for $T_O = 1$ s;
   (c) zero errors for $T_O = 1$ s (equivalent to the concept of error free seconds EFS).

These categories equate to those of Table 1.4. In assessing these objectives, periods of unavailability are excluded.

5. The performance objectives aim to serve two functions:
   (a) to give the user of future national and international digital networks an indication of the expected error performance under real operating conditions, thus facilitating service planning and terminal equipment design;
   (b) to form the basis upon which performance standards are derived for transmission equipment and systems in an ISDN connection.

6. The performance objectives represent a compromise between a desire to meet service needs and a need to realize transmission systems, taking into account economic and technical constraints. The performance objectives, although expressed to suit the needs of different services, are intended to represent a single level of transmission quality.

7. Since the performance objectives intend to satisfy the needs of the future digital network, it must be recognized that such objectives cannot be readily achieved by all of today's digital equipment and systems. The intent, however, is to establish equipment design objectives that are compatible with the objectives in this recommendation. These aspects are currently the subject of discussion within CCITT and CCIR [Kirby, 1.33].

It is further urged that all technologies, wherever they appear in the network, should preferably be designed to better standards than those indicated here in order to minimize the possibility of exceeding the end-to-end objectives on significant numbers of real connections.

8. The objectives relate to a very long connection; recognizing that a large proportion of real international connections will be shorter, it is expected that a significant proportion of real connections will offer a better performance than the limiting value given in Item 2. On the other hand, a small percentage of the connections will be longer and in this case may exceed the allowances outlined in this recommendation [CCITT, 1.21].

Since the objectives defined in Table 1.4 relate to the overall ISDN 27,500-km connection, it is necessary to subdivide this to constituent parts and even to individual equipment and systems. In the draft revision of the **apportionment principles,** the apportionment strategy for the 1-min and error-free second requirements and for services utilizing higher digit rates is presented.

1.4.2 Radio Systems: Performance Objectives

In this section we limit our presentation to only one apportionment, namely, the allowable BER at the output of a 2500-km hypothetical reference digital path for **radio systems** [CCIR, 1.18].

A **2500-km hypothetical reference digital path** (HRDP) may form part of a longer international digital connection. CCITT Recommendation G.821, which defines performance requirements for a hypothetical reference connection forming part of an ISDN over 27,500 km, needs to be taken into account in determining the performance requirements for digital radio systems.

Low Error-Ratio Value

The CCITT has proposed an error-ratio design objective of 1 in 10^{10} per kilometer for the transmission system in a 25,000-km hypothetical reference digital connection. For a 2500-km digital radio-relay path, this gives an error-ratio objective of 2.5×10^{-7} (this figure excludes the contributions due to the multiplexing equipment). In digital radio-relay systems, this error ratio will, according to present design criteria, be achieved at least 99% of the time. Therefore, a low value of error ratio of about 10^{-7} would seem appropriate for a 2500-km HRDP.

High Error-Ratio Value

The proportion of time for which the high error ratio may be exceeded has a very great influence on the design of a system. A requirement that the higher error ratio may not be exceeded for more than 0.01% of any month for a 2500-km circuit would have severe repercussions on the economics of a practical system. On the other hand, a requirement that this error ratio should not be exceeded for more than 0.1% of any month would result in a more economic system but one that may not compare favorably with existing FDM-FM radio-relay systems.

Long-Term EFS Objective

The BER parameter does not give an accurate estimate of the data-circuit performance except when the error distribution is known. The more desirable alternative, block error-ratio criterion, unfortunately requires a knowledge of data block size and data rate, which vary from user to user. The EFS-based objectives are one way in which these limitations can be resolved. The CCITT has proposed that for data transmission, a 95% EFS objective should be attained on the local-exchange-to-local-exchange portion of a 25,000-km hypothetical reference connection. A value of 99.5% EFS for the 2500-km hypothetical reference digital path would seem appropriate.

In general there is no simple relationship between bit error-ratio performance and the long-term EFS performance during periods of multipath fading. In particular, the long-term EFS performance for radio systems during multipath fading is a function of the rate of fading; this contrasts with the bit error probability performance, which may be independent of the rate of fading. Computer-simulation studies have indicated that there is only a slight dependence of EFS performance on the transmitted bit rate during periods of multipath fading.

Short-Term EFS Objective

In addition to the long-term EFS objective just discussed, it may be worthwhile to consider the need for a short-term EFS objective. Its purpose would be to ensure adequate performance over an acceptable percentage of typical data connections.

Performance Objectives for Real Digital Paths

Recommendation 594 gives performance objectives for the HRDP [CCIR, 1.18]. Real paths will differ both in length and composition from the HRDP, and it is therefore desirable to provide planning objectives for the allowable bit error ratios of such paths, particularly those that are shorter in length than the HRDP. At present there is no agreement on the method to be used to derive the performance objectives.

In general for real paths there are two problems:

1. To find a method of apportioning the objectives given in Recommendation 594 for systems that are intended to be connected in tandem in order to form paths similar to the HRDP.

2. To take into account CCITT recommendations in determining objectives for short- and medium-haul systems utilized only in the national part of the hypothetical reference connection.

In addressing the first of these problems (Item (1)), there are a number of assumptions upon which there is general agreement.

Since the performance of digital radio-relay systems is dependent upon fading, it is generally agreed that the behavior of any section of the HRDP will be statistically independent. If this is assumed, then mathematically the HRDP performance could be determined by the convolution of the probability density functions of all sections. This process, however, is not practical, since the probability density function is not known in sufficient detail.

Guidelines to be used for the apportionment of the performance criteria of Recommendation 594 over a section must, therefore, first be clearly established.

The recommended allowable bit error rate for the 2500-km CCIR hypothetical reference digital path (HRDP) stated in CCIR Recommendation 594 is that the bit error ratio should not exceed

$$
\begin{array}{ll}
1 \times 10^{-7} & \text{for more than 1\% of any month;} \\
 & \text{integration time under study} \\
\text{and} & \\
1 \times 10^{-3} & \text{for more than 0.05\% of any month;} \\
 & \text{integration time 1 s}
\end{array}
\tag{1.30}
$$

Report 930 of the CCIR discusses the problem of deriving the performance objectives for so-called real digital paths, by which, in general, is meant paths that are shorter than the 2500-km HDRP. Godier of Bell Northern Research, Canada [Godier, 1.1], has proposed that for paths of length L km, the bit error ratio should not exceed

$$
\begin{array}{ll}
\dfrac{L}{2500} \times 10^{-7} & \text{for more than } \dfrac{L}{2500} \% \text{ of any month} \\
\text{and} & \\
1 \times 10^{-3} & \text{for more than } \dfrac{L}{1500} \times 0.05\% \text{ of any month}
\end{array}
\tag{1.31}
$$

TABLE 1.5 BER and Errored Seconds in the Redrafted CCIR–G.821 Recommendation (May 1984) for a 2500-km (HRDP) [Based on Kirby, 1.33]

BER not to exceed	For more than $X\%$ of any month	Integration time
1×10^{-6}	0.4%	1 min
1×10^{-3}	0.54%	1 s

Errored Seconds Total should not exceed 0.32% of any month.*

*We interpret this as:

No. of errored seconds not to exceed	Per "average" period of
8294	1 month (30 da)
276	1 da
11	1 h

In his paper, Godier demonstrates that this proposal is a reasonable solution to the problem of specifying the performance objectives for paths shorter than the HRDP. Sinnreich of MCI [Sinnreich, 1.11] provides a performance guideline for 1.544-Mb/s transmission over supergroup band FDM systems.

CCIR Recommendation 594, covering digital microwave systems, was re-drafted in May 1984, consistent with revised Recommendation G.821 of the CCITT. It now covers all integrated digital services [Kirby, 1.33]. This revised recommendation is highlighted in Table 1.5.

TABLE 1.6 BER and Errored Seconds of a Satellite Service CCIR Hypothetical Digital Reference Path [Kirby, 1.33]

Extract from Draft Report AA/4, May 1984, "Characteristics of a Fixed Satellite Service Hypothetical Digital Reference Path Forming Part of an Integrated Services Digital Network"

. . . Annex III . . . Possible text of a new Recommendation on Error Performance Objectives . . .[1]

"Either

A:

1. that the performance objectives for the Hypothetical Digital Reference Path (HDRP) should be determined according to the methods outlined in Draft Report AA/4, and should be based on 20% allocation of the ISDN HRX   objectives a) and c) and 15% of objective b) as given in Table I of CCITT Recommendation G.821 . . . These allocations can be stated as:

    (a) at least 98% of the ten-minute intervals in any month to have 38 or fewer errors;

    (b) at least 99.97% of the one-second intervals in any month to have less than 64 errors . . .

    (c) at least 98.4% of the one-second intervals in any month to have zero errors.

Or:

B:

1. that the bit error probability . . . at the output of a satellite Hypothetical Digital Reference Path (HRDP) . . . forming part of a 64 kb/s ISDN connection should not exceed the provisional values given below:

1.1(a) 1×10^{-7} for more than 5% of any month . . .

   (b) 2×10^{-7} for more than 10% of any month, and

       1×10^{-6} for more than 0.5% of any month.

1.2 1×10^{-3} for more than Y% of any month . . .

     Y to be specified."

[1]The performance objectives given above are strictly provisional. Further review is intended in 1985 when account can also be taken of further ISDN work in CCITT. The text is for information and it is not suggested that system design be based on these provisional values at this time.

The draft revision of the error performance objectives for the satellite HRDP for the ISDN is complex; it remains under review until the 1986 Plenary Assembly and is quite dependent on expected CCITT decisions on error performance of overall digital networks (CCITT, 1.21). CCIR Draft Report AA/4 (May 1984) gives the status of this work and the text of a possible draft recommendation, shown in Table 1.6.

1.5 ON-LINE BER PERFORMANCE MEASUREMENTS

Measurements, whether performed in a research laboratory, on the manufacturing floor of a factory, or on an installed digital radio system, are one of the most important tasks of the digital transmission engineer. Measurements may be performed on simple building blocks such as prototype breadboards, but they may also be of extreme importance in more sophisticated systems, as in the case of the continuous performance monitoring of complex long-haul traffic-carrying systems.

Evaluation of the BER performance of nonoperational (out-of-service or off-line) channels is described in Chapter 5. In principle, performance evaluation of such systems is done with a pseudorandom test signal sequence, which is transmitted through the measured channel. The receiver computes the BER by comparing the received bits with a stored replica of the transmitted bit pattern. One of the main problems associated with simple out-of-service BER measurements is that it may not be feasible to evaluate the performance of an operating in-service system carrying the unknown digital data stream of the customer. *The revenue-carrying traffic has to be interrupted in order to perform system tests.*

The probability of error, P_e (also denoted by $P_{(e)}$), and BER terms are frequently used in the literature. For most practical applications, these two terms are identical.

Experimentally the BER is measured and defined by:

$$P_e = \text{BER} = \frac{N_e}{N_t} = \frac{N_e}{f_b \cdot t_o} \qquad (1.32)$$

where N_e = number of bit errors in a time interval t_o

N_t = total number of transmitted bits in t_o

f_b = bit rate of the binary source

t_o = measuring time interval, that is, error-counting time

For a random stationary error-generation process and sufficiently long measurement interval t_o, the measured BER gives an estimate of the true P_e.

The minimum acceptable value of N_e is about 10. In this case, the true P_e is contained in a range of $\pm 50\%$ around N_e/N_t and the confidence coefficient is about 90%. If the steepness of typical $P(e) = f(\text{CNR})$ curves is taken into account, then the $\pm 50\%$ estimate may be considered accurate within 0.5 dB in the corresponding measurement of carrier-to-noise ratio (CNR). In order further to improve the measurement accuracy, a number of instruments accumulate an $N_e = 100$ prior to the BER computation.

The measurement duration and the error-rate estimation for a short t_o might also cause serious difficulties. For example, to evaluate a $P(e) = 10^{-9}$ for a 100-kb/s data stream—assuming that for a meaningful statistical estimate, at least 10 bit errors have to be counted—the measurement has to last for $t_o = 10^5$ s (nearly 30 h). This is not practical in time-variable fading radio channels.

In-service or on-line monitoring can be achieved with

1. Test-sequence interleaving
2. Parity-check coding
3. Code-violation detection
4. Pseudoerror detection

Pseudoerror measurement techniques, also known as *shifted threshold* monitors, enable much faster BER monitoring than off-line methods. These techniques, reviewed in this section, are used in numerous on-line measurement and system diagnostics subsystems [Feher, 1.5; Keelty, 1.29].

For pseudoerror, P_{pe} detection, a *secondary* decision device is connected in parallel with the main data path; see Fig. 1.35. This secondary path has intentionally degraded P_e performance. The output sequence of this path has an error rate much greater than the unknown error rate of the main receiver. This amplified error rate is obtained by taking the main receiver output data as a reference and counting the number of disagreements with respect to the secondary output-data stream. Every disagreement is called a **pseudoerror.** A pseudoerror occurs more frequently than a true error in the main data path, and thus the pseudoerror rate, P_{pe}, is higher than the true P_e.

The controlled degradation of the secondary receiver may be obtained by modifying the baseband decision regions (threshold levels) or the eye-diagram sampling instants with respect to the optimal ones. Alternatively, in the secondary path the postdetection low-pass filter might have a wider band than the postdetection filter in the main demodulator. In this secondary pseudoerror path, more noise will appear than in the main path due to the larger bandwidth. This method is autoadaptive. The SNR difference, and with it the P_e amplification factor in the secondary path, stays nearly constant, a constant that is independent of the time-variable SNR caused by fading.

A pseudoerror monitor conceptual diagram, shown in Fig. 1.35, illustrates the theory of operation of pseudoerror measurement systems. Let us assume that the received QPSK modulated signal has a 10-Mb/s rate and that it is corrupted by AWGN. The I-channel and the Q-channel demodulators feed $f_s = 5$-million-baud-rate binary symbols to the corresponding data regenerator outputs. In this example it is assumed that the filtering at the receive end is performed exclusively by means of post-demodulation low-pass filters. For the 10-Mb/s (5-million-baud) data rate, the minimum bandwidth LPF_1 has a 2.5-MHz bandwidth.

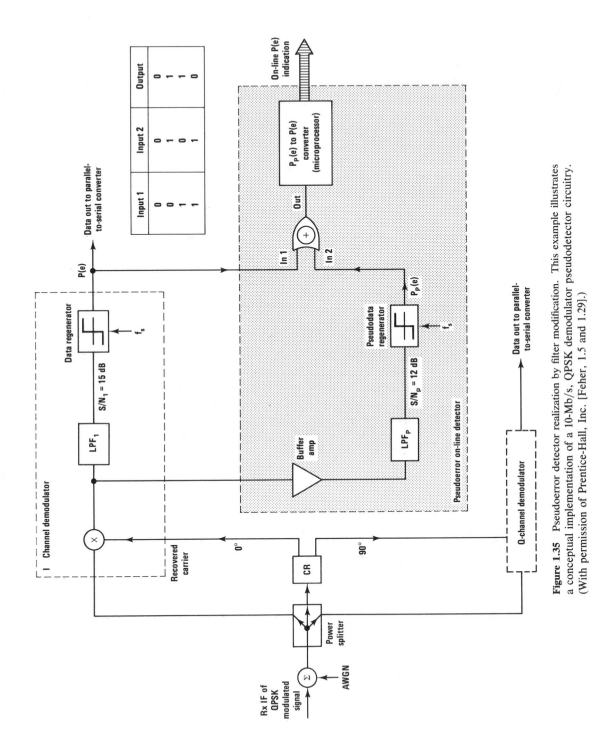

Input 1	Input 2	Output
0	0	0
0	1	1
1	0	1
1	1	0

Figure 1.35 Pseudoerror detector realization by filter modification. This example illustrates a conceptual implementation of a 10-Mb/s, QPSK demodulator pseudodetector circuitry. (With permission of Prentice-Hall, Inc. [Feher, 1.5 and 1.29].)

If the S/N ratio at the data regenerator input is 15 dB, then the error rate of the data path is $P_e = 10^{-8}$; see Fig. 1.20. In the secondary data path, a buffer is followed by a low-pass filter, LPF_p, which is designed to have a noise bandwidth twice as large as the bandwidth of LPF_1. The demodulated signal powers at the outputs of LPF_1 and LPF_p are approximately equal. These signal powers are equal to within 0.6 dB if both filters and the buffer amplifier have the same insertion loss. Assuming that AWGN is the only cause of the error-generating mechanism, then the noise power at the output of the LPF_p filter will be 3 dB higher than at the output of LPF_1. (This is so because the AWGN has a constant-noise spectral density, and the total noise power is directly related to the filter bandwidth). Thus in this case, the $(S/N)_p$ ratio in the pseudoerror path equals 15 dB − 3 dB = 12 dB. This S/N_p ratio corresponds to a probability of error of approximately $P_{pe} = 10^{-5}$.

In an operational system (S/N), $(S/N)_p$, P_e, and P_{pe} are unknown quantities. However, the ingenious idea of the pseudodetection circuitry of using an EXCLUSIVE-OR (EX-OR) gate provides, by means of a simple cancellation trick, a number that is directly proportional to the unknown P_e of the on-line system.

If the P_e and the P_{pe} are both zero ($10^{-\infty}$), then the same data are present at both inputs of the EX-OR gate. Thus in this case the output of the sampled EX-OR gate is continually 0. If the errors in the data path do not occur simultaneously with the errors in the pseudoregenerator path, then at the output of the EX-OR gate, a 1 state appears whenever there is an error in only one of the two paths. In our example the pseudoerror path, having a $P_{pe} = 10^{-5}$, has an error rate of 10^3 larger than the $P_e = 10^{-8}$ of the main data path. Thus the output of the EX-OR gate provides logic 1 states at a frequency that is directly proportional to $P_{pe} - P(e) \approx P_{pe}$. (In our example, $10^{-5} - 10^{-8} \approx 10^{-5}$). The P_{pe}-to-P_e converter is an event counter and a divider that provides the $P(e)$ indication. It can be implemented with hard-wired logic circuits or by microprocessors.

As stated, Fig. 1.35 represents a conceptual pseudoerror monitoring subsystem diagram. Most practical designs have been employing one of the following implementation techniques:

1. Shifted threshold detection
2. Noise addition
3. ISI enhancement
4. Sampling point offset

All these techniques introduce a controlled amount of degradation (increase in the P_{pe}), in the parallel pseudomonitor path. A more detailed description of pseudoerror measurements, hardware implementations, and field measurement results is presented in Keelty [Keelty, 1.29].

REFERENCES

[1.1] Godier, I. "Application of the CCIR Bit Error Ratio Recommendation to Real Digital Radio Path," *IEEE Transactions on Communications,* September, 1984.

[1.2] Wu, K. T., and K. Feher. "256-QAM Modem Performance in Distorted Channels," *IEEE Transactions on Communications,* May, 1985.

[1.3] Newcombe, E. A., and S. Pasupathy. "Error Rate Monitoring for Digital Communications," *Proceedings of the IEEE,* Vol. 70, No. 8, August, 1982.

[1.4] Feher, K. *Digital Communications: Satellite/Earth Station Engineering,* Prentice-Hall, Englewood Cliffs, N.J., 1983.

[1.5] Feher, K. *Digital Communications: Microwave Applications,* Prentice-Hall, Englewood Cliffs, N.J., 1981.

[1.6] Morais, D. H. "Digital Modulation Techniques for Terrestrial Point-to-Point Microwave Systems," Ph.D. Thesis, Department of Electrical Engineering, University of Ottawa, Ottawa, Canada, 1981.

[1.7] Proakis, J. G. *Digital Communications,* McGraw-Hill, New York, 1983.

[1.8] Ritchie, G. R., and P. E. Scheffler. "Projecting the Error Performance of the Bell System Digital Network," IEEE Comsoc Sponsored Conference, 1982.

[1.9] Decina, M., and A. Roveri. "ISDN: Integrated Services Digital Networks: Architecture and Protocol Aspects," Chapter 2 in K. Feher, et al., *Advanced Digital Communications: Systems and Signal Processing Techniques,* Prentice-Hall, Englewood Cliffs, N.J., 1986.

[1.10] Lender, A. "Correlative (Partial Response) Techniques and Applications to Digital Radio Systems," Chapter 7 in K. Feher, *Digital Communications: Microwave Applications,* Prentice-Hall, Englewood Cliffs, N.J., 1981.

[1.11] Sinnreich, H. "Performance Considerations for a T1/SG—1.544 Mb/s Data Modem,"—MCI Technical Memorandum, Washington, D.C., 1984.

[1.12] Sklar, L. B. "Efficiency Factors in Data Communications," *IEEE Communications Magazine,* June, 1984.

[1.13] Prabhu, V. K. "Interference Analysis and Performance of Linear Digital Communication Systems," Chapter 9 in K. Feher, et al., *Advanced Digital Communications and Signal Processing,* Prentice-Hall, Englewood Cliffs, N.J., 1986.

[1.14] Wu, K. T., and K. Feher. "Multilevel PRS/QPRS Above the Nyquist Rate," *IEEE Transactions on Communications,* July, 1985.

[1.15] Kucar, A., and K. Feher. "Performance of Multi-Level Modulation Systems in the Presence of Phase Noise," *Proceedings of the IEEE International Conference on Communications,* ICC-1985. Chicago, June, 1985.

[1.16] Feher, K. *Digital Modulation Techniques in an Interference Environment,* Encyclopedia on EMC, Vol. 9, Don White Consultants, Inc., Gainesville, Va., 1977.

[1.17] CCIR Rec. 556. "Hypothetical Reference Path for Radio-Relay Systems for Telephony: Systems with Capacity Above the Second Hierarchical Level," CCIR Recommendation 556, 1978.

[1.18] CCIR Rec. 594. "Allowable Bit Error Rate Ratios at the Output of the Hypothetical Reference Digital Path For Radio-Relay Systems for Telephony," CCIR Recommendation 594, 1982.

[1.19] Sciulli, A. J., and J. R. Peeler. "Error Analysis on Digital Communications Links," *Telecommunications,* March, 1985.

[1.20] Bell Systems. "Transmission Parameters Affecting Voice-band Transmission Measuring Techniques." *Bell System Technical Reference,* Pub. 41009, May, 1975.

[1.21] CCITT Rec. G.821. "Error Performance of an International Digital Connection Forming Part of an Integrated Services Digital Network." *International Telegraph and Telephone Consultative Committee, Informational Telecommunications Union, Yellow Book,* Fascicle III.3, Geneva, 1981 and Draft Revision of Recommendation G.821, 1983.

[1.22] Brilliant, M. B. "Observations of Errors and Error Rates on T1 Digital Repeater Lines." *Bell System Technical Journal,* Vol. 57, No. 3, March, 1978, pp. 711–746.

[1.23] Rollins, W. W. "Confidence Level in Bit Error Rate Measurement," *Telecommunications,* December, 1977.

[1.24] Johannes, V. I. "Improving on Bit Error Rate," *IEEE Communications Magazine,* Vol. 22, No. 12, pp. 18–20, 1984.

[1.25] Huckett, P. "Performance Evaluation in an ISDN—Digital Transmission Impairments," *The Radio and Electronic Engineer,* Vol. 54, No. 2, February, 1984.

[1.26] McLintock, R. W., and B. N. Kearsey. "Error Performance Objectives for Digital Networks," *The Radio and Electronic Engineer,* Vol. 54, No. 2, February, 1984.

[1.27] Bell Systems. "High Capacity Digital Service Channel Interface Specification," *Bell System Technical Reference,* Pub. 62411, September, 1983.

[1.28] CCIR Report 930. "Performance Objectives for Digital Radio-Relay Systems," *CCIR Recommendation* 930, 1982.

[1.29] Keelty, J. M., and K. Feher. "On-Line Pseudo-Error Monitors for Digital Transmission Systems," *IEEE Transactions on Communications,* August, 1978.

[1.30] Oppenheim, A. V., and R. V. Schafer. *Digital Signal Processing,* Prentice-Hall, Englewood Cliffs, N.J., 1975.

[1.31] Bellamy, J. C. *Digital Telephony,* John Wiley, New York, 1982.

[1.32] Feher, K., et al. *Advanced Digital Communications: Systems and Signal Processing Techniques,* Prentice-Hall, Englewood Cliffs, N.J., 1987.

[1.33] Kirby, C. R. "International Standards in Radio Communications," *IEEE Communications Magazine,* January, 1985.

[1.34] Bell Laboratories, Members of the Technical Staff. "Transmission Systems for Communications," Revised Fourth Edition, *Bell Telephone Laboratories,* 1971.

[1.35] Adoul, J. P. "Speech Coding Algorithms and Vector Quantization," Chapter 3 in K. Feher, et al., *Advanced Digital Communications and Signal Processing,* Prentice-Hall, Englewood Cliffs, N.J., 1986.

[1.36] Qureshi, S. U. H. "Adaptive Equalization," Chapter 12 in K. Feher, et al., *Advanced Digital Communications and Signal Processing,* Prentice-Hall, Englewood Cliffs, N.J., 1986.

2

DIGITAL SIGNAL PROCESSING (DSP) TECHNIQUES

DR. KAMILO FEHER

University of California, Davis
Davis, California 95616
and
Consulting Group, DIGCOM, Inc.

2.1 INTRODUCTION

In this chapter analog-to-digital (A/D) and digital-to-analog (D/A) conversion techniques of audio, television, and frequency division multiplexed (FDM) signals are reviewed. Due to the relatively long equipment and propagation delay of long-distance systems and particularly of satellite systems, the effect of voice echo is annoying to the user. For this reason, recently developed echo suppression and cancellation subsystems are also described.

Signal processing and multiplexing form a large and complex subject. Books have been written that present the theoretical fundamentals and applications of signal processing [Oppenheim, 2.4; Bellamy, 2.5; Feher, 2.6]. In this section we limit our presentation to a brief review of the principles of operation, followed by applications of signal-processing and multiplexing subsystems used in telecommunications systems. This review should facilitate the comprehension of the material presented in Chapters 3 and 4.

The most frequently employed A/D conversion methods used in communication systems are:

PCM: pulse code modulation
DPCM: differential PCM
DM: delta modulation (also abbreviated as DMOD)

The A/D converter, located in the transmitter, is also known as the **encoder,** or simply **coder.** The D/A converter, located in the receiver, is known as the **decoder.** The word *codec* is derived from coder/decoder.

In addition to the basic conversion methods just listed, more-involved codes have been developed. Frequently used acronyms include the following:

APCM:	adaptive PCM
ADPCM:	adaptive DPCM
LDM:	linear (nonadaptive) DM
ADM:	adaptive DM
CDM:	continuous DM
DCDM:	digitally controlled DM

Crucial to the concept of A/D and D/A conversion is the representation of bandlimited signals by sampling. For this reason, before the description of particular voice and video codecs, we present a simple proof of the sampling theorem, followed by a description of measurable imperfections that may be introduced by practical sampling subsystems.

2.2 SAMPLING

The first step in the A/D conversion of an analog waveform is sampling. For many applications it is desirable to have the lowest possible sampling rate, because higher rates lead to higher bandwidth requirements and overall increase in the cost of the system. H. Nyquist derived the minimum sampling rate $(f_{s\min})$ required in order that the samples would contain all of the information of the original signal.

The result of Nyquist's sampling theorem states that the sampling frequency (f_s) has to be larger than the double of the input signal bandwidth, that is,

$$f_s > 2f_{BW} \qquad (2.1)$$

where f_s = sampling frequency
f_{BW} = bandwidth of the input signal prior to A/D conversion

The sampling process may be illustrated as a pulse amplitude modulation (PAM) process, where $m(t)$, the input analog signal, is modulated by a train of impulses $s(t)$ (see the time-domain illustration of Fig. 2.1). The corresponding frequency domain illustration is shown in Fig. 2.2. Note that the input-signal double-sideband amplitude modulates the impulse train. The message signal $m(t)$ is recovered by a low-pass filter (LPF), which removes all higher-order harmonic components.

A simple proof of the sampling theorem, based on the derivation given in [Bell, 2.7] follows:

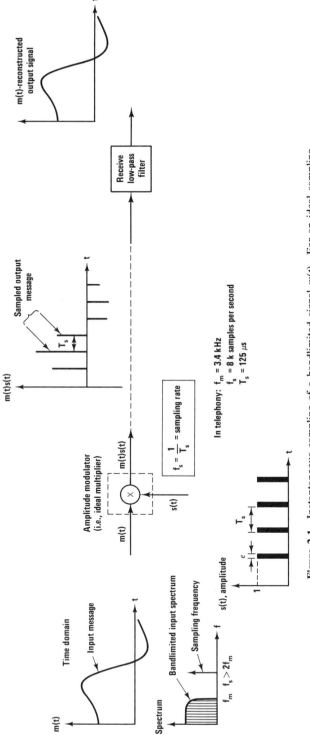

Figure 2.1 Instantaneous sampling of a bandlimited signal $m(t)$. For an ideal sampling process, the width of the pulse train approaches zero, that is $\varepsilon \to 0$.

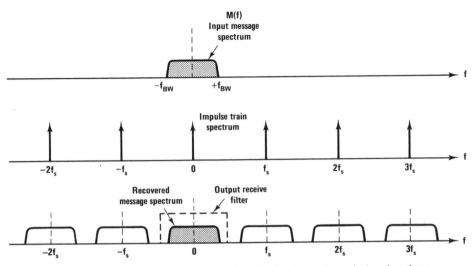

Figure 2.2 Spectrum of a sampled signal; to simplify the mathematical explanations, positive as well as negative frequencies are illustrated.

A signal, $m(t)$, is said to be *bandlimited* if its Fourier transform vanishes outside a finite interval. Therefore, $m(t)$ can be represented as

$$m(t) = \int_{-1/2T}^{1/2T} M(f) e^{j2\pi ft} df \qquad (2.2)$$

where $M(f) = 0$ for $|f| > 1/2T$, and $1/2T$ is the bandwidth in Hertz.

Since the bandlimited input signal $M(f)$ need be defined only between $-1/2T$ and $1/2T$, we use a Fourier series representation over that interval:

$$M(f) = \sum_{n=-\infty}^{\infty} C_n e^{j2n\pi fT} \qquad |f| < \frac{1}{2T} \qquad (2.3)$$

having the following coefficients:

$$C_n = T \int_{-1/2T}^{1/2T} M(f) e^{-j2n\pi fT} df \qquad (2.4)$$

Comparing these equations with (2.2), we note that C_n represents the values of $m(t)$ at discrete points (except proportionality constants), that is,

$$C_n = Tm(-nT) \qquad (2.5)$$

Thus the C_n, the samples of $m(t)$ taken at twice the highest frequency in the band, determine $M(f)$ and hence $m(t)$. Substituting equation (2.4) into equation (2.3) and then into equation (2.2) yields

$$m(t) = \sum_{n=-\infty}^{\infty} m(nT) \frac{\sin \frac{\pi}{T}(t - nT)}{\frac{\pi}{T}(t - nT)} \qquad (2.6)$$

This equation shows how the original signal can be recovered from its samples by using as an interpolation function the familiar impulse response $h(t)$ of an ideal low-pass filter with gain T and bandwidth $1/2T$:

$$h(t) = \frac{\sin \frac{\pi}{T} t}{\frac{\pi}{T} t} \qquad (2.7)$$

In practice, ideal low-pass filters and sampling with pulses of zero width (impulse sampling) can only be approached. The consequences of these imperfections are called **foldover distortion** and **aperture effect**.

If the input waveform $m(t)$ is **undersampled**—that is, $f_s < 2f_{BW}$—the original waveform cannot be recovered without distortion. This output distortion, also known as *foldover distortion,* is caused by the overlap of the original spectrum with the sampled spectrum (see Fig. 2.3). Foldover distortion could be viewed as the generation of undesired or interfering spectral components in the desired frequency band of the original signal. The term *aliasing* is also in use for this distortion.

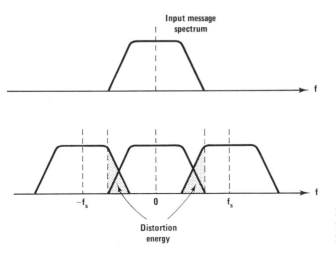

Figure 2.3 Foldover spectrum (distortion) caused by undersampling ($f_s < 2f_{BW}$) of the input message.

Telephony speech signals are typically limited to about 3.4 kHz. However, due to the final steepness of bandlimiting LPFs, a small component of the speech energy is present up to 4 kHz. Experimentally it was found that an $f_s = 8$-kHz (or 8000-sample/s) sampling rate leads to sufficiently low, practically unnoticeable foldover distortion.

In addition to foldover distortion, signal distortion may also be introduced by sampling subsystems having finite-width sample pulses [Bell, 2.7].

2.3 QUANTIZING

We recall that the purpose of the sampler is to convert an input message to a discrete time signal, a signal that is uniquely defined in discrete time instants. Such a discrete time sampled signal may have an infinite number of amplitudes. For synchronous digital baseband transmission it is required that the processed signal be (1) a discrete time signal, and (2) a discrete amplitude signal. That is, it has only a finite number of well-defined amplitude levels, defined in discrete time intervals. These types of PAM signals are described in detail in Chapter 1. To convert a sampled signal having an infinite number of possible amplitude levels into a finite-level discrete PAM signal, a *quantizing* operation has to be introduced. **Quantization** is an essential process that enables a sample to be represented by a finite number of symbols. This is essentially the coding process. Each quantized PAM sample is assigned a unique code word by an encoder. Since the transmission bandwidth has to be kept finite and practically as small as possible, the number of code words available has to be also limited. Therefore, a small range of sampled amplitudes will be assigned the same code word, and thus all are decoded into one particular amplitude at the receiving end. Hence a **quantizing error** is introduced in the quantization process.

In Fig. 2.4 the quantizing error introduced by a *uniform codec* is illustrated. A codec is considered to be uniform if the input amplitude range is divided into N steps of equal width, s, and the output levels are also uniformly spaced.

2.4 PCM: PULSE CODE MODULATION

Another representation of sampling, uniform quantization, and binary encoding is illustrated in Fig. 2.5(a). In this figure the basic steps used in pulse code modulation (PCM) are highlighted.

For simplicity, only eight quantization levels are shown. The continuous signal $m(t)$ has the following sample values: 1.3, 3.6, 2.3, 0.7, . . ., -3.4 V. The quantized signal takes on the value of the nearest quantization level to the sampled value. The eight quantized levels are represented by a 3-bit code number. (Note that with 3 bits, $2^3 = 8$ distinct levels can be identified.) The amplitude difference between the sampled value and the quantized level is called the quantization error. This error is proportional to the step size, d, that is, the difference between consecutive quantization levels. With a higher number of quantization levels (smaller d), a lower quantization error is obtained. Experimentally, it has been found that for toll-grade telephony voice transmission for an acceptable signal-to-noise ratio, 2^8 or 256 quantization levels are required. This represents 8 bits of information per quantized sample.

If the number of quantizer levels is large (greater than 100), we may assume that the quantization error has a uniform probability density function given by

$$p(E) = \frac{1}{d} \qquad -\frac{d}{2} \leqslant E < \frac{d}{2} \tag{2.8}$$

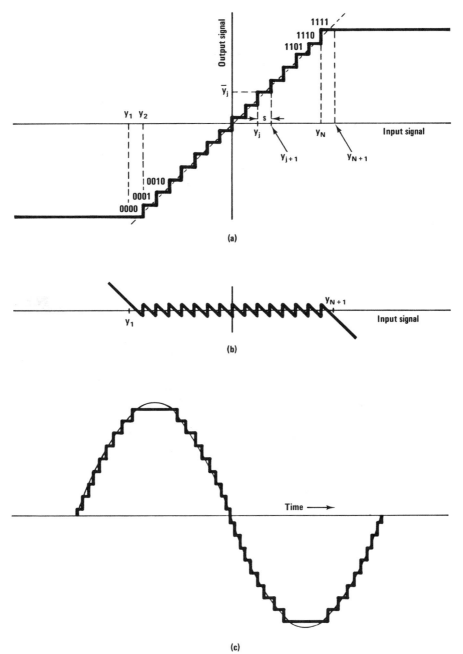

(a)

(b)

(c)

Figure 2.4 Characteristics and quantizing noise of a uniform codec. (a) Uniform codec transfer characteristic. (b) Error characteristic. (c) Quantized full-load sine wave. (With permission of the AT&T-Bell Laboratories [Bell Laboratories 2.7]).

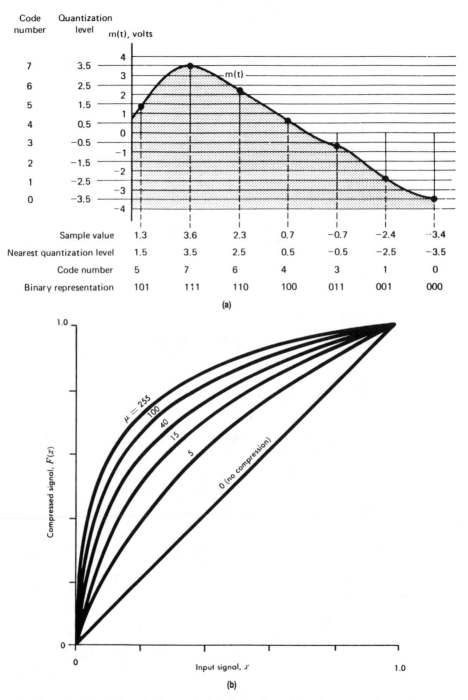

Sample value	1.3	3.6	2.3	0.7	−0.7	−2.4	−3.4
Nearest quantization level	1.5	3.5	2.5	0.5	−0.5	−2.5	−3.5
Code number	5	7	6	4	3	1	0
Binary representation	101	111	110	100	011	001	000

(a)

(b)

Figure 2.5 (a) Quantization and binary encoding for PCM systems. A message signal is regularly sampled. Quantization levels are indicated. For each sample the quantized value is given and its binary representation is indicated. (After [Taub and Schilling, 2.13] with permission from the McGraw-Hill Book Company.) (b) Logarithmic PCM μ-law compression characteristics. (By permission from the AT&T-Bell Laboratories, [Bell Laboratories, 2.7].)

This uniform error distribution is true if the signal $m(t)$ does not overload the quantizer. For example, in a quantizer—such as that shown in Fig. 2.5—the quantizer output might saturate at level 5 for $|m(t)| > 5$. The quantization error during such overload is a linearly increasing function of $m(t)$. In the linear region of operation, the mean-square value of the quantization error is

$$\int_{-d/2}^{d/2} E^2 p(E)\, dE = \int_{-d/2}^{d/2} E^2 \frac{1}{d}\, dE = \frac{d^2}{12} \tag{2.9}$$

If the root-mean-square (rms) value of the input signal $m(t)$ is M_{rms}, then the signal-to-quantization error ratio is

$$\frac{S}{N} = \frac{M_{\text{rms}}^2}{d^2/12} = 12 \frac{M_{\text{rms}}^2}{d^2} \tag{2.10}$$

2.5 COMPRESSION AND EXPANSION

From equation 2.10 we conclude that the signal-to-quantization error ratio is dependent on the rms value of the input signal M_{rms}; that is, for larger input signals, a larger signal-to-quantization-noise ratio (S/N_q) is obtained. This is an undesirable effect in telephony systems, as some people speak at a considerably lower volume than others. For the listener it would be a nuisance if he or she had to listen to a very low volume signal corrupted by a relatively high quantization error (low S/N_q). To achieve the same signal-to-noise ratio for a small-amplitude signal as for a large-amplitude signal, a quantizer with a nonuniform step size is required. To achieve this nonuniform step-size quantization given a uniform step-size quantizer such as that shown in Fig. 2.5, it is necessary to precede it with a nonlinear input-output device known as a **compandor,** or companding device. Note that the compandor followed by the linear quantizer amplifies the low-volume signals more than the high-volume signals.

Companding requirements are different for different signal distributions. For example, voice signals require constant S/N_q performance over a wide dynamic range, which means that the quantizing noise must be proportional to signal amplitude for any signal level. To achieve this, a logarithmic compression law must be used. A truly logarithmic assignment of code words is impossible because it implies both an infinite dynamic range and an infinite number of codes. Two methods for modifying the true logarithmic function have been used. In the first, called the **μ-law,** for the normalized coding range of ± 1,

$$F(x) = \text{sgn}(x) \frac{\ln(1 + \mu |x|)}{\ln(1 + \mu)} \qquad -1 \leqslant x \leqslant 1 \tag{2.11}$$

where x is the input signal (see Fig. 2.5(b)).

A second method of approximating the true logarithmic law is to substitute a linear segment to the logarithmic curve for small signals. This method is called the **A-law** and is represented by

$$F(x) = \text{sgn}(x) \frac{1 + \ln A |x|}{1 + \ln A} \qquad \frac{1}{A} \leq |x| \leq 1$$

$$\qquad\qquad\qquad\qquad\qquad\qquad\qquad\qquad\qquad\qquad (2.12)$$

$$F(x) = \text{sgn}(x) \frac{A |x|}{1 + \ln A} \qquad 0 \leq |x| \leq \frac{1}{A}$$

This curve is smooth at $x = 1/A$. The parameter A determines the dynamic range. Over the intended dynamic range, it has a flatter S/N_q ratio than the μ-law.

At the receiving terminal, signal processing that is the inverse of that performed at the transmitter has to be done in order to recover the transmitted signal. An analysis of the companding improvement is given in references on signal processing, including [Bell Laboratories, 2.7].

2.6 PCM: VOICE- AND VIDEO-TRANSMISSION-RATE REQUIREMENTS

To summarize, we may conclude that in telephony systems the signal is bandlimited to $f_m = 3.4$ kHz. To convert this analog signal into a binary PCM data stream, a sampling rate of $f_s = 8000$ samples/s is used. Each sample is quantized into one of the 256 quantization levels. For this number of quantization levels, 8 information bits are required ($2^8 = 256$). Thus one voice channel being sampled at a rate of 8000 samples/s and requiring 8 b/sample will have a transmission rate of 64 kb/s.

Broadcast-quality color television signals have an analog baseband bandwidth of somewhat less than 5 MHz. For conventional PCM encoding of these video signals, a sampling rate of $f_s = 10$ million samples per second and a 9-b/sample coding scheme is used. Thus the resulting transmission rate is 90 Mb/s. Most television pictures have a large degree of correlation, which can be exploited to reduce the transmission rate. It is feasible to predict the color and brightness of any picture element (pel) based on values of adjacent pels that have already occurred. Digital broadcast-quality color television signals requiring only 20- to 45-Mb/s transmission rates, obtained by means of predictive techniques, have been reported [Feher, 2.6].

2.7 REDUNDANCIES IN THE SPEECH SIGNAL

Removal of redundancies inherent in the nonstationary speech signal is an important part of recently developed digital speech encoders, that is, A/D conversion speech-encoding devices. This redundancy comes from constraints upon the manners and means by which speech is produced. Although speech is a highly nonstationary signal, the means by which it is produced, namely, the larynx and mouth apparatus, are remarkably universal, as is the hearing equipment!

The presence in speech of these statistical invariants makes efficient speech coding a challenging research area. Coding algorithms make varying attempts to account for that redundancy. At high bit rates, there is no need for sophisticated redundancy removal, and therefore algorithms are modestly aimed at modeling long-time average speech properties. At low bit rates, however, coding algorithms must adhere closely to the time-varying properties of speech.

In this section we present a brief review of some conceptually simple voice encoders, namely DPCM and DMOD. In [Adoul, 2.8], a detailed description of the new, more advanced speech-processing principles and applications is given.

2.8 DPCM: DIFFERENTIAL PULSE CODE MODULATION

Differential PCM (DPCM) is a predictive coding scheme that exploits the correlation between neighboring samples of the input signal to reduce statistical redundancy and thus lower the transmission rate. Instead of quantizing and coding the sample value as done in PCM, in DPCM an estimate of the next sample value based on the previous samples is made. This estimate is subtracted from the actual sample value. The difference of these signals is the prediction error, which is quantized, coded, and transmitted to the decoder. The decoder performs the inverse operation; that is, it reconstructs the original signal from the quantized prediction errors.

The basic functional diagram of a DPCM coder is shown in Fig. 2.6. In this diagram, $e(n)$ is the difference between the input sample $x(n)$ and a prediction of $x(n)$ denoted by $\hat{x}(n|n - 1, \ldots, n - p)$. The most successful class of predictors has been the class of linear estimators. In this case, $\hat{x}(n|n - 1, \ldots, n - p)$ is

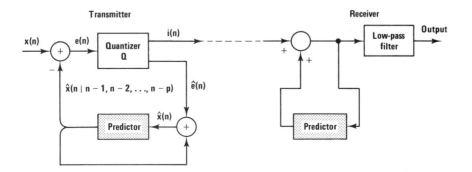

Figure 2.6 Functional diagram of differential PCM coder-DPCM; the predictor computes an estimate $\hat{x}(n|n - 1, n - 2 \ldots n - p)$ of the input sample $x(n)$ from information based on the p previous reconstructed samples, $\hat{x}(n - k)$. The difference $e(n)$ is then quantized as $\hat{e}(n)$ with corresponding index $i(n)$. Note that, since the reconstructed sample $\hat{x}(n)$ is entirely determined by the past sequence of $\hat{e}(n)$, it can be reconstructed in the same manner at the receiver end (providing no transmission error corrupted the sequence of indices). (Adopted from [Adoul, 2.8].)

calculated as a linear combination of the previously transmitted samples as follows:

$$\hat{x}(n|n-1, n-2, \ldots, n-p) = \sum_{k=1}^{p} a_k \hat{x}(n-k) \qquad (2.13)$$

where $\hat{x}(n-k)$ stands for the reconstructed sample at time $n-k$ using the quantized difference $\hat{e}(n-k)$:

$$\hat{x}(n-k) = \hat{x}(n-k|n-k-1, n-k-2, \ldots, n-k-p) + \hat{e}(n-k)$$

$$(2.14)$$

Use of DPCM is predicated on the basis of the fact that the variance of the difference signal $\hat{e}(n)$ is smaller than the variance of the sample $x(n)$. In fact, if we neglect the effect of the quantizer, the contribution of the prediction scheme can be characterized by the energy reduction, expressed in decibels, between the $x(n)$ and the $e(n)$ samples [Adoul, 2.8].

If the quantizer or both the quantizer and predictor adapt themselves to match the signal to be encoded, considerable signal-to-noise improvement can be attained. The dynamic range of the encoder can be extended by adaptive quantization if a nearly optimum step size over a wide variety of input signal conditions is generated.

The two most frequently used quantizer adaptation methods include the **syllabic,** or slow-acting, adaptation and the fast-acting, or **instantaneous,** companding with only one sample memory.

2.9 DM: DELTA MODULATION

The exploitation of signal correlations in DPCM suggests the further possibility of oversampling a signal to increase the adjacent sample correlations and thus to permit a simple quantizing strategy. DM is a 1-bit version of differential PCM. The DM coder approximates an input time function by a series of linear segments of constant slope. Such an A/D converter is therefore referred to as *linear delta modulator*.

At each sample time the difference between the input signal $x(t)$ and the latest staircase approximation is determined. The sign of this difference is multiplied by the step size, and the staircase approximation is incremented in the direction of the input signal. Therefore, the staircase signal $y(t)$ tracks the input signal. The signs of each comparison between $x(t)$ and $y(t)$ are transmitted as pulses to the decoder, which reconstructs $y(t)$, and then to LPFs $y(t)$ to obtain the output signal.

The slope overload distortion region (see Figs. 2.7 and 2.8) occurs for large and fast signal transitions. It is caused by the fact that the maximum slope that can be produced by a linear delta modulator is SS $\cdot f_r$, where SS is the step size and f_r is the sampling rate. The granular noise is introduced by the fact that the staircase is hunting around the input $x(t)$.

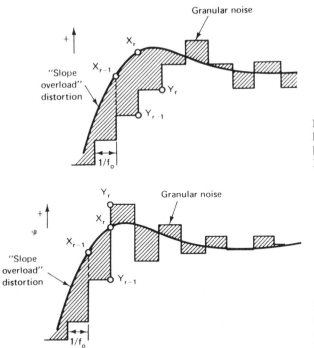

Figure 2.7 Quantization noise in linear delta modulation. (After [Jayant, 2.9] with permission from the IEEE, © 1974.)

Figure 2.8 Quantization noise in adaptive delta modulation, ADM. (After [Jayant, 2.9] with permission from the IEEE, © 1974.)

The use of adaptive techniques reduces the quantization noise and increases the dynamic range of delta modulators. The idea of adaptive step-size modulators is illustrated in Fig. 2.9. A large number of methods for suitable adaptive step-size variation exist. Monolithic integrated-circuit adaptive delta modulators (ADM) using advanced digital algorithms are available from a number of manufacturers.

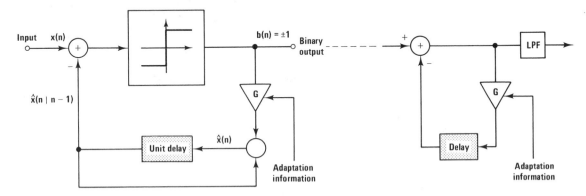

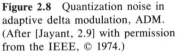

Figure 2.9 Block diagram of adaptive delta modulators. There exists a wide variety of techniques for generating the adaptation information from the signal's past. In fact, techniques based on almost all combinations of short vs. long adaptation-time constant on the one hand and binary ($b_{(n)}$) vs. analog ($\hat{x}(n|n-1)$) representation for the signal's past have been proposed in the technical literature [Adoul, 2.8].

The cost of these high-performance ADM codecs, typically operated in the 10-kb/s to 40-kb/s range, is approximately $10.

Syllabic adaptation refers to schemes that estimate the speech characteristic over a duration of several milliseconds (typically 4 to 25 ms) to accommodate changes in phonemes and syllables. On the contrary, **instantaneous adaptation** schemes have very short time constants of a few samples (typically less than 4 ms).

The frequently used **continuously variable slope-delta** (CVSD) **modulation** technique employs a syllabic adaptation scheme in the form of a first-order digital filter with a time constant greater than 4 ms. It is described by a gain G_n (see Fig. 2.9) of the form

$$G_n = \alpha G_{n-1} + f(b_{n-1}, b_{n-2}, b_{n-3}) \tag{2.15}$$

where the function $f = 1$ or 0 according to whether or not b_{n-1}, b_{n-2}, and b_{n-3} are all of the same sign. Basically, G_n is increased whenever the binary stream b_n exhibits too many consecutive bits of the same sign, that is, when the entropy of the bit stream diminishes.

The **one-word-memory adaptation,** or instantaneous adaptation, scheme of Jayant is also well known and illustrates an instantaneous adaptation. The coding strategy is as follows:

$$G_n = G_{n-1} M(b_{n-1}, b_{n-2}) \tag{2.16}$$

The multiplier M takes one of the two values according to whether or not b_{n-1} and b_{n-2} are of the same sign.

2.10 TRANSMULTIPLEX TECHNIQUES

For interfacing with analog facilities, a means for conversion between analog frequency division multiplex (FDM) and digital time division multiplex (TDM) signals is required. Specially developed FDM/TDM converters called **transmultiplexers** integrate the conversion functions in a compact, efficient form. These converters can be implemented by either analog or digital means.

In Fig. 2.10(a) the conversion between the analog FDM microwave system and the digital PCM satellite system is accomplished with the use of standard TDM and FDM multiplex equipment. Although this method requires no specially developed conversion equipment, it has serious drawbacks in terms of cost, size, and reliability because of the inherent redundancy of circuit functions. The transmultiplexer performs A/D conversion, and this conversion is equivalent to the functions performed by the FDM-DEMUX and PCM-MUX equipment.

An analog transmultiplexer employs conventional channel circuitry as in standard FDM and TDM equipment: analog filters, A/D and D/A converters, and digital multiplexing circuits. A digitally implemented transmultiplexer accomplishes the FDM/TDM conversion using digital processing techniques such as digital recursive filters, nonrecursive filters, and the fast Fourier transform

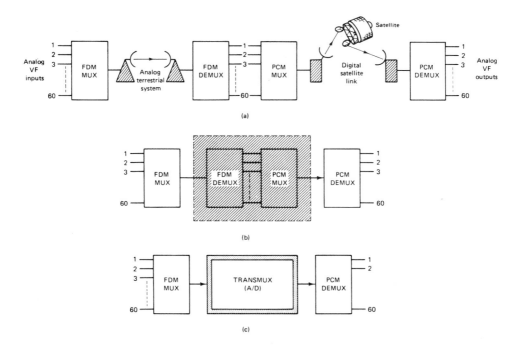

Figure 2.10 Analog and digital system connection for 60 telephony channels using the voice-frequency connection and the transmultiplexer methods. A one-way transmission model is illustrated. (a) Voice-frequency connection between an analog FDM terrestrial and TDM digital PCM satellite system. (b) Equipment required for the interconnection of analog and digital systems if the voice-frequency connection method of (a) is employed. Standard FDM and TDM multiplex set of equipment connected back-to-back is required. (c) Transmultiplexer connection of analog and digital transmission systems. Note that the TRANSMUX equipment is equivalent to the FDM-DEMULTIPLEX-PCM-MULTIPLEX equipment shown in (b) (shaded area). (By permission from Prentice-Hall, Inc. [Feher, 2.1].)

(FFT) and departs substantially from the analog channel processing concepts [Feher, 2.6].

2.11 ECHO CONTROL IN SPEECH AND DATA TRANSMISSION

The telephone network generates echos of the transmitted speech or data signal at points within and near the ends of a telephone connection. The longer the echo is delayed, the more disturbing it is and the more it must be attenuated before it becomes tolerable. Synchronous satellites are placed in an orbit roughly 40,000 km above the earth's surface. Due to this large distance, the round trip of a telephone conversation or data connection relayed via satellite, including the terrestrial segment, is about 500 to 600 ms. In case of a double satellite hop, the round-trip delay can exceed 1 s.

In the terrestrial United States, telephone trunks exceeding 3500 km (approximately 35-ms echo delay), **echo suppressors** are required. In satellite systems echo suppressors or echo cancelers, most frequently one at each end of the long-distance connection, are required. This is known as **split echo control.**

2.11.1 Sources of Echo in the Network

Figure 2.11(a) shows a simplified telephone connection. This connection is typical in that it contains two-wire segments on the ends (the subscriber loops and possibly some portion of the local network), in which both directions of transmission are carried on a single wire pair. The center of the connection is four-wire, in which the two directions of transmission are segregated on physically different facilities. This is necessary where it is desired to insert carrier terminals, amplifiers,

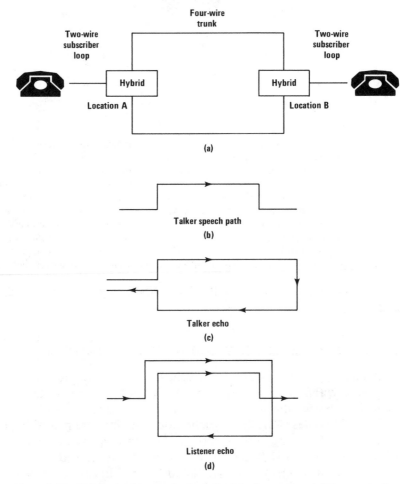

Figure 2.11   Sources and types of echo in the telephone network [Messerschmitt, 2.10].

or digital switches. (In cases where carrier transmission is used, these are not necessarily four physical wires, but the terminology is still useful and descriptive.)

There is a potential feedback loop around the four-wire portion of the connection, and without sufficient loss in this path there is degradation of the transmission or, in extreme cases, oscillation (called singing). The hybrid is a device that provides a large loss around this loop, thereby limiting this impairment. At the same time, the hybrid must not insert significant loss in the speech paths of the two talkers. The remainder of Fig. 2.11 illustrates more graphically the function of this hybrid. One of the two speech paths of the talkers is shown in Fig. 2.11(b). In order that this path not have a large attenuation, it is necessary for the hybrid not to have an appreciable attenuation between its two-wire and either four-wire ports. There are two distinct echo mechanisms, shown in Fig. 2.11(c) and (d). **Talker echo** results in the talker hearing a delayed version of his or her own speech, whereas in **listener echo,** the listener hears a delayed version of the talker's speech. Both these echo mechanisms are mitigated if the echo has significant loss between its two four-wire ports [Messerschmitt, 2.10].

The subjective effect of the echo depends critically on the delay around the loop and the effective transfer function around the loop (which of course incorporates the delay). For short delays, the talker echo represents an insignificant impairment if the echo attenuation is reasonable (6 dB or so), since the talker echo is indistinguishable from the normal sidetone (a version of the talker speech that is deliberately reproduced in the earpiece, making the telephone sound *live*). For longer delays of 40 ms or so, talker echo represents a serious impairment unless the echo is highly attenuated. In this case the echo is disturbing to the talker and in fact can make it very difficult to carry on a conversation.

The situation reverses in the case of listener echo. For short delays in the range of 1 ms or so, the listener echo results in an overall transfer function with peaks and valleys due to the destructive and constructive interference at different frequencies. Subjectively, this has an effect similar to talking into a rain barrel.

The circuit diagram of a hybrid located at B is shown in Fig. 2.12. If the two transformers are identical and the balancing impedance Z_n equals the impedance of the two-wire circuit, the signal originating on the *in* side is transferred to the two-wire circuit of B but produces no response at the *out* terminal. On the other hand, if the signal originates in the two-wire circuit (talker B is active), this signal is transferred to both paths of the four-wire circuit. This signal has no effect on the *in* signal path, since signal on the four-wire connection is amplified only in the opposite direction. Echoes are generated whenever the *in* side is coupled (has a leak-through) to the *out* side. Unfortunately, this occurs in almost all networks, as the Z_n network is not identical to the distributed time-variable impedance of the two-wire circuit. Since Z_n depends on the details of the subscriber loop (such as gauge, length, and the configuration of bridged taps), it varies from one subscriber loop to another. Any fixed balancing impedance can only be chosen to be a compromise. The compromise balancing impedance is usually chosen to be either a parallel or series resistor-capacitor combination. The degree

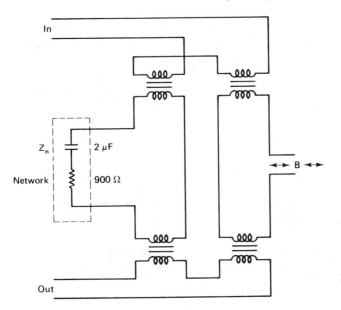

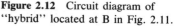

Figure 2.12 Circuit diagram of "hybrid" located at B in Fig. 2.11.

to which the far-end talker signal is attenuated depends on the relationship between the two-wire and the balancing impedance. Also, we note that a four-wire circuit may be connected to a large number of two-wire circuits. Thus the need for echo control (suppression or cancellation) in long-distance systems is imminent.

2.11.2 Echo Suppressors

One conceptually simple approach to controlling echo in the network is to introduce loss in the connection in a controlled fashion. The introduction of loss represents a trade-off between its beneficial effects on the subjective effects of echo and its deleterious effects on the talker's speech path. The effect of the echo increases as the connection round-trip delay increases. Thus to avoid introducing any more loss than is necessary, the loss should be increasing as the connection round-trip delay increases. This is the principle of **via net loss** (VNL), in which a loss is added to a trunk that depends on its length. The objective of the VNL is to achieve a loss between the two local switches in any given connection that approximates

$$dB = 4.0 + 0.4N + 0.102D \qquad (2.17)$$

where dB is the overall loss in decibels of the talker's speech path, exclusive of the loss of the two subscriber loops at the two ends of the connection, N is the number of trunks in the connection, and D is the echo path round-trip delay in milliseconds.

The echo suppressor is a device that inserts a large loss in a connection in one direction or the other. The result is that any echo signal experiences this large loss and is therefore heavily attenuated. The goal is to put the large loss in the direction opposite to the current active speech path, assuming that there is speech in only one direction at a time. This means that there is inevitable clipping during doubletalk (speech in both directions at the same time).

The use of loss to control echo is an inadequate measure for very long delays, since the loss itself becomes an important impairment [Messerschmitt, 2.10]. For this reason, the loss to control echo is used only for terrestrial trunks up to about 1900 km in North America (this number is changing with the introduction of fiber optics systems).

2.11.3 Echo Cancelers

With increasing round-trip delay, the subjective effect of echo becomes more annoying. The echo canceler is a sophisticated form of echo control that can effectively eliminate echo as an impairment on even very long delay channels and without regard to the doubletalk situation.

The basic idea of echo cancellation, illustrated in Fig. 2.13, is to generate a synthetic replica of the echo and subtract it from the leaked-through echo signal returned through hybrid B. If an adaptive filter that perfectly matches the transfer function of the echo path were designed, complete echo cancellation could be achieved. Adaptive filtering is required to match the time-variable distance and device-dependent impedance characteristics of the two-wire circuit.

In practice it is desirable to cancel the echos in both directions of a trunk. For this purpose two adaptive cancelers are necessary, where one cancels the echo from each end of the connection. The near-end talker for one of the cancelers is the far-end talker for the other. In each case, the near-end talker is the "closest" talker, and the far-end talker is the talker generating the echo that is being canceled. It is desirable to position these two halves of the canceler in a split configuration, where the bulk of the delay in the four-wire portion of the connection is in the middle. The reason is that the number of coefficients required

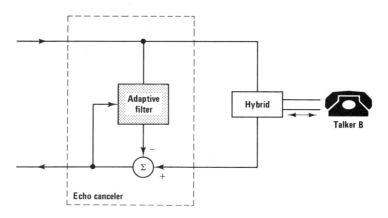

Figure 2.13 Echo canceler: conceptual diagram.

in the echo-cancellation filter is directly related to the delay of the channel between the location of the echo canceler and the hybrid that generates the echo. In the split configuration, the largest delay is not in the echo path of either half of the canceler, and hence the number of coefficients is minimized. Typically, cancelers in this configuration require only 128 or 256 coefficients, whereas the number of coefficients required to accommodate a satellite connection in the end link would be impractically large [Messerschmitt, 2.10].

2.12 DIGITAL SPEECH INTERPOLATION (DSI) SYSTEMS

The digital speech interpolation (DSI) technique exploits the fact that in telephone conversations only one speaker is usually speaking, while the other listens. These speech signals are the product of two-way conversations. It is customary for one talker to pause while the other speaks; thus an active speech signal is present on a transmission channel for only a fraction of the time. In addition, even when only one talker is speaking, pauses occur between utterances, and there are times when the circuit is simply idle. Therefore, it can be expected that, on the average, speech is present for considerably less than 50% of the time. Statistical measurements show that speech is present on a telephone channel approximately 40% of the time, averaged over a large number of trunks.

Digital time-assigned speech interpolation (TASI) without and with overload channels, variable rate delta modulation used with speech interpolation, and speech-predictive encoded communications (SPEC) techniques that exploit these relatively low-voice-activity properties to reduce the information rate needed to handle a multiplicity of speech-carrying telephone channels are being developed and are already in use by a number of organizations [Campanella, 2.12].

For example, application of DSI systems in time division multiple access (TDMA) satellite systems leads to a major capacity advantage of digital TDMA satellite systems. An examination of Table 2.1 indicates that the capacity (expressed in number of telephony channels) of a DSI-TDMA satellite system employing TASI and TDMA is two to three times larger than the capacity of a corresponding frequency division multiple access (FDMA) satellite system. This almost three-

TABLE 2.1 Approximate Capacities for the INTELSAT-V Hemispheric Beam 72-MHz Transponders for INTELSAT Standard A (30-m Antenna Diameter) Earth Stations

Transmission mode	Maximum capacity expressed in terms of number of 3.1-kHz telephony channels
FDMA (analog system)	1100
TDMA (digital system)	
Without DSI	1600
With DSI	3200

to-one capacity advantage warrants the additional complexity required for digital TDMA-DSI satellite systems.

Digital TASI and SPEC have been implemented using the customary 8-b/sample PCM format for 64-kb/s digital telephone transmission with either μ-law companding for T-carrier systems used in North America and Japan or A-law companding for CEPT-32 used in Europe and most other countries outside of North America and Japan. With such systems, the number of channels carried on a transmission facility can be multiplied by between 2:1 and 2.5:1, depending on the number of trunks interpolated and the mean voice spurt activity on them.

Interpolation techniques have also been extended to other methods of digital source coding, such as delta modulation (DM), adaptive differential PCM (ADPCM) and nearly instantaneous companding (NIC). Such source coders are easily capable of providing telephone quality speech reproduction at a rate of 32 kb/s and, when combined with speech interpolation, will yield a net channel multiplication of 4:1 to 5:1 compared with conventional noninterpolated PCM telephony. It is further expected that by 1990, 16-kb/s source coders will be perfected and when combined with interpolation these will yield a net channel multiplication ratio of between 8:1 and 10:1 [Campanella, 2.12].

The original application of speech interpolation was on analog telephone circuits and was intended principally for use on transoceanic telephone cable circuits. Such analog implementation, based on TASI, was subject to impairments caused by the difficulty to realize ideal, click-free analog switches and differences in the quality of the interconnecting transmission channels themselves. TASI-type systems randomly seize any one of the available transmission channels on successive speech spurts and variations due to gain and noise differences on the transmission links can become noticeable. Digital implementation of speech interpolation techniques applied to digital speech transmission channels eliminates both of these impairments. When interpolation is implemented by digital means on digital transmission facilities, the method is referred to as **digital speech interpolation.** The TASI technique, when implemented by digital techniques, is frequently referred to as **digital TASI,** or TASI-D. An alternative digital implementation using sample by sample prediction of PCM speech has been called **speech predictive encoding** (SPEC).

The principle of compression of a larger number of incoming (source) channels into a smaller number of outgoing (transmit) channels is illustrated in Fig. 2.14. The PCM encoder accepts N analog telephony inputs and converts them into N binary bit streams. The output of the DSI module provides M binary channels for transmission. (*Note: $M < N$.*) One of these channels, the assignment channel, is used to transmit the transmit channel assignments to the receiving earth station for proper reconnection to the outgoing channels. The transmit channel assignments are transmitted by means of assignment messages, which consist of the incoming channel numbers and associated transmit channel numbers.

For the INTELSAT-V TDMA satellite network, digital speech interpolation using the TASI technique has been adopted. Probably the single most important

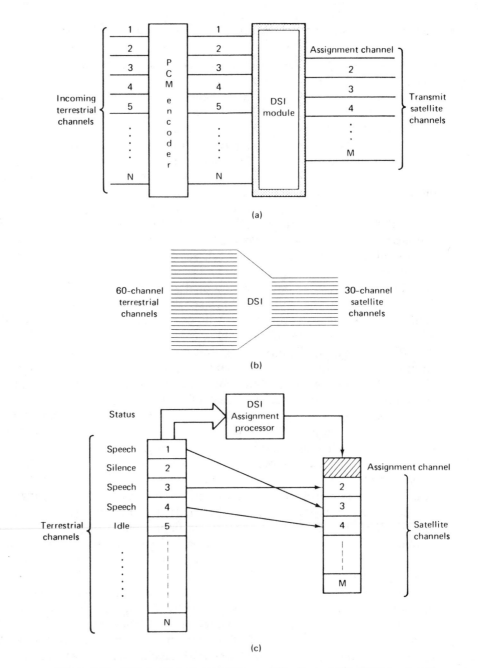

Figure 2.14 DSI channel-compression capability in a satellite TDMA network: (a) DSI transmit equipment: *N* incoming terrestrial channels are compressed into *M* outgoing satellite channels through DSI operation, where the ratio *N/M* is the DSI gain. (b) 60 incoming channels are compressed into 30 satellite channels using DSI with a gain of 2.0. (c) Mapping of terrestrial channels into satellite channels.

reason for the choice of TASI (instead of SPEC) is the more elaborate operational experience and almost off-the-shelf availability of the TASI-DSI equipment.

Theoretically, the DSI gain can be adjusted over a fairly wide range. If the number of transmit channels for a given number of incoming channels is reduced, the DSI gain increases. However, the gain increase is limited by increased clippings of the incoming channels, which cannot be transmitted for given periods of time due to nonavailability of transmit channels. Thus the incoming channels that do not have a transmit channel assignment are *frozen out* until one of the terrestrial channels that was previously assigned a satellite channel becomes inactive. Thus front-end clipping of speech is evident during freeze out. The development of the INTELSAT-V freeze-out satellite system specifications was preceded by numerous subjective tests. These specifications require that the percentage of clipping be less than 2% of the speech spurts that experience clips greater than 50 ms. (Note that each period of time occupied by a caller's speech is called a **speech spurt**.)

In the SPEC digital speech interpolation system, the competitive clip problem experienced by the TASI-DSI system is completely avoided. The speech predictor algorithm used in the SPEC system eliminates unnecessary samples in the instantaneous speech waveform or in the short intersyllabic pauses. The SPEC predictor is similar to an adaptive DPCM system. In this system more than 25% of the PCM samples are removed during voice spurts. On the other hand, the major performance degradation in the SPEC digital speech interpolation system is the production of prediction distortion.

2.13 DIGITAL TELEVISION-PROCESSING TECHNIQUES

Data-compression coding combined with digital transmission techniques have been viewed for a long time as promising and powerful means to achieve efficient television transmission. Recent progress in LSI and digital technologies has made complex signal processing a technically feasible reality and has led to progress in digital television encoding, particularly in interframe coding, by which the transmission bit rate can greatly be reduced [Kaneko and Ishiguro, 2.11].

During the 1980s the demand for video teleconferencing services has been growing at a rapid rate. Visual communication is a key to teleconferencing, a full-motion video signal being thought to be very helpful and useful in accomplishing interactive communications. However, a full-motion television signal requires a thousand times wider bandwidth than a voice telephone channel. Bandwidth compression is, therefore, a powerful means to provide economical teleconferencing systems with compression ratios of 1:40 or even more. Such high data compression can only be achieved by digital video processing techniques.

Digital television encoding schemes are generally categorized into three classes (see Fig. 2.15):

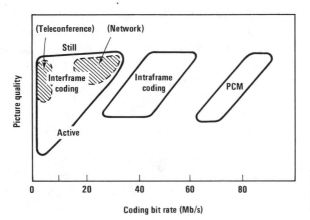

Figure 2.15 Television (picture) quality as a function of the coding bit rate (Mb/s) [Kaneko and Ishiguro, 2.11].

1. Conventional PCM
2. Intraframe coding
3. Interframe coding

Standard color television signals (video part) are bandlimited to about 4.5 MHz. Sound (audio) subcarriers may be added in the 4.5-MHz to 6.5-MHz frequency range. Conventional PCM or straight A/D conversion provides high-quality encoding with 7- or 8-b/sample at about a 10.7 MHz sampling, resulting in a 75-Mb/s to 90-Mb/s transmission rate. Intraframe coding enables a reduction of the transmission bit rate by intraframe processing such as differential PCM or orthogonal transform coding. By these intraframe coding methods, the transmission bit rate can be reduced to about 30 through 60 Mb/s, depending on quality requirements and technique employed. There are also certain trade-offs between picture quality, bit rate, and hardware complexity.

Much greater reduction in transmission bit rate can be achieved by use of interframe coding. The general concept of bit rate reduction by interframe coding is to transmit the difference information of the two successive frames instead of transmitting the entire frame information. The information to be transmitted is dependent on picture object movement: The more active the movement, the greater the information becomes. With recently developed television processors, *broadcast-quality network television* signals can be transmitted at a bit rate around 6 through 30 Mb/s. Relatively still pictures such as those encountered in conference-room scenes can be transmitted at 1.5 Mb/s or even lower bit rates [Kaneko and Ishiguro, 2.11].

REFERENCES

[2.1] Feher, K. *Digital Communications: Satellite/Earth Station Engineering*, Prentice-Hall, Englewood Cliffs, N.J., 1983.

[2.2] Feher, K. *Digital Communications: Microwave Applications*, Prentice-Hall, Englewood Cliffs, N.J., 1981.

[2.3] Decina, M., and A. Roveri. "ISDN: Integrated Services Digital Networks: Architecture and Protocol Aspects," Chapter 2 in K. Feher, et al., *Advanced Digital Communications: Systems and Signal Processing Techniques*, Prentice-Hall, Englewood Cliffs, N.J., 1987.

[2.4] Oppenheim, A. V., and R. V. Schafer. *Digital Signal Processing*, Prentice-Hall, Englewood Cliffs, N.J., 1975.

[2.5] Bellamy, J. C. *Digital Telephony*, John Wiley, New York, 1982.

[2.6] Feher, K., et al. *Advanced Digital Communications: Systems and Signal Processing Techniques*, Prentice-Hall, Englewood Cliffs, N.J., 1987.

[2.7] Bell Laboratories, Members of the Technical Staff. *Transmission Systems for Communications*, Revised 4th ed., Bell Telephone Laboratories, 1971.

[2.8] Adoul, J. P. "Speech Coding Algorithms and Vector Quantization," Chapter 3 in K. Feher, et al., *Advanced Digital Communications: Systems and Signal Processing Techniques*, Prentice-Hall, Englewood Cliffs, N.J., 1987.

[2.9] Jayant, S. N. "Digital Coding of Speech Waveforms: PCM, DPCM and DM Quantizers," *Proceedings IEEE*, May, 1975.

[2.10] Messerschmitt, D. G. "Echo Cancellation in Speech and Data Transmission," Chapter 4 in K. Feher, et al., *Advanced Digital Communications: Systems and Signal Processing Techniques*, Prentice-Hall, Englewood Cliffs, N.J., 1987.

[2.11] Kaneko, H., and T. Ishiguro. "Digital Television Processing Techniques," Chapter 6 in K. Feher, et al., *Advanced Digital Communications: Systems and Signal Processing Techniques*, Prentice-Hall, Englewood Cliffs, N.J., 1987.

[2.12] Campanella, S. J. "Digital Speech Interpolation Systems," Chapter 5 in K. Feher, et al., *Advanced Digital Communications: Systems and Signal Processing Techniques*, Prentice-Hall, Englewood Cliffs, N.J., 1987.

[2.13] Taub, H., and D. L. Schilling. *Principles of Communication Systems*, McGraw-Hill, N.Y., 1971.

3

DIGITAL SIGNAL PROCESSING IN TELEPHONE CHANNEL MEASUREMENTS AND INSTRUMENTATION

DAVID DACK and BOB COACKLEY

Hewlett-Packard Ltd.

3.1 INTRODUCTION

The history of telecommunications is one of continuous advance both in type and quality of service. Not only has the quality of voice communication improved steadily over the years, but the same channels used for speech are now being called upon to carry data, a service for which they were never originally intended. Data signals are, of course, digital in origin, linking computer or terminal to computer, where it would be ideal if there were a direct digital link to carry them.

Looking ahead we can foresee a time when each telephone will have a digital output, and a digital transmission path will be provided with direct dialing between all subscribers. The technology is certainly available to achieve this today, but to appreciate the difficulties involved in implementation, we must remember that the present equipment and plant investment is enormous. Something like one-half of the capital cost of the world's telecommunications equipment is in the subscriber loop and its associated local exchange equipment. This is analog, voice-optimized equipment that must live out its useful life of many tens of years before it can economically be replaced. However, the demand for data communications equipment is far outstripping the provision of digital switching and transmission equipment. This is why we need devices such as modems, which convert digital data into analog signals suitable for transmission over existing voice channels.

The present telecommunications networks and systems are extremely complex, since they accommodate past methods together with modern services and technology. Indeed, tribute must be paid to the many network planners and designers who have achieved a fully integrated system when faced with so many requirements and constraints.

3.1.1 Implications for Test Equipment

As each new service requirement developed, an appropriate test was introduced to aid in characterizing voice channels for the application. A good example of this is the development of noise-measurement practices, which we review in this chapter. This particular measurement problem provides a good case study, showing how measurement methods have developed and how modern technology can make additional contributions even when applied to conventional measurements. Although digital signal-processing techniques can now bring significant contributions to modern test equipment, many of the transforms used have been known in mathematical form for many years. The real change that has occurred is that many mathematical ideas can now be implemented economically by using low-cost digital electronics. Computing power and memory are available with price and performance characteristics that now give an opportunity to introduce new methods or to replace analog methods with digital. An added advantage is that the need for factory adjustments can be often minimized, and servicing facilities can be enhanced.

We have chosen to illustrate the power of digital signal processing by describing how the digital filter is applied to voice-channel testing and, in particular, to the measurement of noise.

3.2 NOISE MEASUREMENTS

If a subscriber complains of noise on a line, it is not enough simply to say that the line is noisy. The repair person needs to know just how noisy the line is and how to measure the line at several points to find out where the noise is at its worst and thus where the noise originates. Clearly, a simple noise-measuring instrument is of great help, and a set of standard calibrated noise meters available across the country enable reliable comparative measurements to be made at different locations by different people.

The first noise-level meters measured noise of different bandwidths equally, but it was soon realized that the human ear responds differently to different frequencies. It was reasoned that if a noise-level meter had a defined frequency response that weighted noises of different frequencies in a similar way to the ear, then a measurement could be made that would correspond to a human assessment of how objectionable the noise was. In this way, the **C-message curve** was created, as shown in Fig. 3.1. It was not possible to abandon the flat level meter, however, since the C-message curve severely attenuates the low frequency (60 Hz) that would have to be measured when investigating power induction.

It was therefore necessary for a new power meter to have two filters. Not everyone agrees on the frequency response of the human ear, and an alternative weighting curve is required for use with CCITT standard circuits. This curve is known as the **psophometric curve;** it is also shown in Fig. 3.1. We now have three filters, but the worst is yet to come. The introduction of pulse code modulation (PCM) voice channels brings about an improvement in quality when the system is working well, together with some potentially nasty noise on the rare occasions when bits get dropped.

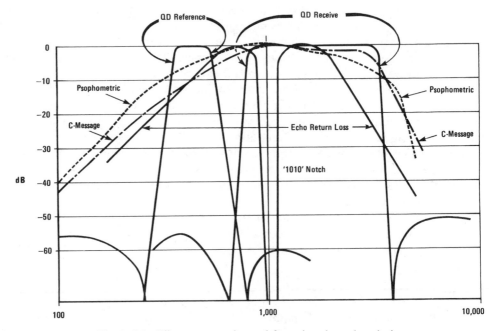

Figure 3.1   Filters commonly used for voice channel analysis.

This **digitally generated noise** is much more difficult to measure, however, since when there is no traffic, the digital system is not exercised and there is no noise. This is quite unlike the analog case, where we need a condition of no traffic so that the residual noise can be measured. The solution is to stimulate the voice channel with a tone and to remove the tone at the measurement point with a notch filter before measuring the noise that was stirred up by the tone. This noise with tone measurement uses as a stimulus a tone that does not, of course, represent normal traffic, so the CCITT specifies as a stimulus a band of noise from 350 Hz to 550 Hz for its equivalent measurement. This is the **quantization**

distortion (QD) **reference bandwidth.** At the other end of the link, a receiving filter is used with a bandwidth of 2600 Hz centered on 2100 Hz, known as the **QD receive filter.**

In an ideal system, of course, this receiving filter will pick up no signal at all, but any nonlinearity, quantizing noise, thermal noise, or crosstalk will result in a signal in the receiving filter that can be measured as an indication of the quality of the channel. We can now add two more filters to our collection, and we have not yet concerned ourselves with anything other than noise measurements.

A further important measurement concerns echoes on a telephone line that interfere with communication and, in extreme cases, can cause oscillation or ringing. The method of testing for this condition is to stimulate the line with a defined bandwidth of noise and to measure the returned, or echoed, signal.

Of course, this required yet another filter, the echo return loss filter of Fig. 3.1. This figure shows the superimposed frequency responses of most of the filters required for a universal voice channel set, and Table 3.1 lists the measurements to be made and the filters required. Full details of the measurements referred to are contained in the IEEE and CCITT standards [IEEE, 3.7; CCITT, 3.8].

TABLE 3.1   Measurements Commonly Made on Voice Channels and the Filters Required. Taken from IEEE and CCITT Standards.

Measurement	Filter
Level and frequency response	High pass to eliminate 60 Hz. Low pass above 10 kHz to eliminate interference from AM transmission.
Noise	C-message, psophometric. Sharp notch. 3 kHz flat, 15 kHz, flat (Butterworth).
Transients	Notch, C-message with specified phase characteristics.
Phase and amplitude jitter	A variety of weighting filters to discriminate between jitter in various bands.
Return loss, ERL, and SEL	Weighting filters for the stimulating noise source.
Intermodulation distortion	Narrow-band filters at 1900 Hz, 520 Hz, and 2240 Hz.
Selective level	Tunable narrow-band filters of 3, 10, 30, and 100 Hz are suggested.
Quantization distortion (CCITT)	Flat-topped sharp cutoff filters of 200 Hz and 2600 Hz bandwidth.

3.3 DIGITAL MEASUREMENTS

Just as digital technology is in the process of transforming the telecommunications network, it can also transform the way in which measurements are made. In theory, a totally new "do everything" measurement could be devised with a random noise stimulus applied to a channel and a sophisticated computer-based analyzer deducing everything we need to know from the response. However, in an industry where capital equipment has a 30- to 40-year life span and experience in interpreting results has a similar longevity, old measurements die hard. In fact, as we have seen, a new measurement never supersedes the old measurements; it simply adds one more to the growing list. Digital technology can come to our aid in another way, however, by giving us the means to implement the old measurements with cheaper, more stable circuitry, requiring no adjustments.

We have seen that all of the measurements defined in the IEEE and CCITT standards require filters, each measurement requiring a different set. These filters, when implemented in traditional analog technology, occupy a great deal of space, and many of them require adjustment in the factory and subsequent periodic calibration to maintain their shapes. What we really need is a single digital filter that can be programmed to perform whichever filter is required, when it is required. The space saved, together with the ease of manufacturing and the stability of the resulting measurements—with no adjustments or calibration—provide cost benefits for the manufacturer and user alike.

3.3.1 Sampling

The signals that we have to filter and measure are continuously varying analog waveforms. On the other hand, the microprocessors and digital integrated circuits that we wish to use to analyze these signals handle only sequences of binary numbers occurring at a fixed rate.

Obviously, we need some means of converting the analog signals into digital form before they are processed—that is, an analog-to-digital (A/D) converter. Such an A/D converter will sample the incoming signal at a fixed rate and will output a series of (usually) binary numbers at that rate. It is fairly easy to see that if we do not sample the signal often enough, we will lose some of the fine detail of the waveform. We do not want to sample too fast, however, since the faster we sample, the faster the following digital circuitry will have to go and the more expensive it and the ADC will become. The phenomenon that occurs if we do not sample fast enough is called **aliasing**; we have all seen samples of aliasing on television. If a television film shows a car or a stagecoach in motion, as the wheels rotate faster and faster, they appear to slow down, stop, or even go backward. This is because the film consists of a sequence of still shots or samples of the scene, and the sample rate is often not fast enough to capture the rotation of the wheels at high speed.

The same phenomenon occurs if we use a stroboscope to view a piece of rotating machinery. If the stroboscope frequency exactly matches the rotation speed, then the machinery appears to be stationary. If the stroboscope speed is slightly less than that of the machinery, then the machinery appears to be rotating slowly. Figure 3.2 shows what happens when we sample a sine wave at various rates. If we sample too slowly, we get a low frequency output even though the original waveform was at a higher frequency. The situation is best summed up in the frequency domain as in Fig. 3.3. A signal that is f hertz away from the sampling frequency f_s appears as if it were, in fact, a frequency of f hertz.

To avoid this aliasing problem, we have to make sure that the A/D converter never sees any signals that are too high in frequency, so we have to precede it with an analog filter, which (not surprisingly) is known as the **anti-aliasing filter**. If the anti-aliasing filter had a perfectly sharp cutoff, we could handle frequencies of up to one-half of the sampling rate. In practice, however, to allow a realizable filter to be used, its cutoff frequency needs to be lower than this.

For example, the voice channels in a PCM system contain frequencies of up to around 3 kHz, so with a perfectly rectangular anti-aliasing filter, a sampling rate of 6 kHz could be used. In practice, a realizable nonrectangular filter is used, and so the sampling frequency must be increased to 8 kHz. It seems then

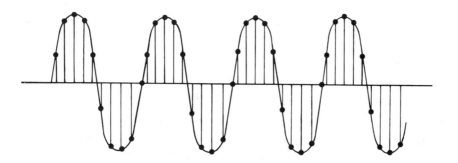

A sinewave sampled fast enough to represent it properly

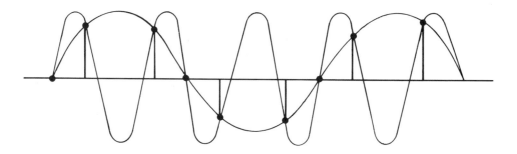

A sinewave sampled too slowly looks like a lower frequency sinewave

Figure 3.2   Sampling rate.

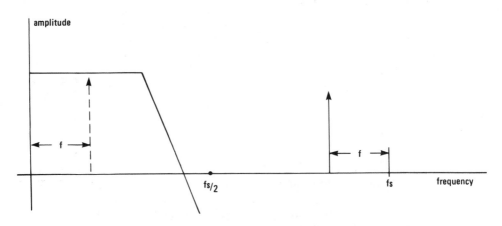

Figure 3.3 Signals above $f_s/2$ are "aliased" and appear to be lower frequency. Signals up to $f_s/2$ are sampled correctly. An anti-aliasing filter must be used, therefore, to limit the ADC input to frequencies with which it can cope.

that the first step toward our goal of replacing analog filters with digital filters is to make an analog anti-aliasing filter! Such a course is sensible because only one anti-aliasing filter is required, and this is a fairly simple low-pass filter. The following digital filter will perform all the complicated filter shapes that would be much more difficult to realize in analog technology.

3.4 ANALOG AND DIGITAL TECHNIQUES COMPARED

All filter designers should be aware of the terminology of analog filter design. Figure 3.4 lists these concepts and compares them with their equivalents in the digital world. First, as we explained earlier, analog filters operate on continuous waveforms, whereas their digital counterparts operate on discrete sequences of numbers that have been produced from continuous waveforms.

One way of describing an analog filter completely is to use its impulse response. If the input to an analog filter is a short, sharp pulse (impulse), the resulting output will have enough information in its waveform—the impulse response of the filter—to calculate what the response would be for any input [Oppenheim and Schafer, 3.5]. We can see that this is so by considering any arbitrary input signal as the sum of lots of individual impulses. The weight of each impulse represents the value of the signal at that time. We know the response of the filter to each of these impulses. It is the filter's impulse response. To work out the filter's response to the arbitrary signal, we add the impulse responses from all earlier inputs, spaced out in time with scaling factors corresponding to the values of the inputs at those times. In the case of an analog filter, the impulses are infinitesimally narrow and close together, so that the result is a smooth waveform. The above assumes that the filter is linear so that superposition

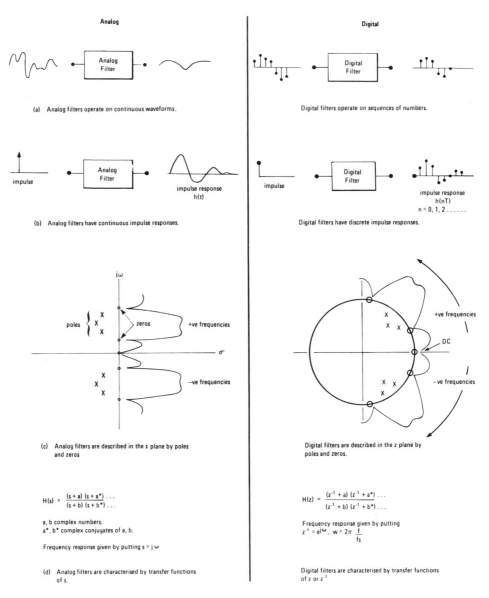

Figure 3.4 Review of analog and digital filter concepts.

applies. Mathematically, the output of the filter is the convolution of the input and the impulse response.

In the digital case, of course, the impulse response has discrete values spaced out by the sampling interval. In Fig. 3.4 we see a digital impulse response, which looks like a sampled version of the adjacent analog impulse response.

3.5 FINITE IMPULSE RESPONSE (FIR) FILTERS

Here then is one way of designing a digital filter to give similar results to an analog filter. First, calculate or measure the impulse response of the analog filter. Then choose a sampling rate that is more than twice the maximum frequency with which the filter is required to deal, and sample the analog impulse response at this rate. The samples of the impulse response become the coefficients of the digital filter, as in Fig. 3.5.

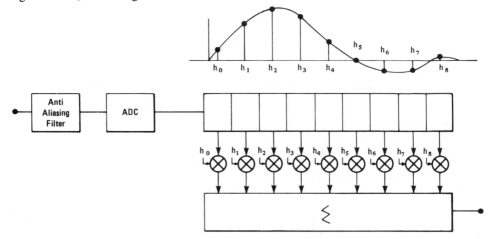

Figure 3.5 Finite impulse response (FIR) digital filter.

Having chosen the sampling rate, we can design a suitable anti-aliasing filter to remove all frequencies that might have been aliased. The ADC turns this filtered input into a series of numbers, which are then fed into a shift register. This produces in parallel the successively delayed versions of the input ready to be multiplied by the sampled impulse coefficients. The results of these multiplications are added to produce each digital output. This type of filter is known as a **finite impulse response filter,** or **FIR filter,** because the impulse response only lasts as long as the length of the shift register, and in the practical world this must be finite. As a result, the tail end of the analog impulse response must be omitted; this distorts the frequency response somewhat, but in practice this is usually a satisfactory method of designing digital filters.

3.6 TRANSFER FUNCTIONS

Analog filters can also be described by their transfer functions. We talk about the poles and zeros of a filter and their effects on its frequency response. The transfer function $H(s)$ is a ratio of two polynomials; for example

$$H(s) = \frac{s^2 + A_1 s + A_2}{s^2 + B_1 s + B_2}$$

(3.1)

represents a second-order filter. The transfer function is usually factored as

$$H(s) = \frac{(s - C)(s - C^*)}{(s - D)(s - D^*)} \tag{3.2}$$

where C and D are complex numbers of the form $a + jb$ and C^* and D^* are the complex conjugates of the form $a - jb$. The numerator of $H(s)$ is defined by the **zeros** of the filter at C and C^*, which are called zeros because if s were to equal C or C^*, the numerator and $H(s)$ would be equal to zero. The denominator of $H(s)$ is defined by the poles of the filter D and D^*, because if s were to equal D or D^*, the denominator would equal zero and $H(s)$ would "climb a pole" toward infinity.

The example shows a second-order transfer function with the maximum power of s being 2. More-complex filters with more poles and zeros can, of course, be found by cascading these second-order sections.

The evaluation of a frequency response is depicted graphically in Fig. 3.4(c), which illustrates the poles and zeros of the filter in the s-plane. The frequency response of the filter can be evaluated by setting $s = jw$, that is, by evaluating $H(s)$ along the vertical axis. As we can see, the proximity of a zero to the frequency of interest makes the response dip toward zero, whereas the proximity of a pole increases the gain of the filter at that frequency. Clearly, this is a very useful way of visualizing the relationship between the mathematical description of a filter and the result obtained in real life when the filter response is plotted on a network analyzer.

There is a similar, graphical way of describing digital filters in what is known as the z-plane. The mathematical reasoning behind this method is explained fully in the references [Gold, 3.3; Cappellini, 3.4; Oppenheim and Schafer, 3.5]. There is space here to list only the results. First, the transfer function is presented in terms of the variable z. A delay of one sampling period is represented by z^{-1}, since z is regarded as a unit advance in time. The transfer function of a digital filter takes the form

$$H(z) = \frac{L_2 z^{-2} + L_1 z^{-1} + 1}{L_2 z^{-2} + k_1 z^{-1} + 1} \tag{3.3}$$

As in the analog plane description, we have poles and zeros, but the big difference lies in the evaluation of the frequency response. Instead of evaluating the response at points on the jw-axis, we must evaluate at points on a circle of unit radius. This should not be too surprising, since we are already used to the idea of aliasing making high frequencies look like low ones. Zero frequency, or DC, is represented by the point $(1, 0)$ shown on the right-hand side of Fig. 3.4(c), and as the frequency is increased, the point rotates around the unit circle until it reaches one-half of the sampling frequency at 180° from the start point. Now, just as in the s-plane, the portion of the unit circle above the horizontal axis represents positive frequencies; that below the axis represents negative frequencies, which are indistinguishable. Any frequency above one-half of the sampling frequency is thus indistinguishable from a frequency positioned symmetrically below one-half of the sampling frequency, which is exactly the phenomenon of aliasing. When calculating the frequency response of the filter, we can imagine it as being curved around the unit circle, as in Fig. 3.4(c), or we can imagine the unit circle from 0° to 180° being unwrapped and flattened out to a straight line, taking the frequency response with it. In either case, the same intuitive feel as in the s-plane for the effect of poles and zeros on the frequency response can be gained.

3.7 INFINITE IMPULSE RESPONSE (IIR) OR RECURSIVE DIGITAL FILTERS

As we have seen, a second-order digital filter can be represented by the transfer function

$$H(z) = \frac{y(nT)}{x(nT)} = \frac{1 + L_1 z^{-1} + L_2 z^{-2}}{1 + K_1 z^{-1} + K_2 z^{-2}} \qquad (3.4)$$

where the input and output of the filter are

$$x(nT) \quad \text{and} \quad y(nT) \qquad \text{for } n = 0, 1, 2, 3, \dots$$

and T is the sampling period.

$$y(nT) = x(nT) + L_1 z^{-1} x(nT) + L_2 z^{-2} x(nT) + K_1 z^{-1} y(nT) + K_2 z^{-2} y(nT) \qquad (3.5)$$

However, since z^{-1} is a unit delay,

$$y(nT) = x(nT) + L_1 x[(n - 1)T] + L_2 x[(n - 2)T] + K_1 y[(n - 1)T] \qquad (3.6)$$
$$+ K_2 y[(n - 2)T]$$

The equation simply tells us how to add delayed versions of x and y, suitably weighted by the coefficients L and K to produce a filtered output. The block diagram of a digital system to perform this operation would look like Fig. 3.6.

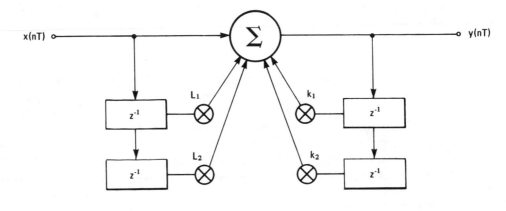

z^{-1} represents a unit delay

Figure 3.6 Block diagram of a recursive infinite impulse response (IIR) digital filter.

The big difference between this filter and the FIR filter is that we now have feedback terms K_1 and K_2. Because of this feedback, this type of filter is a *recursive* filter and, again because of feedback, the impulse response is no longer restricted in length by the number of registers.

In fact, the impulse response of a filter of this type can continue indefinitely, so it is called an infinite impulse response (IIR) filter. One benefit of this approach is that very sharp cutoff filters can be designed with fewer registers and many fewer multipliers than would be required for a similar FIR filter.

3.8 METHODS OF DESIGNING FILTERS

We have now seen how simple the hardware for a digital filter can be. The response can be varied by varying the coefficients L_1, L_2, K_1, K_2, and so on, for each section in a cascade of second-order sections to achieve any desired response. The problem is, of course, how to work out what the Ls and Ks should be.

Fortunately, a great many of the filters required for telecommunications are simply specified in terms of straight-line masks, as in Fig. 3.7. As long as the filter response falls within the mask, then we do not care about precisely how the response ripples between the mask boundaries. For this case, computer programs have been written that calculate the Ls and Ks for us. All we have to do is specify the ripple, cutoff frequency, and stop band of the filter, and the computer can calculate the coefficients and even program them into a ROM to be plugged into the digital filter hardware [IEEE, 3.1; Hewlett-Packard, 3.2]. Obviously no adjustments are required to tune the filter, as might be required in the analog case, and the filter is completely insensitive to drift through temperature changes. What is more, the same hardware can implement any other filter by simply changing the ROM.

Filters like the C-message are a little more difficult, however, since they are not of a regular shape. In general, a complex hill-climbing program would be required to optimize the coefficients to meet the required shape. If, however, a frequency response plotting program is available on a general-purpose digital computer with good graphics capability, then the coefficients can be changed manually until the observed response is acceptable.

3.9 THE IMPACT OF DIGITAL FILTERING ON PCM TEST EQUIPMENT

Figure 3.8 shows a very crude representation of a PCM system. The voice signals are encoded into digital form at the telephone exchange nearest to the subscriber, and thereafter all transmission is handled digitally until the decoding is performed at the other end of the link to present an analog signal to the far-end subscriber.

Traditional analog test equipment can, of course, make end-to-end analog measurements, but when *fault finding* is carried out, we may wish to stimulate

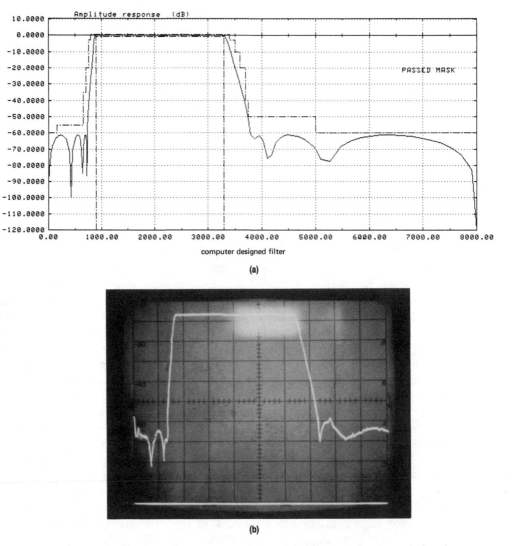

Figure 3.7   Example of a computer-designed digital filter. (a) Computer-designed filter. (b) Actual digital filter results.

the system with an analog signal but measure the result at a digital test point. Clearly, if a basically analog test instrument is to make measurements from an analog point to a digital point and also from a digital point to an analog point, then it requires A/D and D/A converters. Why then is one measurement preferred over the other? Consider the A/D converters. The output of the analog test set has to be converted accurately to digital form if its harmonic purity is to be maintained.

Similarly, the digital input to the digital test set must be an accurate rep-

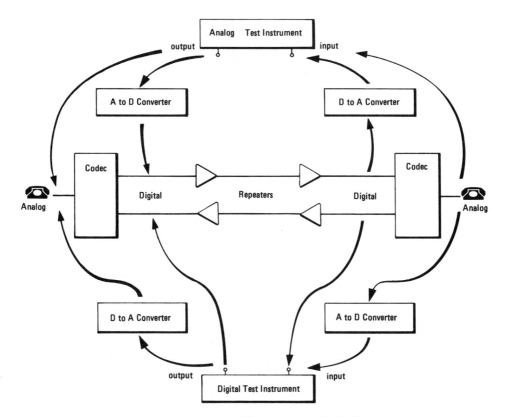

Figure 3.8   Simplified PCM test system block diagram.

resentation of the unknown analog signal. All the distortions of the analog signal introduced by the telecommunications network must be preserved, since it is these distortions that the test set is designed to measure. The two A/D converters must then be of comparable quality.

The same reasoning applies to the D/A converter used at the input to the analog test set. It has to operate on the expanded digital output of the PCM system, and this has to handle 13 bits. The D/A converter at the output of the digital test set, however, is operating on a known signal, usually a sine wave of precisely calculated digital form. The harmonic properties of the signal can be calculated exactly, and the D/A has to be good enough only to preserve this. There are some slight savings to be made, then, in the A/D converter of the *digital instrument*. The real benefits of the digital approach are as follows:

1. All required filters can be implemented by only one set of hardware.
2. No adjustments or periodic calibrations of filters are required.
3. Digital filters cannot drift with temperature changes.
4. The digital hardware lends itself to rapid automatic production testing.

These benefits can be realized, however, only if the digital approach is also cost effective. The large-scale integration of digital circuitry that is possible today holds out the promise of all the telecommunications filters we shall ever need being available on a single chip. Obviously, the large semiconductor manufacturers have not missed this opportunity to tackle a new market.

A good example of a commercially available digital filter chip is the INTEL 2920 shown in Fig. 3.9. Not only does this chip implement digital filters, but it also contains its own coefficient ROM and its own A/D and D/A converters. It is obviously intended as a direct replacement for an analog filter in applications where its 9-bit A/D capability is adequate. This includes a wide range of tele-communications applications—for example, codecs and modems. The telecommunications test equipment about which we are talking will be testing networks containing chips like this, so it is not surprising that we have to design test equipment to be an order of magnitude better than these chips. For example, a 12-bit A/D converter is frequently required for test equipment. Similarly, more bits are often required in the arithmetic portion of a digital filter for test equipment.

Finally, the ROM of the Intel 2920, having been designed to hold the coefficients for only one filter at a time, is not big enough for the wide variety of filters of Fig. 3.1. Thus we conclude that, at the moment, commercially available chips—although ideal for practical telecommunications functions—are not suitable for use in instruments that test these functions, so special-purpose chips must be used.

The photograph of Fig. 3.10 shows the amount of hardware required to implement any or all of the filters of Fig. 3.1. As an intermediate stage, it is just about cost effective compared to analog filters, but the compression of all that circuitry into the chip pictured alongside signals the *end of analog filters for telephone-channel test instruments*.

3.10 SUMMARY

We showed how digital signal processing can be applied to voice-channel measurements and that many advantages can be gained, both for the main telecommunications equipment and for the test gear. The test-equipment industry is currently implementing these techniques in modern products in two basic forms.

First, there is the replacement of traditional analog functions by digital implementations. This can give many advantages and allows engineers to adhere to well-proven measurement methods while simultaneously gaining several of the advantages that digital electronics offer. Second, new measurement methods are now made possible and are being studied. Progress along this front is, however, somewhat slower due to the need to agree on national and international standards.

Our view is that test-equipment technology will continue to advance along both of these paths and that new test-equipment products will be provided to help engineers deal with the ever-increasing demands upon their networks. Some of these products will contribute significantly to the need for monitoring and

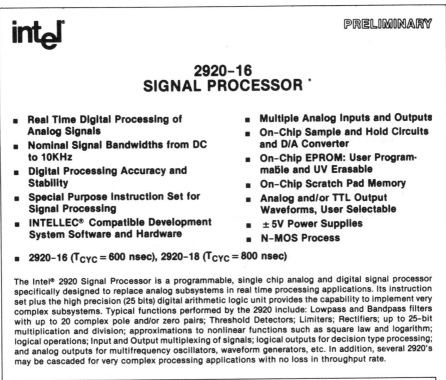

PRELIMINARY

2920-16
SIGNAL PROCESSOR *

- **Real Time Digital Processing of Analog Signals**
- **Nominal Signal Bandwidths from DC to 10KHz**
- **Digital Processing Accuracy and Stability**
- **Special Purpose Instruction Set for Signal Processing**
- **INTELLEC® Compatible Development System Software and Hardware**

- **2920-16 (T_{CYC} = 600 nsec), 2920-18 (T_{CYC} = 800 nsec)**

- **Multiple Analog Inputs and Outputs**
- **On-Chip Sample and Hold Circuits and D/A Converter**
- **On-Chip EPROM: User Programmable and UV Erasable**
- **On-Chip Scratch Pad Memory**
- **Analog and/or TTL Output Waveforms, User Selectable**
- **± 5V Power Supplies**
- **N-MOS Process**

The Intel® 2920 Signal Processor is a programmable, single chip analog and digital signal processor specifically designed to replace analog subsystems in real time processing applications. Its instruction set plus the high precision (25 bits) digital arithmetic logic unit provides the capability to implement very complex subsystems. Typical functions performed by the 2920 include: Lowpass and Bandpass filters with up to 20 complex pole and/or zero pairs; Threshold Detectors; Limiters; Rectifiers; up to 25-bit multiplication and division; approximations to nonlinear functions such as square law and logarithm; logical operations; Input and Output multiplexing of signals; logical outputs for decision type processing; and analog outputs for multifrequency oscillators, waveform generators, etc. In addition, several 2920's may be cascaded for very complex processing applications with no loss in throughput rate.

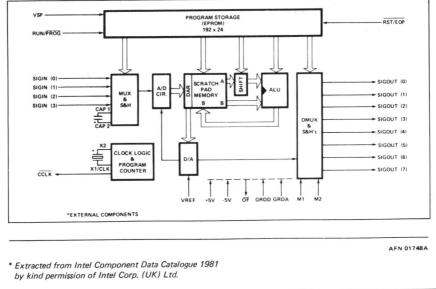

AFN 01748A

** Extracted from Intel Component Data Catalogue 1981
by kind permission of Intel Corp. (UK) Ltd.*

Figure 3.9 The INTEL 2920 signal-processing chip.

Figure 3.10 Photograph of digital filter discrete hardware and digital filter chip.

analysis of systems, thus making it possible for maintenance practices to be modified and provide a higher grade of service to end users.

REFERENCES

[3.1] IEEE. "Programs for Digital Signal Processing," Edited by the Digital Signal Processing Committee of the IEEE Acoustics Speech and Signal Processing Society, IEEE Press, New York, 1975.

[3.2] Hewlett-Packard 92835 Signal 1000 Software. This software contains the programs of [3.1] for use on the HP1000 F-series computers.

[3.3] Gold, X., and Y. Rader. *Digital Processing of Signals*, McGraw-Hill, N.Y., 1975.

[3.4] Cappellini, Constantinides and Emiliiana Cappellini. *Digital Processing of Signals*, Academic Press, London, 1975.

[3.5] Oppenheim, A. V., and R. V. Schafer. *Digital Signal Processing*, Prentice-Hall, Englewood Cliffs, N.J., 1975.

[3.6] Hamming, R. W. *Digital Filters*, Prentice-Hall, Englewood Cliffs, N.J., 1981.

[3.7] IEEE Standard P743. "IEEE Standard Methods and Equipment for Measuring the

Transmission Characteristics of Analog Voice Frequency Circuits,'' New York, N.Y., 1980.

[3.8] CCITT. "Specifications of Measuring Equipment," *Yellow Book*, Vol. 4, Recommendations of the 0 series, Geneva, Switzerland, 1981.

Note. References 3.3, 3.4, and 3.5 all provide good introductions to their subjects. Cappellini includes chapters on hardware implementation and applications. Hamming is particularly good on FIR filters and deals extensively with windowing techniques to improve these filters.

4

OVERCOMING INTRINSIC UNCERTAINTIES IN PCM CHANNEL MEASUREMENTS

M. B. DYKES

Hewlett-Packard Ltd.

4.1 INTRODUCTION

The move toward the integrated digital network and the development of single-chip pulse code modulation (PCM) codecs has called for separate performance tests on transmit and receive portions of PCM terminals and on subsystems within them. The resulting requirement for increasingly accurate measurements has highlighted the intrinsic uncertainties due to the coding method itself, as distinct from errors in either the channel or the test instrument.

The interaction of coding laws with the amplitude distributions of standard test signals results in quantizing noise and apparent gain even in an ideal PCM channel and sets limits to the accuracy of measurements. Careful choice of test signal and averaging time can minimize the effects. Additional impairments affect direct digital-to-digital (D-D) measurements and crosstalk measurements on PCM coders. To understand how test equipment operates and to choose good test parameter values, it is first necessary to understand the quirks of PCM and the ways in which conventional tests can be invalidated.

4.2 PCM QUANTIZING

4.2.1 The Quantizing Mechanism

Digital coding of analog signals quantizes the samples to discrete values; each sample, on reconstruction, suffers an error up to half a quantum step in magnitude. Figure 4.1 shows a simple coding law, with three equally sized level quanta. Considering the encoding of a single amplitude x, we can plot the quantizing error for any value of x.

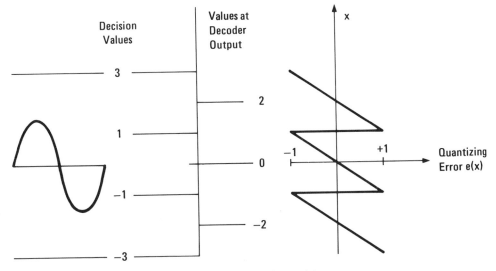

Figure 4.1   Distribution of quantizing error.

As the amplitude x is increased from zero and passes through a decision value from below, its own value is first understated and then overstated at the decoder output. A continuous signal will have samples at all values, and as the signal level is varied, so each of these samples will encounter decision values. The global effect on the level of the signal depends on the way the errors average together: At some levels, the signal will experience gain and at others, loss as the signal is transmitted through the channel.

4.2.2 Signal Statistics

The impact of quantizing on any particular signals depends on its amplitude distribution. Figure 4.2 shows the *rectified* probability density functions (pdf) of some common signals, all normalized to have a root-mean-square (rms) value of 1. The area under all the curves is 1 unit.

The pdf $p(x)$ of a continuous variable is such that the probability of a value in the small interval $x \pm \delta x/2$ is $p(x) \cdot \delta x$. The pdf of a sinusoid is particularly

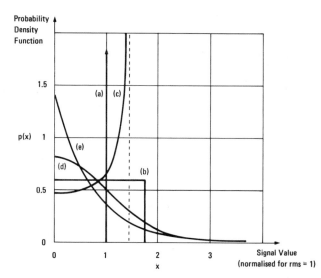

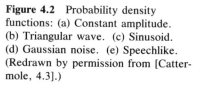

Figure 4.2 Probability density functions: (a) Constant amplitude. (b) Triangular wave. (c) Sinusoid. (d) Gaussian noise. (e) Speechlike. (Redrawn by permission from [Cattermole, 4.3].)

interesting, in that the waveform is concentrated at its extremes. As a result, the averaging-out of quantizing errors is relatively poor. Based on the simple uniform coding law, illustrated in Fig. 4.1, Fig. 4.3 shows the **quantizing gain** of an ideal coder-decoder (codec) pair for sinusoidal signals with peak amplitudes close to a decision value. The gain has a sharp minimum each time the peak of the sinusoid passes through a decision value; it then rises quickly to a maximum before falling away again. For a uniform coding law, the gain excursions decrease as the level of the sinusoid rises.

Figure 4.2 shows that the pdf of a sinusoid is very unlike that for speech. Gaussian noise is a much better approximation, and having no singularity in its

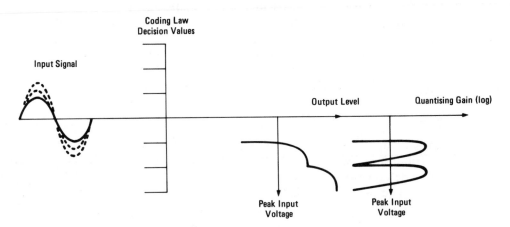

Figure 4.3   Quantizing gain: Uniform coding of a sinusoid.

pdf, it would be expected to have much better gain linearity performance—that is, to have quantizing gain values near zero.

4.2.3 Quantizing Noise Generation

The coding law illustrated in Fig. 4.1 has a uniform step size. It can easily be shown that such a law, when exercised by a signal with equal probability of all amplitudes (e.g., a triangular wave), gives rise to a constant quantizing noise power of $\Delta^2/_{12}$, where Δ is the size of the quantizing steps. [Cattermole, 4.3] (This is true if the peak-to-peak signal amplitude is any multiple of Δ.) Clearly, the signal-to-noise ratio (S/N) deteriorates as level decreases.

4.2.4 Companding

The coding laws actually in use for PCM telephony are not uniform but are approximately logarithmic, having progressively finer quantization expected at lower levels; such laws give approximately constant S/N over a wide dynamic range—a property of dominant importance in telephony. See Fig. 4.4.

The nonuniformity of the practical coding laws means that gain and noise impairments due to quantizing are reasonably bounded at higher levels but increase

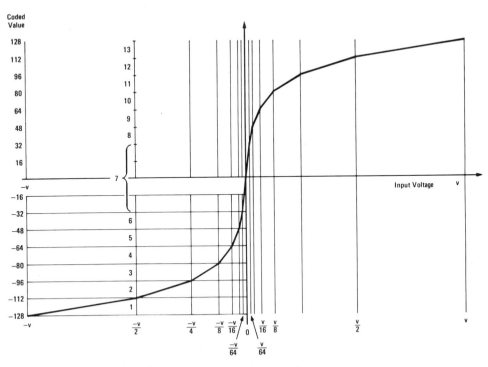

Figure 4.4   The A-law: A 13-segment piecewise-linear approximation to a logarithmic companding law.

markedly below a level of about 45 dB lower than the standard test level (abbreviated as −45 dBm0) for tones.

4.2.5 Half-channel Measurements

It is important to realize that quantizing impairments apply to half-channel measurements in exactly the same way as to full-channel measurements: In each case, there is a codec pair or equivalent (see Fig. 4.5). For example, an analog-to-digital (A/D) measurement, stimulating the analog input of a transmit terminal and analyzing its digital output, involves the encoding process of the terminal and an actual or simulated decoding process in the test instrument. Hence the result, on the transmit side alone, includes the intrinsic impairments of a *complete* ideal channel. One example of the practical significance of this is that the measured gains of the transmit and receive ends of a channel will not, in general, add up to the gain measured overall. Figure 4.6 illustrates the point for a test tone of 0 dBm0, at which the quantizing gain q in a practical system is about 0.02 dB.

For each result, the departure from ideal of the test channel is found by subtracting the quantizing gain from the measured value. Once this is done, the half-channel results combine as expected. In algebraic terms,

$$g_T + g_R = g_C \tag{4.1}$$

However, the quantities *measured* will be $g_T + q$ and so on, and

$$g_T + q + g_R + q = g_C + 2q \tag{4.2}$$

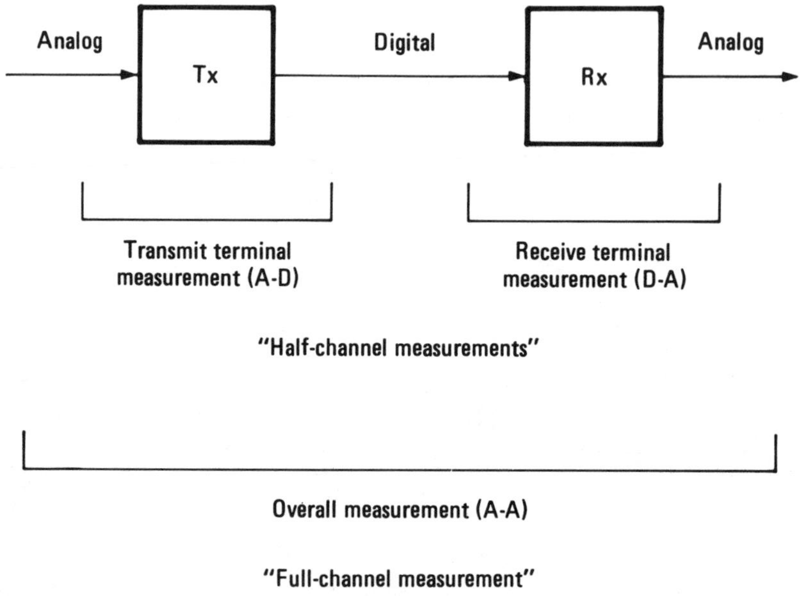

"Half-channel measurements"

"Full-channel measurement"

Figure 4.5 Measurement mode terminology.

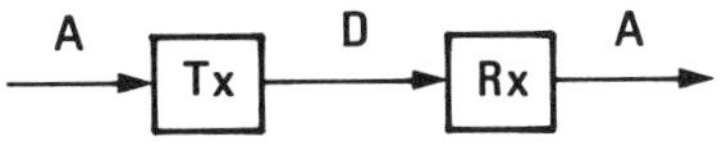

	Test Channel		Overall Results
	A → Tx → D → Rx → A		
	Half-channel Results		
Measured Gain (dB)	$m_T = +0.05$	$m_R = -0.07$	$m_C = -0.04$
Quantizing Gain (dB)	$+0.02$	$+0.02$	$q = +0.02$
Error of UUT (dB)	$g_T = +0.03$	$g_R = -0.09$	$g_C = -0.06$

Figure 4.6   The influence of quantizing gain on the summation of half-channel results.

Expressing the equation directly in measured quantities,

$$m_T + m_R = m_C + q \qquad (4.3)$$

Summation of half-channel results must, therefore, be approached with caution.

The intrinsic error of 0.02 dB, though small, is significant when test equipment accuracies approach ± 0.05 dB and codec chip gain specs are ± 0.15 dB. At lower levels, the quantizing gain can be much worse: For example, a tone at -54.3 dBm0 applied to an ideal μ-law channel suffers a loss of 0.2 dB, whereas with the A-law, a tone of -53.6 dBm0 experiences a loss of 0.6 dB. Further, the steepness of the error slope at these levels means that a calculation like the one in equation (4.3) is more complicated because the signal level seen by the Rx terminal will be different, having experienced gain in the Tx terminal; and so the values of quantizing gain could be different for the two sides! Figure 4.7 shows a configuration capable of full- and half-channel measurements without recabling.

4.3 QUANTIZING GAIN

4.3.1 Definition

The long-run **average quantizing gain** for a signal whose samples never repeat (q_{av}) is

$$10 \log_{10} \frac{1}{r^2} \cdot \sum_i p_i y_i^2 \quad \text{(dB)} \qquad (4.4)$$

where p_i is the probability of occurrence of PCM code i, y_i is the decoder output value associated with code i, and r is the rms value of the input signal. Also, p_i is derived from the pdf of the input signal. This calculation fails to eliminate the quantizing noise power and therefore overstates the level of the signal at the output, yielding the result that would be obtained by broadband measurement.

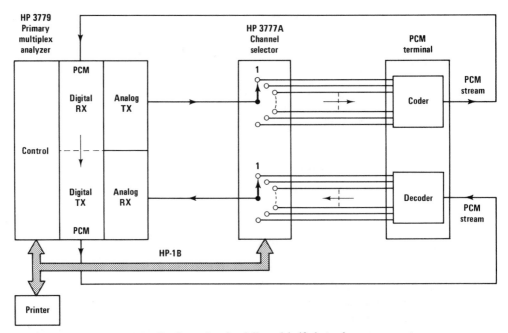

Figure 4.7   Configuration for full- and half-channel measurement.

If the signal pdf is such that $p(x)$ is reasonably constant over any one quantum step, then the quantizing noise may be removed in the formula

$$10 \log_{10}\frac{1}{r^2} \cdot \sum_i p_i \left(y_i^2 - \frac{\delta_i^2}{12} \right) \quad \text{(dB)} \qquad (4.5)$$

where δ_i is the size of quantum step i and the rest is as before. By this means, one can easily plot the quantizing gain of an ideal PCM channel for any signal of known pdf.

4.3.2 Quantizing Gain for Test Signals

Figure 4.8 shows the result for the μ-law exercised by a tone. The piecewise-linear sections of the coding law are referred to as segments, and on this diagram, the segment boundaries and decision values can be clearly seen, with the increase in error at low levels. The BSTR 43801 mask for half-channel gain tracking is shown on the same scale for comparison. [AT&T, 4.1].

Figure 4.9 presents similar information for the A-law, which has fewer segments and consequently greater error at low levels. [CCITT, 4.5] Figure 4.10 shows the performance of Gaussian noise—a great improvement in intrinsic error, which allows much better assessment and control of actual errors due to the UUT. This improvement is because of the pdf properties of Gaussian noise and explains why some gain-tracking measurements are made with a noise stimulus.

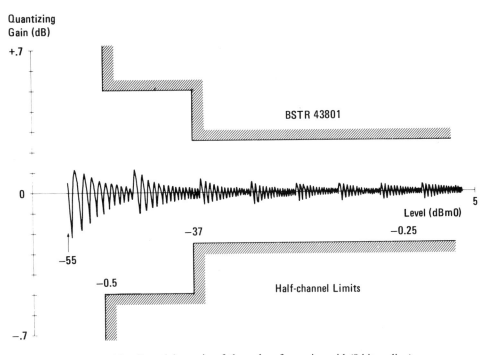

Figure 4.8   Quantizing gain of the μ-law for a sinusoid (8-bit coding).

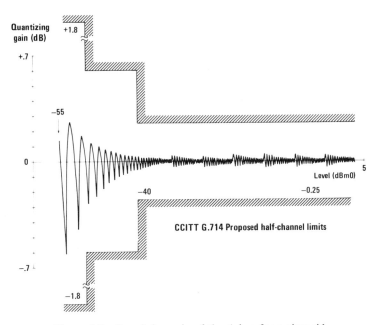

Figure 4.9   Quantizing gain of the A-law for a sinusoid.

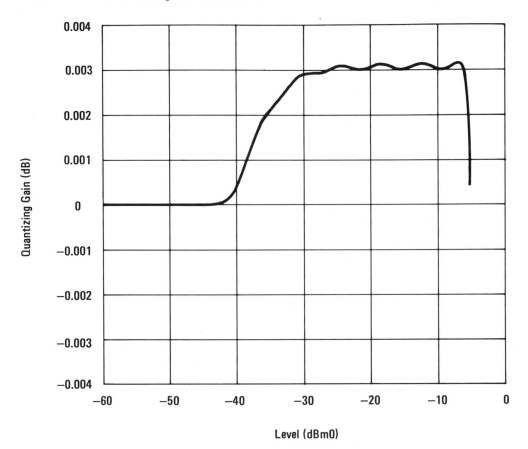

Figure 4.10 Quantizing gain of the A-law for Gaussian noise.

4.3.3 Effect of Bit Stealing

So far in this chapter, when the μ-law has been mentioned, 8-bit coding has been assumed. But most of the currently installed μ-law systems in fact employ $7\frac{5}{6}$-bit coding, using the least significant audio bit of every sixth sample for signaling purposes. This is known as **bit stealing.** Being a coarsening of the quantization, bit stealing brings with it a worsening of quantizing gain and noise.

4.4 QUANTIZING NOISE

4.4.1 Effect of Quantizing Noise on Measurements

The S/N characteristic of a logarithmic companding law has the general form shown in Fig. 4.11. At high levels, S/N is virtually constant; at low levels, any

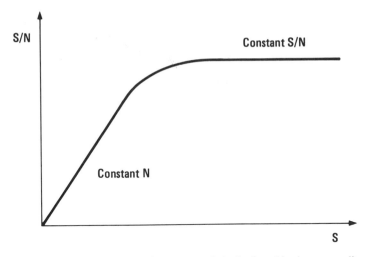

Figure 4.11   General form of S/N characteristic for logarithmic companding.

practical law approximates uniform quantization, noise becomes independent of signal level, and S/N falls off with level. Measurements made in the presence of significant noise suffer from level error and instability. To overcome these difficulties, it is useful to reject the quantizing noise with a selective filter. Keeping the error due to this cause alone, less than 0.05 dB for tones down to -55 dBm0 on a worst-case channel, requires a receiver bandwidth less than 350 Hz—allowing some margin for the nonuniform frequency spectrum of quantizing noise.

4.4.2 S/N *Characteristics of Standard Test Signals*

The S/N characteristics for practical coding of tones contain undulations at high levels. As the pdf singularity crosses coding-law segment boundaries, the dominant step size contributing to the quantizing noise changes by a factor of two. At the same time, whenever the singularity crosses any decision value, the quantizing error is maximum and the S/N is minimum. There are short-range and long-range variations, and the overall effect is as shown in Figs. 4.12 and 4.13.

The graphs also show masks for half-channel tests of S/N. In designing tests of quantizing distortion (QD) performance, it is worth bearing in mind that the intrinsic QD varies in level by about 4 dB peak-to-peak at higher levels and that small changes in input level can cause considerable changes in intrinsic QD. It is, therefore, important that test equipment has good control of transmitted level. Figure 4.14 shows the S/N characteristic for the A-law as applied to Gaussian noise. The improved pdf results in a smooth curve with much smaller variations. In Europe, QD measurements are usually made with Gaussian noise.

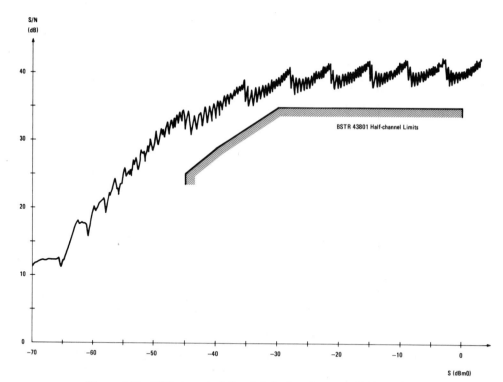

Figure 4.12 S/N characteristic of the μ-law for tones (8-bit coding).

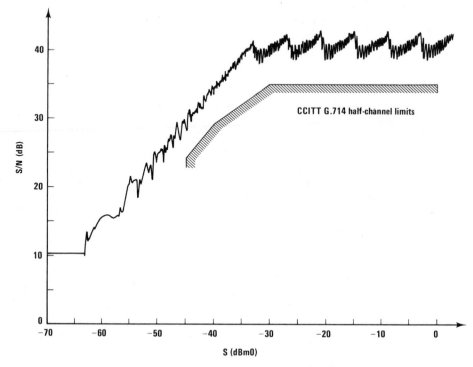

Figure 4.13 S/N characteristic of the A-law for tones.

118

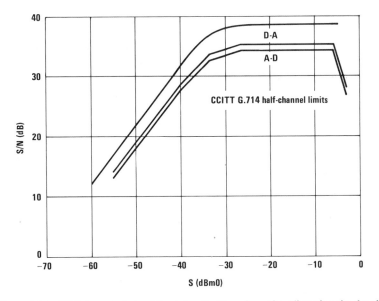

Figure 4.14   *S/N* characteristic of the A-law for Gaussian noise (distortion simulated in 3.1-kHz bandwidth).

4.5 AVERAGING

4.5.1 Test Signals

We have described the uncertainties due to quantizing in the PCM channel. Further uncertainties arise when practical measurements are attempted.

Test signals aim to exercise the channel in a representative and repeatable way: Ideally, the same set of PCM codes should be generated every time a test is run, and these codes should correspond to a representative time period of the waveform. The Gaussian noise signal recommended by the CCITT is a pseudorandom sequence generated, in principle, by repeating a set of sample values. Averaging over an integral number of periods therefore satisfies our requirements. How can the same be achieved for tone signals?

Clearly, any arbitrary frequency tone can be averaged simply by waiting long enough: The measured value oscillates about the average but gradually converges on it. Analysis of an unquantized system indicates less than 0.01 dB error in the measured rms after 34 periods of a sinusoid. However, this length of time is not constant, rising to 170 ms at 200 Hz but including only 80 samples around 3400 Hz.

4.5.2 Number of Samples and Start Phase

Including the effect of quantizing, we might intuitively expect that convergence would be slower and that there would be a constraint on the minimum number of independent samples to be averaged. Consider again a quantized sine wave: In Section 4.3 quantizing gain is described as the long-run average gain suffered by an unsynchronized signal—that is, one whose periodicity is unrelated to the PCM sampling rate. See Fig. 4.15. For such a tone, an infinite number of independent sampling phases exist, and the pdf of the sample values (before quantization) tends to that of the tone itself. Consequently, the averaged quantizing gain is exactly that calculated by the method in Section 4.3. By contrast, consider as an extreme case a synchronized 2-kHz tone with exactly four samples per cycle. No matter how long the averaging, only four phases are ever sampled, and the pdf of the unquantized samples consists of four impulses—a very poor approximation to the pdf of a tone and one that would only by accident have the same rms value after quantization, even if the average was over an integral number of periods. Since the sampling phases do not drift, the set of samples varies with the actual start phase of the measurement; so, too, will the quantizing gain, and the smaller the number of independent samples, the larger will be the limits of the variation.

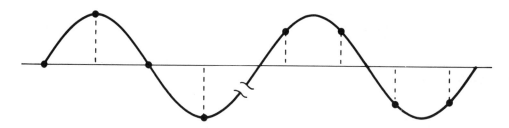

Figure 4.15   Sampling phases of a synchronous 2-kHz tone.

Figure 4.16 illustrates the mechanism. If the tone exercises the full range of the coding law of Fig. 4.1, the resultant waveforms are as shown in Fig. 4.17. Fourier analysis shows the level of the fundamental to differ by 3 dB between the two cases.

The conclusion is that quantizing gain is a function of

1. Start phase, with the size of the variation depending upon the strength of the relationship between the signal period and the sampling rate; and
2. the coarseness of quantization at the signal level in view.

Some practical values will give perspective. For a signal with 800 independent samples, the variation of quantizing gain with start phase at various signal levels is shown in Table 4.1 for three types of coding. This level of error is acceptable compared with other measuring errors.

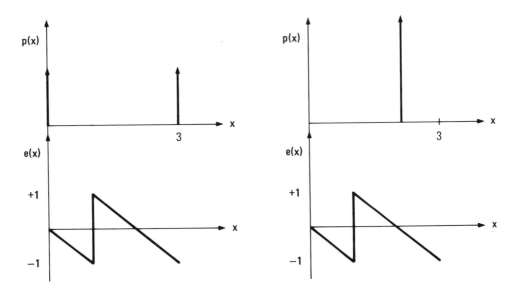

Figure 4.16 Pdf of unquantized samples of 2 kHz for extreme cases of Fig. 4.15, together with quantizing error function for simple law of Fig. 4.1.

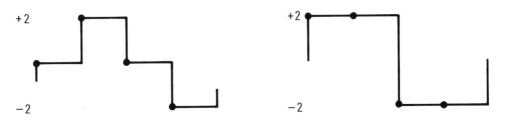

Figure 4.17 Waveforms of Fig. 4.15 after coding and decoding using the law of Fig. 4.1.

TABLE 4.1 Variation with Phase of Quantizing Gain for Tones Yielding 800 Independent Samples: Estimates Based on Simulations of Spot Levels. The Level of Error Is Acceptable Compared with Other Measuring Errors.

| (800 samples) | Gain variation with phase ($\pm$ dB) | | |
Level (dBm0)	A-law	μ-law (8-bit)	μ-law (7⅚-bit)
$\geqslant -40$	0.005	0.005	0.01
$\geqslant -60$	0.015	0.01	0.03
$\geqslant -75$	0.03	0.03	0.05

4.5.3 Independent Samples

The relationship between the signal frequency f and the number of independent samples n is simple. For a sampling frequency of 8000, the phase difference between adjacent samples is $2\pi f/8000$. The condition for averaging over a whole number of periods is

$$n\frac{2\pi f}{8000} = K2\pi$$

where K is an integer, or

$$n = \frac{8000K}{f}$$

For a given frequency, the number of independent samples is given by the lowest possible value of n and occurs when K is such that $n = 8000/h$, where h is the greatest common factor of f and 8000.

4.5.4 Integral-period Averaging

Averaging over a fixed number of samples yields a measurement time independent of frequency and with bounds on the error due to gain variation with phase. For $n = 800$, the averaging time is 100 ms, and there is a very high probability that every PCM code is exercised, even at the highest signal levels. Frequencies such as 210, 230, 270, 290, 310, . . . , 810, . . . , 1010, . . . , 3390 Hz have 800 independent samples contained in an integral number of periods and constitute a set adequate for test purposes.

A sinusoid may be correctly averaged over any multiple of a quarter-period only if the start phase is itself a multiple of $\pi/4$. For arbitrary start phase, multiples of a half-period are needed; even in this case errors can arise if, for example, the coder is offset. For this reason and to be equally representative of *all* the relevant PCM codes, it is desirable to average over an integral number of whole periods.

In this case, it is possible to visualize all the sampled phases mapped onto one cycle of the input and separated in phase by $2\pi/n$. It follows that the quantizing gain versus phase characteristic is independent of frequency for synchronized signals and that it is periodic in the spacing of the sampled phases. In the 800-sample case, the same set of samples will recur if the start phase is moved by a multiple of $2\pi/800$.

The quantizing gain (q) versus phase function for a synchronized signal can be normalized by reference to the long-run average quantizing gain for an unsynchronized signal of the same level (q_{av}). Figure 4.18 illustrates a typical normalized characteristic: $q_n = q - q_{av}$.

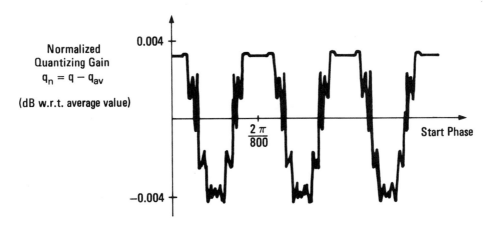

Figure 4.18 Variation with phase of quantizing gain as measured over 800 samples, normalized to the average value for a tone at -40 dBm0 coded with the μ-law using 8 bits and measured broadband.

4.5.5 Variation of Quantizing Gain with Phase

An average taken over an integral number of periods can be based on a different set of samples, depending on the start phase. This set is exactly repeated in every n sample, where n is $8000/h$. Averaging instead over an arbitrary interval yields a value that converges on the exact value for the particular start phase: Figure 4.19 illustrates this for two arbitrary phases. For lower values of i, the envelopes overlap.

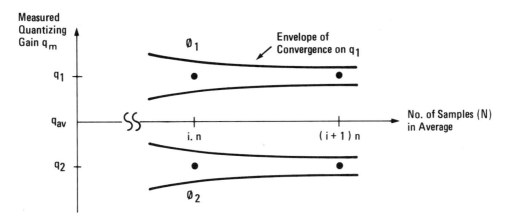

Figure 4.19 Notional quantizing gain vs. phase characteristics for various start phases, normalized to the average value and showing exact values when $N\Delta\phi = K \cdot 2\pi$ together with long-run convergence on the exact value.

We have discussed averaging over an integral number of signal periods; this may not be achieved in practice if the signal generator is not phase-locked to the PCM sampling rate. Such a situation is termed **skew**.

4.5.6 Skew Error

Figure 4.19 is correct even when the sampling rate and the test frequency are totally unrelated. The result measured for any given start phase will oscillate about the exact q_n value, and if the test signal is unsynchronized, the number of independent samples, n, becomes effectively very large, q_n values become very small, and all start phases tend to yield the long-run unsynchronized average quantizing gain, q_{av}. However, if both test frequency and sample rate are crystal-derived, the asynchronism can be controlled to ± 100 ppm, for example, and the situation is best treated as a slight departure from the synchronized case.

The effect is that the averaging interval is no longer an exact multiple of the signal period—that is, $800 \cdot \Delta\phi \neq K \cdot 2\pi$, where $\Delta\phi$ is the (actual) phase difference between adjacent samples. This extra source of error widens the tolerance on any single result. But because the start phase will slowly drift through all possible values, the start phase error (and skew error) can be averaged out of results taken with a skewed test tone, and q_{av} can be found. This is not so easy with a zero-skew tone, since the start phase error is constant and cannot be removed by averaging unless the tone generator is stopped and restarted between each measurement.

Figure 4.20 suggests the convergence of the measured value toward the exact value (q_0), together with the exact values at multiples of 2π. For slight deviations from $K \cdot 2\pi$, the values incur a small error. Figure 4.21 shows the skew error $q_s - q_0$ for a typical case.

As expected, the skew error is of the same order of magnitude as the phase variation, since, as just described, it tends to change the sample set in the same

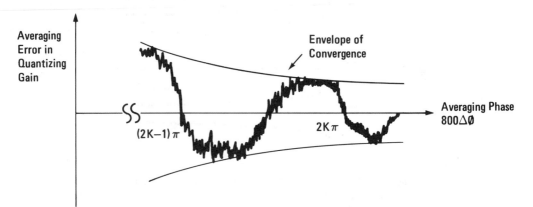

Figure 4.20 Notional form of averaging error characteristic for arbitrary start phase.

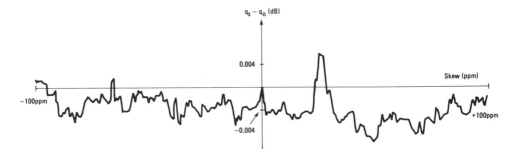

Figure 4.21 Gain error vs. skew characteristic for 800 samples of a 0.81 kHz -42 dBm0 signal starting at phase 0° with μ-law coding using 8 bits, measured broadband and normalized to the quantizing gain at zero skew.

sort of way. Similarly, the magnitude of the effect depends on the coarseness of the quantization—that is, on the level of test signal. Skew error in an unquantized system is frequency-dependent. However, below 3400 Hz the effects of quantization dominate skew error, and frequency effects may be ignored.

4.5.7 Effects of Bit Stealing

What is the effect of bit stealing on skew error? Since every sixth sample is affected, there are six possible results for each start phase; the more coarsely quantized samples add a further component to the quantizing gain versus phase relationship. A slower variation with phase is superimposed on the 8-bit characteristic (see Fig. 4.22). The magnitude of the slow variation depends upon the

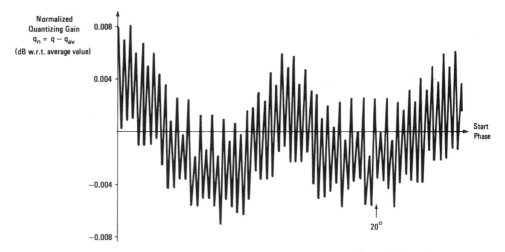

Figure 4.22 Variation with phase of quantizing gain as measured over 800 samples, normalized to the average value, for a 0.81-kHz tone at -40 dBm0 coded with the μ-law using $7\frac{5}{8}$ bits and measured broadband.

coarseness of the quantization and so is level-dependent and similar to that for the 8-bit variation itself.

As in the 8-bit case, the addition of skew superimposes several further components—varying with the signal frequency—and the total characteristic is quite complex. Measurement tolerances due to skew and gain variation with phase approximately double when bit stealing is present in a μ-law system.

4.5.8 Phase Shift Due to Quantizing

We have examined the effects of quantizing as they would appear to a level-detecting device. Clearly, though, a phase detector would also be affected: The errors in the reconstructed samples at the output of an ideal PCM channel can appear as phase shift. The phase detected depends on how the samples in the measurement set average together in the phase domain.

An ideal PCM channel includes delays due to the digital transmission path and phase shifts in filters. However, there is no long-run average phase effect for an unsynchronized sinusoidal signal due to the coding law itself, since all phases of the signal are sampled, and any effect on the first and third quadrants must by symmetry be cancelled by an opposite effect on the second and fourth quadrants. However, there will be phase jitter, whose magnitude and frequency will depend on the signal frequency.

By contrast, the set of phases sampled from a synchronized signal is finite, and the averaged phase error seen at the output is unlikely to equal zero, but it is constant for any given start phase of measurement. The greater the number of independent samples n, the lower the maximum phase error, and this depends on the strength of the relationship between the signal frequency and the sampling rate. Skew—that is, a slight departure from the synchronized case—gives rise to slow jitter.

Envelope delay measurements using DSB-AM techniques on PCM channels must take account of these effects. If unsynchronized signals are used, then long averaging times are needed, whereas if synchronized signals are used, large enough numbers of independent samples for both carrier and modulation frequencies should be chosen so that quantizing phase shifts are kept small.

4.6 D/D TESTS

Ease of interfacing, the advent of time division multiplex-frequency division multiplex TDM-FDM transmultiplexes (TMUX), and the desire to test hybrid loss directly have all contributed to the growth in transmission testing between digital points or **D/D** tests, [Feher, 4.9]. Based on the reasoning in Section 4.2, the signal path for a D/D test contains two codec pairs. Quantizing impairments, both gain and noise, are therefore compounded in an involved manner, depending, for example, on the presence of gain errors in Side A of the TMUX (Fig. 4.23). For good channels, the overall D/D quantizing gain is predictable, and the quantizing noise might be expected to be 3 dB worse.

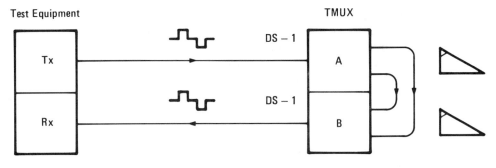

Figure 4.23 D/D test configuration for a TDM-FDM TMUX.

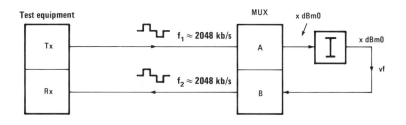

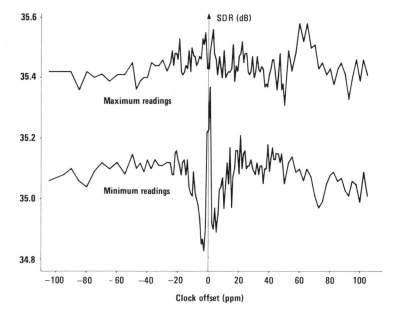

Figure 4.24 Effect of sampling clock offset on the noise-based SDR of an A-law multiplex measured D/D using the method of CCITT Rec. 0.131 and a level of −10 dBm0.

However, an extra source of uncertainty is provided by the relationship of the PCM sampling clocks in the two sides of the TMUX. If they are identical in frequency, then each sample encoded in Side A corresponds to one decoded in Side B; otherwise, the sampling instants drift through each other. This effect is particularly visible in the QD performance D/D. Figure 4.24 indicates the results of measurements on a looped-back PCM multiplex, which suggest the desirability of performing D/D tests with the PCM sampling clocks synchronized.

4.7 MEASUREMENT METHODS

Test equipment must satisfy requirements for speed, accuracy, and convenience. In addition, for PCM channel tests, the methods chosen must minimize errors due to intrinsic impairments.

4.7.1 Feasibility of Eliminating Uncertainty Due to Quantizing Gain

A test method that focuses attention on the actual errors in the test channel by removing the effects of intrinsic impairments is conceptually available. Consider a D/A measurement: If the actual set of PCM codes represented in the result is passed through an ideal decoder (easily simulated digitally) as well as the test channel, then the intrinsic quantizing gain for the actual test signal can be measured and subtracted from the test result. Such "corrected" results give a good indication of the quality of the decoder. But a graphical plot of results obtained in this way could be harder to interpret—for example, in the case of a faulty device—than if the characteristic of the coding law were visible.

The method relies upon knowing the ideally sampled values of the test signal, which is simple in the case of a digital stimulus, as used for D/A and D/D testing. However, to find the ideally sampled values of an analog test signal requires an accurate measurement (or prediction) of its absolute level. In addition, the influence of the start phase must be measured or ignored. If it is to be measured, we must know the actual sampling instants of the coder under test so that the sampled phases can be duplicated in a reference coder. This can be done using synchronized sampling, dc coupling and a stair-step waveform (tantamount to a *dynamic dc* test). However, few coders accept dc—even single-chip codecs now include preceding filters—so this method is limited in application. What is more, in all circumstances if the start phase error can be reduced, a more accurate final result is obtained by ignoring it rather than attempting to measure and eliminate it. The start phase error is reduced by keeping high the number of independent samples in the test signal.

In summary, then, to attempt the elimination of intrinsic error is straightforward in D/A and D/D measurements but can be error-prone in A/D and A/A measurements. Its area of application is in codec testing: In a transmission environment, "uncorrected" results should be used, because no assumptions can be made about the incoming signal and compatibility with results from existing test equipment must be maintained.

4.7.2 Multitone Test Signals

Measurement speed is a prime concern. Time can be saved by the use of a broadband test signal consisting of a comb of frequencies; signal processing in a receiver can then simultaneously yield the results corresponding to several conventional measurements. The difficulties with such techniques are their complexity and lack of standardization. A sinusoidal signal is well understood; it is easily generated. Interworking is possible, and measurements represent physical realities. It seems likely that complex test signals will gain acceptance in the long run, but at present the recognized test standards throughout the world rely for the most part on tone-based measurements.

4.7.3 Number of Samples and Representativeness

One way to save time while using a tone is to cut down the number of independent samples in the measurement set. The uncertainty in the quantizing gain is thereby increased; if it can be removed from the result, as described earlier, then accuracy is maintained. However, apart from the problems of eliminating the intrinsic quantizing gain that have already been mentioned, this method has a drawback that is not immediately obvious. Consider the case of a tone that is to be represented by a set of n samples and whose level is 0 dBm0 (we should expect similar results at most levels). The pdf of a sinusoid lets us calculate the occurrences of each PCM code. If the channel is then to be representatively tested for any signal level, all codes capable of contributing significant power should be exercised— for example, down to a quarter of peak voltage (voltages below this level account for only about 0.3% of total power).

Table 4.2 shows the probability of occurrence of codes by segment of either the A-law or the μ-law and, for segments 6 and 7, three figures that point to the number of samples n needed in the measurement. These figures assume

TABLE 4.2   Occurrence of PCM Codes for a 0 dBm0 Tone Using the A-law or the μ-law.

Segment number	± Code range exercised	p (segment) p_s	No. of ± codes (n_s) in range	$\dfrac{n_s}{p_s}$	Av. occurrence of codes	
					$n = 50$	$n = 800$
8	113 - 119	0.49	14	—	—	—
7	97 - 112	0.28	32	(116)	(0.4)	(7)
6	87 - 96	0.07	20	270	0.2	3
1-6	1 - 86	0.17	172	—	—	—

Note: Voltages between the peak and a quarter of the peak excite only 14 of the codes in segment 8 and 20 of the codes in segment 6. Codes are in the range ± 1 – 128.

equal probability of codes within one segment, which is correct to 5% for segment 6. The first shows the minimum value of n that yields an average occurrence greater than 1 for codes in that segment. The others show the average occurrence of these codes in sets of 50 or 800 samples.

The indications are that a minimum of around 270 independent samples is needed to ensure that all significant codes are represented; with only 50 samples, the majority are not. Relating this to the sampling frequency, we find that putting $n \geqslant 400$ allows the use of convenient 10-Hz multiples: <u>210</u>, 220, <u>230</u>, 260, <u>270</u>, <u>290</u>, <u>310</u>, <u>330</u>, 340, <u>370</u>, 380, <u>390</u>, ...

4.7.4 800 Samples, Selected 10-Hz Multiples

A consideration of quantizing gain and phase tolerances suggests that the use of 800-sample frequencies—underlined in the last paragraph above—yields a useful accuracy enhancement. The resultant averaging time of 100 ms is not unreasonable in relation to other components of the total measurement length, such as filter-settling time in both the test instrument and the signal path.

4.7.5 Limitation of the Digital Milliwatt

It is worth noting that the standard 1-kHz **digital milliwatt** sequence of 8 samples is itself subject to the foregoing objection. Quantizing gain is not a problem: The signal was chosen as the definition of 0 dBm0 precisely because it experiences a gain very close to the average of a sinusoid of the same peak voltage: $q_n \approx -0.017$ dB; because the sampling phases are fixed, there is no variation on that score. However, because it exercises only four PCM codes, the digital milliwatt does not provide a very good means of measuring the average gain of a decoder.

4.7.6 Advantages of Gaussian Noise as a Test Signal

Another special test signal in use for PCM channels is, of course, pseudorandom Gaussian noise [CCITT, 4.6]. Two important parameters remain inadequately specified: The maximum departure from a Gaussian pdf (which determines the quantizing gain) and the sequence length (which determines the feasibility of interworking).

However, the reasons for its adoption have been made clear: The lower intrinsic uncertainties associated with it allow much closer specification of both transmission equipment and test instrumentation in the digital domain. For example, the CCITT are specifying an instrument to make A/D and D/A measurements using tone and noise signals, and Table 4.3 shows limits that have been discussed for its digital transmitter [CCITT, 4.7].

4.7.7 Measurement Accuracies on PCM Channels

Finally, then, what kind of accuracies may we expect for tone-based measurements on PCM channels? The sources of error are skew and quantizing gain variation

TABLE 4.3 D/A Tester: Proposed Tx Level Accuracy Spec (A-law) [CCITT, 4.7].

Level (dBm0)	Level tolerance (dB): tone	Level tolerance (dB): noise
≥ − 30	± 0.05	
≥ − 49	+ 0.13 - 0.2	} ± 0.1
≥ − 60	+ 0.5 - 1.2	

with phase; intrinsic uncertainty is due to average quantizing gain. The effect of quantizing noise can be eliminated by selective filtering.

Table 4.4 indicates that using low-skew tones and averaging over 800 independent samples can result in measurement tolerances that are reasonably balanced between modes and, taken together with the quantizing gain, are well within the applicable specifications for transmission equipment.

TABLE 4.4 Comparison of Error Sources, Quantizing Gain, and PCM Multiplex Specification (CCITT) for the Main Tone-Based Level Measurements in Three Modes, Assuming 800-sample Averaging, Skew Better than ± 100 ppm and Selective Measurement.

Measurement	Mode	Errors due to measuring equipment			Intrinsic uncertainty	
		"Analog" errors	"Digital" errors	Total measurement tolerance	Quantizing gain	UUT "spec"
Gain 0dBm0	A/A	0.045	—	0.05	+0.02 −0.03	0.5
	A/D	0.08	0.04	0.12	+0.02 −0.03	0.3
	D/A	0.09	0.01	0.10	0.02	0.3
Gain vs. Frequency 0 dBm0 300–3400 Hz	A/A	0.08	—	0.08		0.3
	A/D	0.01	0.07	0.08	NIL	0.15
	D/A	0.08	—	0.08		0.15
Gain vs. Level 0.81 or 1.01 kHz −10−−55 dBm0	A/A	0.18	—	0.18	} +0.12 +0.03 μ, { } +0.25 +0.16 A {	3.0
	A/D	0.07	0.13	0.20		(1.5)
	D/A	0.11	0.04	0.15	0.09μ, 0.18A	(1.5)

Notes: "Analog" errors refer to analog Tx and Rx.

"Digital" errors refer to digital Tx and Rx.

Quantizing gain variation with phase is associated with the Rx error.

UUT "spec" is the tighter of BSTR 43801 and CCITT G712, etc. [AT&T, 4.1], [CCITT, 4.4], [CCITT, 4.5]

Tolerances on quantizing gain for A/A and A/D modes allow for ± 0.1-dB gain error before the coder.

Total measurement tolerance is for 7⅝-bit μ-law (the worst case).

4.8 CODER ENHANCEMENT OF CROSSTALK

A further uncertainty due to quantizing arises in the measurement of crosstalk contributions, which disturb the coder of a channel. This uncertainty is because the apparent gain of a coder to a small perturbation, such as a crosstalk signal, depends on the instantaneous voltage present at the coder input: If it is midway between decision values, then perturbations of less than half a code level step have no effect at all, whereas an idle coder with a steady input voltage close to a decision value changes its output code for even the least perturbation. The latter effect is known as **coder gain enhancement** and accounts for the variability of crosstalk readings even on idle coders of similar design. Clearly of interest is the crosstalk performance under traffic conditions, when the coder has at its input a range of voltages and the gain to perturbations has some average value. For this reason, it is becoming common to stimulate the coder of a disturbed channel with an auxiliary signal to sweep the coder through a representative range of input voltages. The auxiliary signal must not contain energy within the crosstalk measurement frequency band. The validity of the crosstalk measurement depends upon the statistics of the auxiliary signal, so a Gaussian noise signal is recommended to minimize uncertainty in the results. The level of the signal is chosen to provide a good averaging effect while remaining within the lowest segment of the coding law; the finite rejection of the instrument receiver filter provides another upper limit. For noise signals, the useful range is about −66

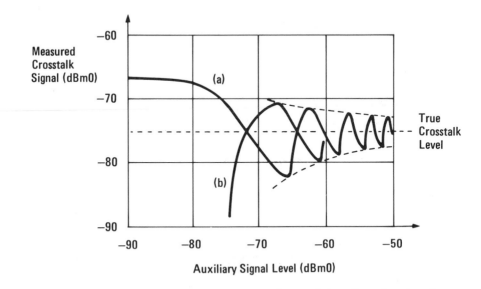

Figure 4.25 Crosstalk measurement uncertainty and the effect of a sinusoidal auxiliary signal for the A-law; (a) Idle input voltage on a decision value. (b) Idle input voltage midway between decision values. (Redrawn by permission from [Kühne, 4.8].)

to -41 dBm0 (A-law) or -72 to -53 dBm0 (μ-law). Furthermore, the presence of the auxiliary signal generates quantizing noise within the receiver bandwidth, which provides an effective noise floor to crosstalk measurements.

The effect of the idle input voltage on the measurement of a -75-dBm0 crosstalk signal at the input of an A-law coder may be read from the y-axis intercepts of Fig. 4.25. It can be seen that the measured values converge as the auxiliary signal level is increased. However, the pdf of a sinusoidal auxiliary signal itself interacts with the decision values, and considerable uncertainty remains. Figure 4.26 shows the performance of a Gaussian auxiliary signal: Measurement uncertainty is eliminated for levels greater than about -68 dBm0.

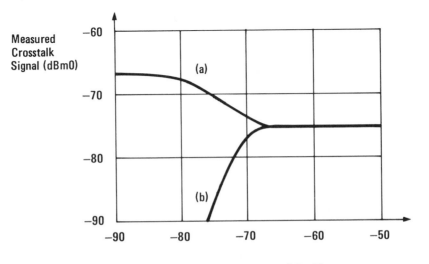

Figure 4.26 Crosstalk measurement uncertainty and the effect of a Gaussian auxiliary signal for the A-law: (a) Idle input voltage on a decision value. (b) Idle input voltage midway between decision values. (Redrawn by permission from [Kühne, 4.8].)

It is fairly simple to connect test equipment so that disturbed channels are presented with an auxiliary signal at the correct time (see Fig. 4.27). Using Gaussian noise as the auxiliary signal can completely eliminate uncertainty, but worthwhile improvements in accuracy are still obtainable using a tone.

4.9 ACKNOWLEDGMENTS

I am indebted to Andrew Batham for graphs of quantizing gain variation with phase and skew, to Mike Smith for ideas on phase impairments, and to Malcolm Paterson for work on D/D sampling skew and the behavior of ideal Gaussian noise.

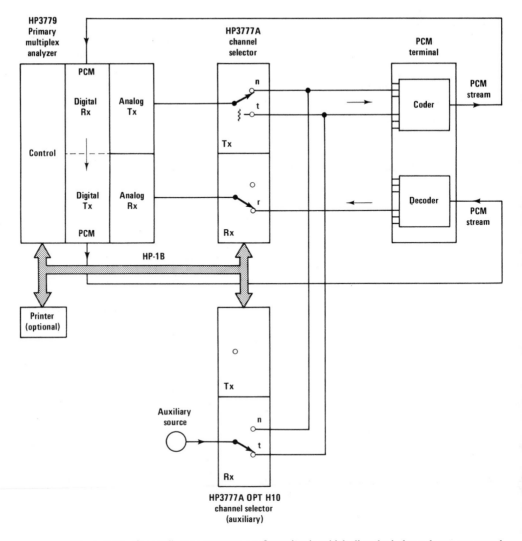

Figure 4.27 Crosstalk measurement configuration in which disturbed channels are presented with an auxiliary signal at the correct time.

PROBLEMS

4.1 A maintenance technician is suspicious of the A-law PCM test gear. The technician loops its digital output to its digital input and sends a tone at -51.1 dBm0. The resulting measurement is -51.48 ± 0.01, and the instrument is sent for repair, with a note that the measurement was out of specifications and in spite of all-digital processing, the results varied! Describe how the repair engineer replied to the technician's complaints.

4.2 The end-to-end QD performance of a point-to-point PCM link is observed to vary. In an effort to locate the source of the variation, the QD is measured A/A on each of the terminals looped back on itself at the digital side. Both results are stable. Comment on each of the following.

(a) Possible reasons for these observations.

(b) The validity of tests made on looped channel banks to qualify them for point-to-point operation.

4.3 The report of a codec chip designer is under review by his boss. She reads that the send side gain, the receive side gain, and the overall gain of the chip at a particular level are all 0.6 dB, yet the report concludes that the chip design is working well. Which of the following should the supervisor do?

(a) Fire the designer.

(b) Scrap the test equipment.

(c) Go to a "PCM basics" course.

REFERENCES

[4.1] AT&T. "Digital Channel Bank Requirements and Objectives" (BSTR 43801), Facilities Systems Engineering, AT&T Co., Basking Ridge, N.J., 1982.

[4.2] Bell Telephone Laboratories. *Transmission Systems for Communications*, Western Electric Co., Winston-Salem, N.C., 1971.

[4.3] Cattermole, K. W. *Principles of Pulse Code Modulation*, Iliffe, London, 1969.

[4.4] CCITT Rec. G.712. "Performance Characteristics of PCM Channels at Audio Frequencies," *Yellow Book*, Vol. 3, No. 3, ITU, Geneva, 1981.

[4.5] CCITT Draft Rec. G.714. Study Group XVIII. "Final Report to the VIIIth CCITT Plenary Assembly (Part IV)" including "Separate performance characteristics for the send and receive sides of PCM channels applicable to 4-wire voice-frequency interfaces," Draft Recommendation G.714, COM XVIII, No. R31, June, 1984.

[4.6] CCITT Rec. O.131. "Specification for a Quantizing Distortion Measuring Apparatus Using a Pseudo-Random Noise Stimulus," *Yellow Book*, Vol. 4, No. 4, ITU, Geneva, 1981.

[4.7] CCITT. "Draft Specification for Equipment to Measure the Performance of the Send and Receive Sides of a Primary Multiplex," COM IV, No. 138-E, February, 1979.

[4.8] Kühne, F. "Use of Auxiliary Signal in Measuring Intelligible Crosstalk in PCM Multiplex Systems," *Siemens Forschungs- und Entwicklungsberichte*, Vol. 7, No. 4, Springer-Verlag, Munich, 1978.

[4.9] Feher, K. *Digital Communications: Satellite/Earth Station Engineering*, Prentice-Hall, Englewood Cliffs, N.J., 1983.

5

ERROR PERFORMANCE ANALYSIS OF DIGITAL TRANSMISSION SYSTEMS

PETER HUCKETT and GEOFF THOW

Hewlett-Packard Ltd.

5.1 SUMMARY

Evolving integrated digital networks (IDNs) and integrated services digital networks (ISDNs) require definition and analysis of error performance of digital line sections comprising the network. This chapter describes some of the methods currently in use, development of international standards, and appropriate measurement equipment for digital muldex (multiplex-demultiplex), digital radio, and digital line systems (over copper or glass).

5.2 INTRODUCTION

The basic measure of performance of any digital transmission system is the probability of any transmitted bit being received in error. In general, these errors occur in the regenerative repeaters of digital line sections (be they cable, fiber or radio links), which are used to achieve the most significant advantage of digital transmission over analog transmission, namely, distortion in the service carried almost independently of distance. This can be achieved only if the spacing of the regenerators and other contributing sources of error are controlled in such a way to give a "low enough" overall error rate [CCITT, 5.1]. For example, line-induced errors may be modified or magnified by the operation of terminal equipment such as descramblers. Several other parameters of the system may contribute to errors—for example, signal level, noise level, and timing jitter—but the most important criterion on which a system is judged is its error performance and the effect of those errors on the service being carried. The implementation of an ISDN, where the integrated functions of switching, terminals, and transmission of an IDN are used to provide digital transmission of all customer services over a common network, can be realized only if the error performance of individual line sections and overall digital connections is defined and measured [CCITT, 5.2]. The concept of common transmission path irrespective of service carried may in fact be difficult to achieve in practice because of the following reasons:

1. Many of the terminals and transmission links already in use have been designed and their performance optimized for conventional telephone voice.
2. The main reason for (1) is that over 90% of the traffic carried by most national and international networks is speech, and this balance is likely to be maintained for some time to come.

With the increasing introduction of new digitized services within the framework of a partial ISDN, it is becoming increasingly important to characterize both existing and new digital transmission links in terms of *error distribution* as opposed to just total error count or error ratio measured at spot times.

One other parameter, timing jitter, also has a significant effect on the service being carried by a digital transmission link, particularly if that service is a PCM-encoded wideband signal, such as television. Chapter 6 deals with the fundamental limits of jitter tolerance in digital transmission systems and the formation of jitter specifications for digital line sections within an ISDN [Thow, 5.3]. We restrict ourselves here to considering the question of error performance and methods of measuring it.

5.3 ERROR MEASUREMENTS

There are two general approaches to error measurements:

1. **Out-of-service testing,** where the transmission equipment is stimulated by a test pattern, usually a pseudorandom binary sequence (PRBS) to simulate traffic and the output of the link compared bit-by-bit with a locally generated, error-free reference pattern in an error detector. Note that this type of test requires that live traffic be removed from the link, implying a loss of revenue. It is therefore more suited to production testing and installation situations, particularly in the case of high-capacity links (see Fig. 5.1).
2. **In-service testing,** where some property of the traffic being carried, the line transmission code being used, or some added pattern (to the traffic) is monitored continuously for error conditions. This is particularly relevant to maintenance situations, where it is important to have at any instant an indication of the state of a link, and, if possible—particularly for high-

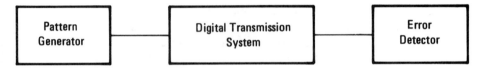

Figure 5.1 Basic out-of-service test configuration.

capacity routes—to have the link be self-supervising with automatic protection switching (see Fig. 5.2).

Before considering each case in more detail, it is perhaps worth reviewing some basic concepts of error measurements. The probability of error in any transmitted bit is a statistical property and has to be treated as such. Any attempt to measure this over a given time period can be expressed in various ways, the most common of which is:

$$\text{Bit error rate (BER)} = \frac{\begin{array}{c}\text{Number of errors}\\ \text{counted in the}\\ \text{averaging interval}\end{array}}{\begin{array}{c}\text{total number of}\\ \text{transmitted bits}\\ \text{in the}\\ \text{averaging interval}\end{array}} \qquad (5.1)$$

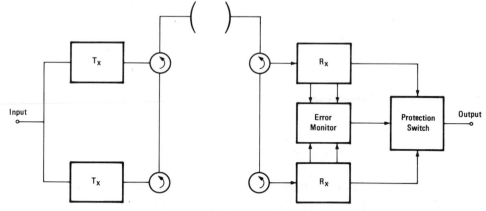

Figure 5.2 Example of in-service error monitoring.

Clearly, the result will have a statistical variance from the long-term mean error rate dependent upon the size of the sample taken from the population—in this case, the number of errors counted. Three methods of computing BER are in general use and are illustrated in Fig. 5.3.

The most common of these is Method 2, counting errors after a fixed, **repetitive gating period,** in which case the variance in the result will be continuously

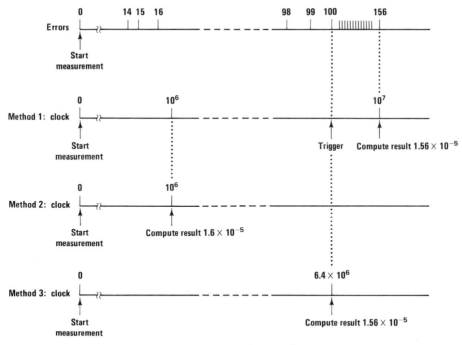

Figure 5.3   Methods of computing BER.

changing, so it is normal to give some kind of warning if the variance exceeds generally acceptable levels. As shown in Fig. 5.3, the most widely accepted level is 10%—that is, an error count of at least 100 errors is required. As will be shown later, the gating period, or averaging interval, is usually related to the service being carried by a particular link and the discernible effect errors have over that period.

5.3.1 Out-of-Service Testing

The choice of test pattern is usually made between a PRBS, to simulate traffic or specific word patterns to examine the transmission system for pattern-dependent tendencies or critical timing effects, and ISDN, the concepts of which require bit-sequence independence of any transmission path within the network. See Fig. 5.4.

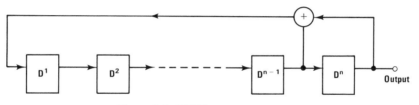

Figure 5.4   PRBS pattern generator.

The test pattern is normally encoded in some interface code to allow single port connections of the data signal between terminal and transmission equipment. Apart from the lower-speed transmission links, which also use this interface code as a line transmission code, the interface code is not particularly relevant to the choice of test pattern and resulting observed error performance of the link as a whole. What is more important is the choice of binary sequence and the resulting spectral and run properties in the particular system under test.

These properties may be summarized as follows:

- Sequence length in bits.
- Shift register configuration, which defines binary run properties.
- Spectral line spacing, which depends on bit rate (see Fig. 5.5).

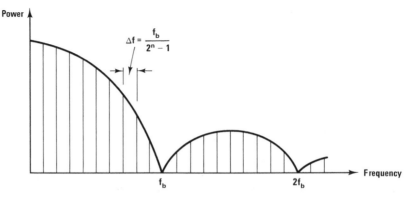

Figure 5.5   Spectrum of a binary NRZ PRBS.

Two PRBS patterns have been standardized by the International Telegraph and Telephone Consultative Committee (CCITT) for testing digital transmission systems [CCITT, 5.4]. These are based on 15-stage and 23-stage shift registers and are summarized in Table 5.1.

TABLE 5.1   Test Patterns

Bit rate (f_b)	Sequence length (n)	Shift register configuration (polynomial)	Spectral line spacing $(\Delta f = f_b/n)$
1,544 kb/s	$2^{15} - 1$ bits	$D^{15} + D^{-14} + 1 = 0$	47.1 Hz
2,048 kb/s	$2^{15} - 1$ bits	$D^{15} + D^{-14} + 1 = 0$	62.5 Hz
6,312 kb/s	$2^{15} - 1$ bits	$D^{15} + D^{-14} + 1 = 0$	192.6 Hz
8,448 kb/s	$2^{15} - 1$ bits	$D^{15} + D^{-14} + 1 = 0$	257.8 Hz
32,064 kb/s	$2^{15} - 1$ bits	$D^{15} + D^{-14} + 1 = 0$	978.6 Hz
34,368 kb/s	$2^{23} - 1$	$D^{23} + D^{-18} + 1 = 0$	4.1 Hz
44,736 kb/s	$2^{15} - 1$	$D^{15} + D^{-14} + 1 = 0$	1,365.3 Hz
139,264 kb/s	$2^{23} - 1$	$D^{23} + D^{-18} + 1 = 0$	16.6 Hz

Inspection of Table 5.1 leads to the following observations:

1. Since the low frequency performance of digital transmission links tested with these patterns is a function of both bit rate and sequence length, it might be argued that $2^{15} - 1$ is not a long-enough pattern for systems operating at bit rates greater than 2 Mb/s. This is probably more true for systems that operate on essentially binary information—for example, fiber and radio—than those that use true ternary line codes, which may modify the spectrum of the transmitted signal. Since more and more systems are employing scrambling to randomize the signal to be transmitted, it is difficult to generalize.

2. Adequate (i.e., close enough) spectral line spacing is important when testing systems containing relatively narrow band (high Q) clock timing recovery circuits in order to see the jitter contribution of these and its effect on error performance [Thow, 5.3; Bell Northern Research, 5.6].

3. The choice of shift-register configuration affects the run properties of the PRBS, which, in turn, affects the jitter performance in terms of the length of zero blocks over which phase error is accumulated by the timing recovery circuits [Thow, 5.3]. This leads to pattern-dependent jitter and, if not controlled, to regenerator errors. Line coding or scrambling are two frequently used methods to control this. However, studies have shown that jitter observed on these digital test patterns may be different to that observed on live traffic [Netherlands Administration, 5.5; Bell Northern Research, 5.6]. This may be due to the unfortunate choice of shift register configuration for $2^{15} - 1$ leading to peculiar run properties of the binary PRBS (see Fig. 5.6) or a combination of this with the repetition rate of the PRBS approaching that of the jitter bandwidth of the system under test (6) (see Item (2.)).

Before moving on, it is perhaps worth considering the error detector in more detail. The normal method of error detection, which is now internationally standardized [CCITT, 5.4], is closed-loop bit-by-bit comparison of the input against a local reference pattern, both in binary form—that is, after any interface coding has been removed. Optionally, some error detectors also provide bit-by-bit comparison at the code level to allow error polarity distinction and the separation of errors of *commission* from those of *omission*. Either case raises the question of synchronization of the two patterns to be compared before error detection can commence. In order to synchronize rapidly yet remain in synchronism at high error rates or during large error bursts, an error detector **sync criterion** has to be established that has variable gate times over which a test for sync is made.

The sync criterion may be expressed as

$$\text{Sync gain BER} = \text{sync loss BER}$$

$$\frac{x}{n} = \frac{X}{N} \tag{5.2}$$

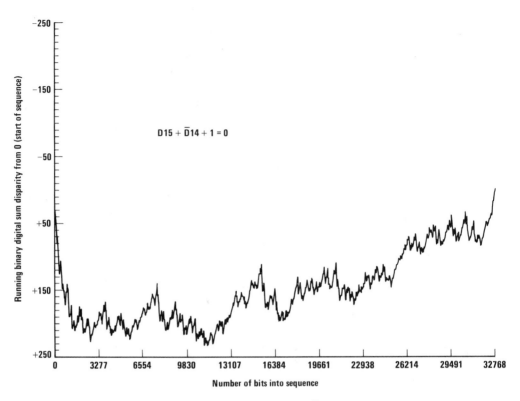

Figure 5.6   Run properties of CCITT $2^{15} - 1$ binary PRBS.

where $x \ll X$ (number of errors counted in the sync test interval)

$n \ll N$ (number of bits over which a test for sync is made)

(To avoid oscillation, it is normal to make the sync loss BER equal to the sync gain BER.)

The normal method of achieving synchronization (see Fig. 5.7) is to open the feedback loop in the reference pattern shift register and feed the input data signal into the register until it is full, close the feedback loop, and test for sync. Clearly, two PRBS patterns out of sync have a BER of approximately 0.5, so the sync criterion must be lower than this.

Other types of error detector have been proposed and used in the past, notably the *open-loop* detector first proposed by R. J. Westcott of the British Post Office (now British Telecom) [Dieckman, 5.7]. This error detector has the feedback loop broken (see Fig. 5.8).

This type of error detector does not need synchronization but does not give correct results at high error rates or for bursts of errors because of cancellation effects. It does operate correctly at low error rates and for errors having a normal distribution, but—as discussed in the next section—it is now generally agreed that for most digital transmission systems, errors do not always occur at

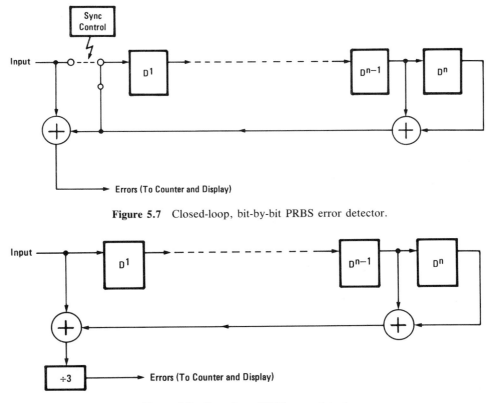

Figure 5.7   Closed-loop, bit-by-bit PRBS error detector.

Figure 5.8   Open-loop PRBS error detector.

randomly spaced intervals according to a normal distribution and are often *bunched,* or *clustered.* Thus an open-loop error detector is not to be recommended and will not be considered further.

5.3.2 In-Service Testing

The most common method of in-service error monitoring is line code violation detection, and a specification for one instrument to do this is now standardized by CCITT [CCITT, 5.8]. This particular instrument is for low-speed bit rates where the interface codes used, AMI, HDB3, B6ZS, and B8ZS, are the same as the line code used on paired cable or coaxial cable line systems. It should be noted that this type of error monitoring is relevant only where there are no code-conversion processes over the line section being monitored. Some systems have attempted to avoid this problem by monitoring the line code for errors and reinserting those errors in the interface code after code conversion. This allows conventional code-violation monitoring and automatic protection switching based on the violation error rate. An example of this is the Western Electric T1D line system [Anderson, 5.9].

In attempting to relate code error rate to the resulting binary bit error rate, we must take into account the terminal decoder error-extension factor. The relationship between true line error rate (bit-by-bit comparison at the line code level), observed code violation error rate and resulting binary error rate depends on the code violation and decoding rules used. For low-speed bit rates and simple pseudoternary codes such as AMI, HDB3, B6ZS and B8ZS, the standards in use are [CCITT, 5.8]:

Code Violations

AMI:	Two consecutive marks of the same polarity.
HDB3:	Two consecutive bipolar violations of the same polarity.
B6ZS and B8ZS:	Two consecutive marks of the same polarity excluding violations caused by the zero substitution code itself.

Decoding Rule

AMI:	Simple rectification.
HDB3:	Detection of a bipolar violation preceded by two consecutive zeros and substitution of four consecutive zeros in place of the bipolar violation and its three preceding digits.
B6ZS:	Detection of the sequence 0V10V1 and substitution of it by six consecutive zeros (where V = bipolar violation).
B8ZS:	Detection of the sequence 00V100V1 and substitution of it by eight consecutive zeros (where V = bipolar violation).

For higher-order, true ternary line codes such as 4B3T, MS43, FOMOT, and 6B4T, it is common to monitor the digital sum variation (DSV)—that is, the running polarity disparity—as a means of in-service error detection. Line errors will cause "forbidden" states in the ternary words, which push the DSV outside its normal bounds.

An alternative to line code-violation monitoring, which gets over some of the limitations discussed, is monitoring of a fixed pattern always present in the transmitted signal. This can take many forms, the most common of which is the frame alignment signal of the terminal multiplex originating the digital traffic. An alternative is to use some added *housekeeping* bits or *trace* signal, which can be monitored. An example of this is the parity bit monitoring possible in DS-3 level systems based on checking the parity of each preceding frame against the transmitted parity bits in the next frame. Both of these approaches have the attraction that, in an overall digital connection consisting of multiplexing-demultiplexing and a number of digital line sections at various bit rates, a fixed pattern within the original signal can be checked at various points along the route. This can be achieved wherever the original appears or via demultiplexing from system monitor points at any interface along the route (a sort of "digital SLMS with stored TDM plan" analogous to the checking of pilots in FDM analog transmission systems).

As an example, consider the 2-Mb/s frame alignment signal (FAS) monitor for systems conforming to CCITT Rec. G.732 now standardized and in use in Europe [CCITT, 5.10]. This instrument detects the presence of errors within the 7-bit FAS contained in time-slot zero (TSO) of every other frame of the 32-TS frame structure and predicts the average error rate in the overall traffic signal from a count of FAS errors. The prediction is based on two key assumptions:

1. The errors occur in a random manner and thus follow a Poisson distribution.

2. The average error rate is present for the whole of the measuring period.

The normal method is to check the number of errors x seen in a given averaging period t, where Poisson statistics lead to a high probability that a given error rate exists or is exceeded. Take the case of a long-term mean error rate of $p = 10^{-3}$.

$$\text{Poisson probability } P(x) = \frac{e^{-\mu} \cdot \mu^x}{x!} \tag{5.3}$$

where x = number of actual errors in the sample
  p = probability of error in the bit stream, that is, the long-term mean error rate
  n = number of bits in the sample
  $= pn$ = mean number of errors in the sample

If $t = 0.3$ s is chosen as a reasonable time to indicate an average error rate of 10^{-3} on a 2-Mb/s path carrying speech,

$$n = t f_b \times \frac{\text{number of bits in FAS}}{\text{number of bits in frame}} \qquad t = \text{indication time,}$$
$$\text{where } f_b = \text{bit rate}$$

$$n = \frac{(0.3)(2048)(10^3)(7)}{(512)} = 8400 \text{ bits} \tag{5.4}$$

$$\mu = 8.4$$

Then by working out the individual probabilities of seeing 0, 1, 2, . . ., 8 errors, we obtain

$$\sum_{x=0}^{x=8} P(x) = 0.53$$

Thus in $t = 0.3$ s there is 47% chance of the cumulative error count exceeding 8, indicating an error rate of 10^{-3}. A simple FAS error counter can, therefore, indicate the probable traffic error rate.

Although this type of error monitoring may be valid at 2 Mb/s on line systems that are essentially crosstalk-limited and the two key assumptions just given can be made, it is doubtful if those assumptions can be made on higher-order systems because of the nature of error occurrences. Also, it is unclear how errors in higher order systems are translated down to the 2-Mb/s level after line terminals and demultiplexing.

A third method of in-service error monitoring, which is really a mixture of both in-service and out-of-service testing, is to measure the errors on a pattern stimulated 64-kb/s or 1544/2048-kb/s path contained within the higher-order system. This implies demultiplexing capability in the error detector to extract the test stream from the traffic signal at system monitor points. Although this method does involve taking one or more 64-kb/s circuits out of service with consequent loss of revenue, it is perhaps a better method of gaining a closer estimate of the average error rate in the traffic, based as it can be on a bit-by-bit comparison of the decoded binary signal, including any error extension that is seen at the system terminal.

The **pseudoerror on-line** (in-service) **monitoring technique** has been gaining increased popularity in digital microwave systems. By means of a second data detector in the pseudopath, an estimate of the in-service BER is obtained. A detailed description of the principles and a number of applications of pseudoerror detectors is presented in [Feher, 5.16].

5.4 ERROR ANALYSIS

Considering the following:

1. Error ratio is a time- and statistically varying parameter prone to large fluctuations if the errors are not randomly distributed but are bunched or clustered.
2. There is a trend away from specifying the performance of digital transmission paths in terms of long-term mean bit-error rate.
3. An ISDN needs to have one network performance objective that is appropriate to and compatible with each type of service carried.

In light of these facts, it seems likely that the next decade will see an increasing emphasis placed on characterizing error performance of digital equipment, line sections, and overall connections in terms of **error distribution** with time. Indeed, considerable study of error-performance objectives for the ISDN and resulting equipment-design objectives is already underway in CCITT and elsewhere.

As a first step along this road, the concept of an **error-free second**—a second in time that is error free—was established in the data transmission world as a parameter that is convenient and relevant to data terminal equipment. Conversely, we may define an **errored second** as a second in time that contains one or more errors. Unfortunately, these definitions are not specific enough to avoid the situation that currently exists, namely, that there is more than one definition of errored second already in use in the absence of any international standardization. We shall return to this problem later.

Another more recent approach is to measure the percentage of **averaging periods,** T_0, during which the BER exceeds a certain threshold value, over a much longer time interval, T_L. To take account initially of the differing needs of different services such as voice and data, one may state the performance objective for different values of T_0, T_L, and BER threshold. This is, in fact, the approach taken in CCITT Rec. G.821 for a 64-kb/s ISDN connection [CCITT, 5.2] and is summarized in Table 5.2.

TABLE 5.2   Error Performance Objectives for International 64-kb/s ISDN Connections

$T_0 = 1$ min (voice)		$T_0 = 1$ s (data)	
BER in 1 min	% of available minutes	BER in 1 s	% of available minutes
$>10^{-6}$	$<10\%$	>0	$<8\%$ (ES)
$<10^{-6}$	$>90\%$	0	$>92\%$ (EFS)

Note that international ISDN connections should meet both requirements. The value of T_L, the **total time interval of measurement,** is under review—a month has been suggested as a reference. Not included in this table is the BER threshold at which the connection is said to be unavailable. This has been provisionally set at 10^{-3} for periods exceeding 1 to 10 s and is subject to review because, although all right for speech, it may not be suitable for other services such as facsimile. Clearly, further study is required, including practical measurements before a single performance objective can be chosen for all services carried on a 64-kb/s path. For wideband services such as television, other values will have to be chosen for T_0 and BER threshold because of the severe subjective effects of loss of television synchronization.

Let us now consider the question of definition of errored second and error-free second. The definition in CCITT Recommendation G.821 is that discrete time intervals of 1 s are checked for errors. A second in real time containing one or more errors is defined as an *errored second* and, conversely, all 1-s intervals T_0 within T_L that are free of errors are defined as *error-free seconds*. This definition has been internationally standardized by CCITT. An alternative definition in current use, particularly in North America, is to define a 1-s interval following the occurrence of the first error and call this an errored second (a *synchronous errored second* because the time interval T_0 is synchronized to the error occurrences and thence to the transmitted bit rate).

Consider the examples shown in Fig. 5.9. These illustrate the possible variations in results for a particular error distribution and measurement period.

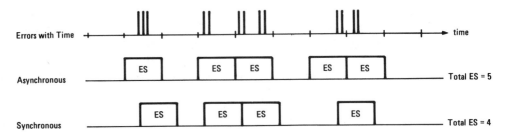

Figure 5.9   Asynchronous and synchronous errored seconds.

Using the asynchronous technique, pairs and groups of errors will be classified differently depending upon the starting time of the 1-s real-time clock. On the other hand, the synchronous technique will count all errors in bursts of less than 1 s as single errored seconds, yielding the lowest number of errored seconds over a given measurement period.

The advantages of the synchronous errored-second detection method may be summarized as follows:

1. Measurements made with different instruments simultaneously on the same link will yield consistent answers over a given measurement period.
2. Measurements made simultaneously at different points on a link will yield constant answers over a given measurement period unless the error rate increases along the route. This can be used to sectionalize faulty digital line sections. (However, this can also be achieved by simultaneous measurements of total error count with asynchronous errored-second detectors).
3. As already noted, measurements made on an error distribution will yield the lowest total number of errored seconds. This may be a mixed blessing, because although it shows up the transmission equipment in its best light, thus favoring the manufacturer, it may not relate to the service carried and the tariffs applied, which may concern the network operator or end user of the service.

The main disadvantage of the synchronous errored-second detection method is that it is difficult to obtain the converse parameter, namely, error-free seconds, which is the key parameter of interest to the operator and user of the network. Only the error-free time can be calculated directly.

The two measurement methods yield the same results where error occurrences are spaced far apart in time and are singular in occurrence. Under high-error-rate conditions, the measurement results are also similar. In the middle of these two extremes, the percentage difference between the methods vary, dependent upon error distribution and, in particular, upon error-burst length. It has been shown [CCITT, 5.1] that the difference reaches a maximum of approximately 18% at average error rates of between one and two errors per second. In the case of transmission links, which operate free of errors for most of the time and experience only occasional error bursts, the difference becomes greater, especially for error-burst durations approaching an integer number of seconds. Clearly, it would be advantageous to all if one or the other method could always be used.

Considerable study into error distributions and extensive measurement programs are underway in an attempt to establish network-performance objectives for the ISDN. Some of these programs have shown that most systems experience burst or clustered errors, and that the assumption of equiprobable bit errors is invalid. An attempt [Jones, 5.12] to describe the behavior of the error-clustering exhibited by real digital networks led to a mathematical model based on a statistical distribution proposed by J. Neyman in 1939 [Neyman, 5.13]. This is known as Neyman's **Type A contagious distribution** and is so called because it was used to describe the spread of contagious diseases and, in particular, the distribution of larvae in experimental field plots. In summary, this distribution describes error occurrences in terms of clusters that have a *Poisson distribution* over the measurement interval, and errors within each cluster also have a Poisson distribution. Each distribution is described by a combination of parameters m_1 and m_2,

where m_1 = mean number of clusters per sample n bits or per unit time

m_2 = mean number of errors within a cluster

$m_1 m_2$ = mean of the combined distributions

It has been shown [Jones, 5.12] that the probability of observing exactly r events per unit time is given by

$$P(r) = e^{-m_1} \cdot \frac{m_2^r}{r!} \sum_{t=0}^{\infty} \frac{Z^t}{t!} \cdot t^r \tag{5.5}$$

where $Z = m_1 e^{-m_2}$ and m_1 and $m_2 > 0$

For an error-free second,

$$P(0) = \exp(m_1 e^{-m_2} - m_1) \tag{5.6}$$

Jones and Pullum [Jones, 5.12] go on to relate the Neyman Type A distribution to the CCITT Recommendation G.821 network performance objectives for a 64-kb/s connection for both intervals of 1 s (EFS objective for data) and 1 min (BER threshold objective for voice) and show that to some extent these objectives are conflicting. For a given value of long-term mean-error rate, as the mean number of errors per cluster m_2 increases, so too does the probability of finding an error-free period. However, when one considers the BER threshold objective over a short averaging period of 1 min, the percentage of minutes below the threshold first decreases and then increases as the cluster size increases. Here, m_2 may be thought similar to the "nominal mean signal power during the busy hour" of an analog network [CCITT, 5.14].

Although this analysis goes a long way toward proposing a method of characterizing the error performance of a 25000-km **hypothetical reference digital connection** (HRX), it does not appear to cover the case of burst binary errors caused by single line errors at the input to a line system terminal containing a line code decoder and descrambler. It is doubtful whether the errors within this type of burst or cluster have a Poisson distribution described by m_2. Thus, an attempt to specify a 64-kb/s HRX in terms of clustering indices m_2 for the individual digital line sections probably needs further study. In any event, the values of m_2 would need regular confirmation by measurements to monitor network error performance. If the Neyman distribution approach is taken, the most relevant averaging periods to use in describing the clustering index would be 1 min and 1 s to relate them to the major services being carried in the ISDN, voice and data.

5.5 MEASUREMENT EQUIPMENT

Having described the trends in error measurements and analysis of error performance, it is appropriate to summarize the required main features needed in test equipment. Such equipment is necessary to evaluate the behavior of individual transmission links in establishing the availability and reliability of overall digital connections in an ISDN. To facilitate both field trial and remote maintenance measurements, it is desirable that the instruments feature IEEE-488 interfacing and control capability.

5.5.1 Pattern Generator Features

- At least two bit rates corresponding to TDM muldex tributary and multiplexed levels in a digital hierarchy.
- Clock jitter capability to investigate jitter tolerance [Thow, 5.3].
- PRBS patterns as specified by CCITT Rec. 0.151 [CCITT, 5.4] with optional longer patterns.
- Optional programmable word patterns for investigation purposes.
- Data outputs in accordance with CCITT Rec. G.703 [CCITT, 5.15].

5.5.2 Error Detector Features

- Data inputs in accordance with CCITT Rec. G.703 [CCITT, 5.15] including cable equalization and monitoring capability.
- Clock recovery at two bit rates (corresponding to the generator), which has maximum tolerance to input jitter [Thow, 5.3].
- Reference pattern capability (corresponding to the generator) with automatic and manual synchronization.
- Error counter averaging periods of 10^6 to 10^{10} bits and 1 s, 10 s, and 1 min to 24-h, or continuous, time.
- Minimal dead time on repetitive measurements and zero dead time on manual continuous gating to detect and count, if possible, all the errors occurring in the system under test.

- Various displays or means of analyzing the detected errors in terms of total error count, bit error ratio, errored seconds, error-free seconds, percentage of error-free seconds, and percentage of periods for which BER is below a presettable threshold.
- Optional error-burst analysis capability for detailed investigation into error occurrence and link behavior (*Note:* IEEE-488 interface bus is not fast enough for most digital transmission systems in use).

5.6 CONCLUSIONS

We reviewed the trends in error performance analysis of digital transmission systems and, in particular, the moves towards characterizing error distribution with time. Considerable further study and extensive measurements are required before administrations can build and operate ISDNs with confidence.

REFERENCES

[5.1] CCITT Rec. G.104. *Yellow Book* Vol. 3 Fascicle III.1 Recommendation G.104, pp. 28–30.

[5.2] CCITT Rec. G.821. *Yellow Book* Vol. 3 Fascicle III.3 Recommendation G.821, pp. 193–195.

[5.3] Thow, G., T. Crawford, and P. Scott. "Fundamental Limits of Jitter Tolerance in Digital Transmission Systems," HP Communications Symposium 1982 and in a revised, updated form. See Chapter 6 in this book.

[5.4] CCITT Rec. 0151. *Yellow Book,* Vol. 4 Fascicle IV.4 Recommendation 0.151, pp. 87–91.

[5.5] Netherlands Administration. "Live Traffic Jitter Measurements," CEPT/T/TR/SG3, Temporary document No. 16, Florence, October, 1979.

[5.6] Bell Northern Research. "Test Sequences for Jitter Measurements on Digital Line Sections," CCITT COM XVIII Contribution in 1980–1984 study period, Geneva.

[5.7] Dieckman, D. J. "Pseudo Random Sequence Binary Digit Generators and Error Detectors," *Post Office Electrical Engineers Journal,* Vol. 64, 1972, p. 245.

[5.8] CCITT Rec. 0.161. *Yellow Book* Vol. 4 Fascicle IV.4 Recommendation 0.161, pp. 91–93.

[5.9] Anderson, D. V., and R. A. Demers. "T1D Office Repeater Design," *IEEE National Telecommunications Conference Record,* Section 39.4.1, 1980.

[5.10] CCITT Rec. 0162. *Yellow Book* Vol. 4 Fascicle IV.4 Recommendation 0.162, pp. 93–97.

[5.11] Rollins, W. W. "Error-Second Measurements as Performance Indicators for Digital Transmission Systems," Telecommunications, September, 1980, pp. 80, 82, 132.

[5.12] Jones, W. T., and G. G. Pullum. "Error Performance Objectives for Integrated Services Digital Networks Exhibiting Error Clustering," IEE Conference, Pub. No. 193, pp. 27–32.

[5.13] Neyman, J. "On a New Class of Contagious Distribution, Applicable in Entomology and Bacteriology," *Annual of Mathematics & Statistics,* Vol. 10, No. 35.

[5.14] CCITT Rec. G.223. *Yellow Book* Vol. 3 Fascicle III.2 Recommendation G.223, pp. 23–31.

[5.15] CCITT Rec. G.703. *Yellow Book* Vol. 3 Fascicle III.3 Recommendation G.703, pp. 34–62.

[5.16] Feher, K. *Digital Communications: Microwave Applications,* Prentice-Hall, Englewood Cliffs, N.J., 1981.

6

LIMITS AND MEASUREMENT OF JITTER TOLERANCE IN DIGITAL TRANSMISSION SYSTEMS

Dr. TOM CRAWFORD, GEOFF THOW, and PETER SCOTT

Hewlett-Packard Ltd.

6.1 INTRODUCTION

The error-rate performance of digital transmission links is frequently degraded by the following factors:

1. Additive system noise (Gaussian and interference), which can cause erroneous decisions within regenerators.
2. Time displacement of the sampling point within the regenerator caused by phase modulation of the data stream, called **timing jitter.**
3. Intersymbol interference (ISI), which may be caused by hardware imperfections, selective fading, or both.

In this chapter the fundamental jitter tolerance in regenerative digital transmission systems is defined and analyzed. Following this, jitter in asynchronous multiplex-demultiplex (muldex) equipment is described. A description of intrinsic jitter and jitter tolerance measurement techniques follows. The characteristics of an instrument are highlighted.

For error rate, the overall accumulation law in a chain of digital line sections is sufficiently well understood so that each regenerator section can be specified in order to meet the quality requirement on the overall connection, [CCITT, 6.1].

However, the jitter behavior in a composite digital path, even though widely investigated by many authors [Byrne, 6.2; Duttweiler, 6.3], is not completely known. There are some difficulties in determining specifications on jitter performance, and further investigations, including experimental tests, are needed. The problem is becoming more urgent with the widespread installation of digital transmission and switching systems, which is leading to the introduction of integrated digital networks.

6.1.1 Effects of Jitter

Timing jitter introduces transmission impairments in two ways:

1. In the case of a digitally encoded analog signal, jitter causes the decoded analog samples to be irregularly spaced, thus introducing a distortion into the baseband filtered signal.

This distortion is particularly harmful in wideband coded signals, like FDM assemblies or television signals.

Jitter can be classified as *systematic* or *nonsystematic,* according to whether or not jitter sources degrade the pulse train in the same way. **Systematic jitter** is introduced in the timing-recovery process of a nonideal regenerator by the irregularity of pattern density in the signal when particular parts of the data sequence produce similar jitter effects in each regenerator. The jitter is then linearly additive in a well-designed system. **Nonsystematic** jitter occurs when there is no correlation between jitter contributions added at different points in the system, as in the case of jitter contributed by thermal noise. In this case, the jitter will tend to accumulate on a power basis.

Since systematic jitter accumulates more rapidly, it will be predominant in sufficiently long digital paths.

2. In the transmission of a digital pulse stream, timing irregularities introduce two effects:
   (a) In the regeneration process, the decision instant can be displaced from the center of the signal eye, thus leading to a reduction in the noise margin and a degradation in the error rate performance.
   (b) At the output of asynchronous demultiplexers (or at the input of digital exchanges), slips can occur due to overflow of the elastic store, which causes losses of frame alignment in the tributary signals.

6.2 FUNDAMENTAL JITTER TOLERANCE IN DATA REGENERATION

6.2.1 Definition of Peak-to-Peak Jitter (J_{pp})

We may define peak-to-peak **jitter** (J_{pp}) (in seconds, bits, or unit intervals) as the maximum peak-to-peak displacement of the ith data bit, symbol, or clock pulse with respect to its position in a hypothetical unjittered reference stream.

More specifically, for periodic jitter terms we may define it with the aid of Fig. 6.1 as the maximum displacement of the nth bit occurring in one-half cycle of the modulating function. In the second half of the cycle, the phase will be lost relative to the reference, to give a mean phase displacement of zero.

6.2.2 Derivation of J_{pp}

As an introduction, let us derive an expression for the J_{pp} in seconds and in bits for the triangular case in terms of

$$f_j = \text{modulating rate}$$

$$f_b = \text{bit rate}$$

$$\Delta f = \text{peak frequency deviation}$$

$$T_j = \text{modulating period} = \frac{1}{f_j}$$

Let n be the number of bits occurring in the positive half of the jitter modulating period $1/2f_j$.

$$n = \frac{1}{2f_j} \times \text{mean bit rate} \tag{6.1}$$

$$= \frac{1}{2f_j} \cdot \frac{(f_b + \Delta f) + f_b}{2} \quad \text{bits} \tag{6.2}$$

Now n bits without jitter take $n \cdot (1/f_b)$ seconds; that is,

$$\frac{1}{2f_j} \cdot \frac{(f_b + \Delta f) + f_b}{2} \cdot \frac{1}{f_b} \quad \text{seconds} \tag{6.3}$$

Thus the J_{pp} is given by

$$J_{pp} = \frac{1}{2f_j} \cdot \frac{(f_b + \Delta f) + f_b}{2} \cdot \frac{1}{f_b} - \frac{1}{2f_j} \tag{6.4}$$

$$= \frac{1}{4f_j} \cdot \frac{\Delta f}{f_b} \quad \text{seconds} \tag{6.5}$$

Normalizing to the period of a bit at f_b

$$\boxed{J_{pp} = \frac{1}{4} \cdot \frac{\Delta f}{f_j} \quad \text{bits}} \tag{6.6}$$

6.2.3 Recovering a Sampling Clock from Jittering Data

Figure 6.2 shows a typical configuration found in pulse code modulation (PCM) line regenerators, demultiplexers, or transmission test equipment.

For digital transmission-measuring instruments in general, the tolerance to

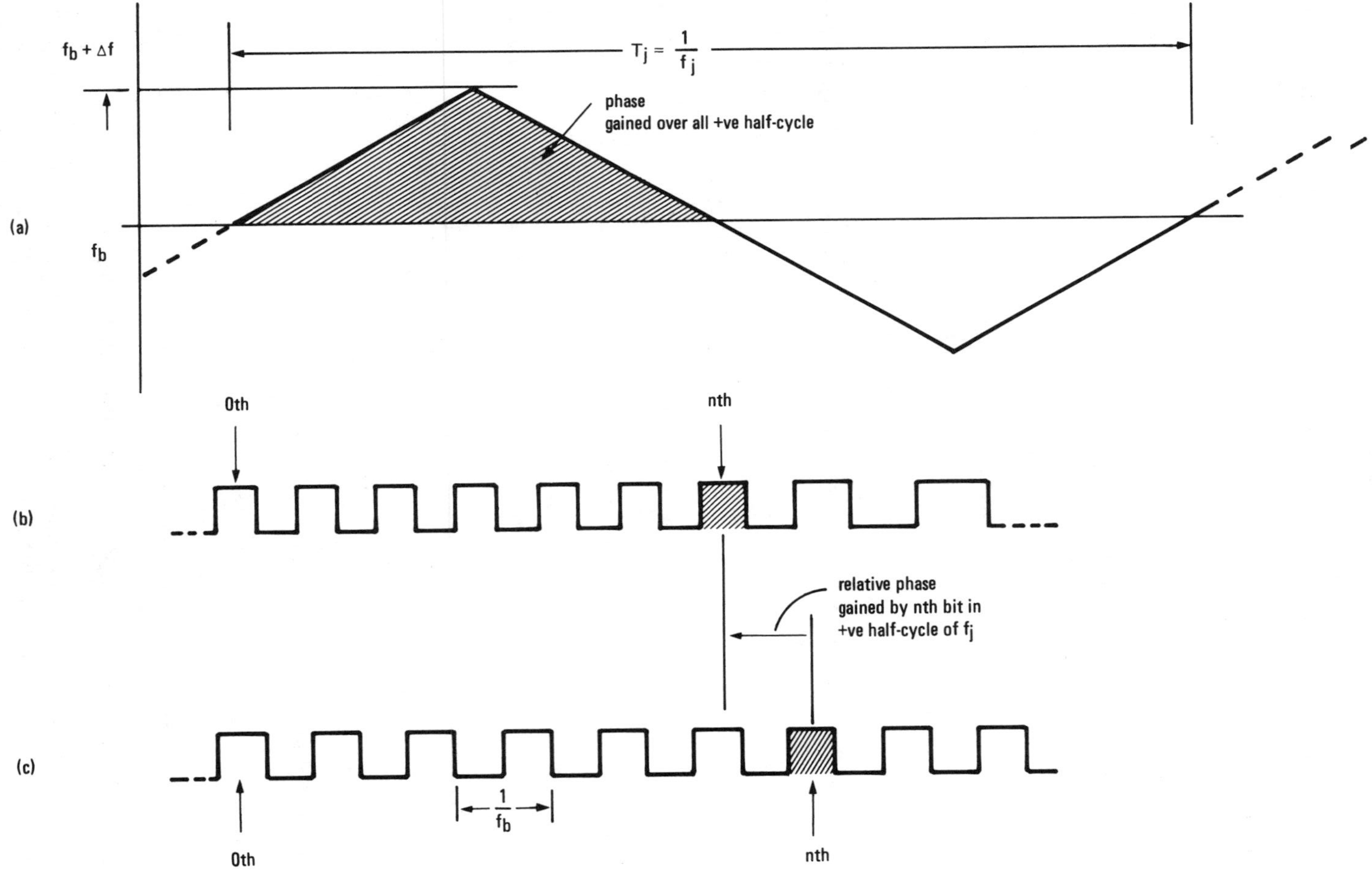

(a)
(b)
(c)
fb + Δf
fb
Tj = 1/fj
phase
gained over all +ve half-cycle
0th
nth
relative phase
gained by nth bit in
+ve half-cycle of fj
1/fb
0th
nth

Figure 6.1 Triangular jitter-modulation waveform.

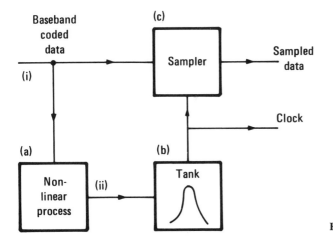

Figure 6.2   Regenerator block diagram.

jitter should be high, and for instruments measuring timing jitter in particular, wide tolerance and flat jitter-transfer functions are essential.

We now derive the *fundamental limit of a clock recovery circuit* appropriate for a jitter-measuring instrument. The analysis applies to data codes such as AMI, B3ZS, and HDB3. The only parameter of interest to us is in the length of the maximum run of zeros allowed in the code. If the code has no upper bound, then it is assumed that some other control exists, such as the one offered by the 14-zero-limit rule employed at the DS-1 level (North American standard).

If the data is binary return-to-zero (RZ) coded, then Item (a) of Fig. 6.2 is obviously not required; its purpose is to convert the spectrum of the ternary line code to which we just referred (which has no clock spectral components) to a binary signal that has a discrete spectral line at the system clock rate. Full wave rectification is frequently used, and Fig. 6.3 illustrates this process.

Item (b) is some oscillatory device, which resonates at its natural frequency in the absence of data marks but is updated in phase during the presence of data marks. Often a simple tuned circuit is used that has sufficient Q to ring for the period of the longest permissible zero block. Alternatively, (b) may be a narrow-band phase-locked loop. For purposes where the best possible jitter acceptance is demanded, we will show that a combination of two normally conflicting parameters is required: a circuit that will (1) not decay in the presence of zeros and (2) be instantaneously updated by a single data mark. We use a simple form of injection locked oscillator to achieve these requirements.

In our jitter analysis, we derive limits assuming the above two properties and then comment on the reduction in tolerance that results from the conventional tank circuit approach.

Figures 6.4 and 6.5 serve to show the asymmetrical nature of the failure mechanism if instantaneous phase updating by a single mark is allowed. We take the negative Δf limit as being the most significant, as noncentral sampling or mistuning is required to gain extra tolerance from the asymmetry.

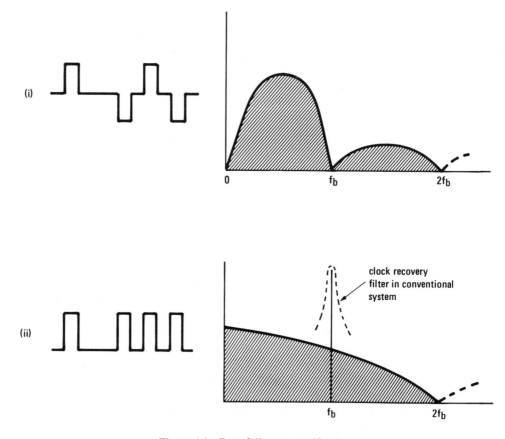

Figure 6.3 Data full-wave rectification.

Referring to Fig. 6.6, the time-domain limitation to the process of clock recovery is determined by the maximum phase error that can be allowed to accumulate during the run of zeros. This phase error (between the sampling clock and the first data mark to occur at the end of the zero block) occurs because the free-running oscillator or tank runs at its natural frequency (nominally, the bit rate) during the data zeros.

The worst case occurs when the zero block straddles the point of peak frequency deviation of the modulation. Since the zero block can accumulate phase only in one direction, the maximum allowable phase deviation over the n bits is one-half of the data pulse width for centrally sampled data; that is, for RZ data the maximum phase deviation is $\frac{1}{4}$ bit.

From Fig. 6.6, we relate the average frequency of the data during the zero block f_{av} to the number of bits (x) occurring in the complete negative half of the

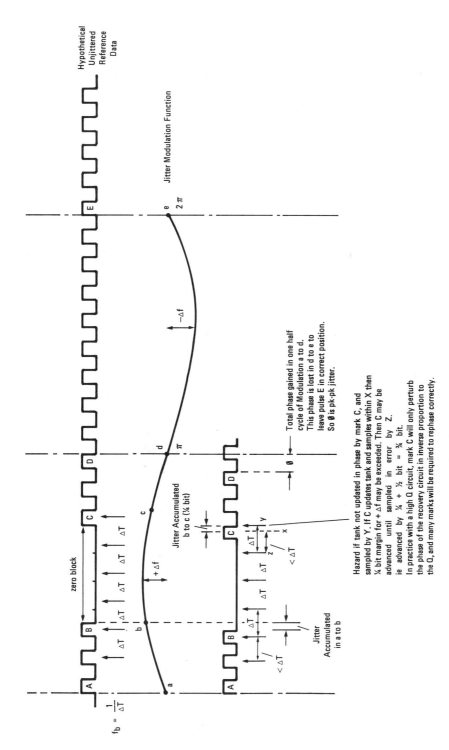

Figure 6.4 Zero block in $+\Delta f$ half of modulation period.

159

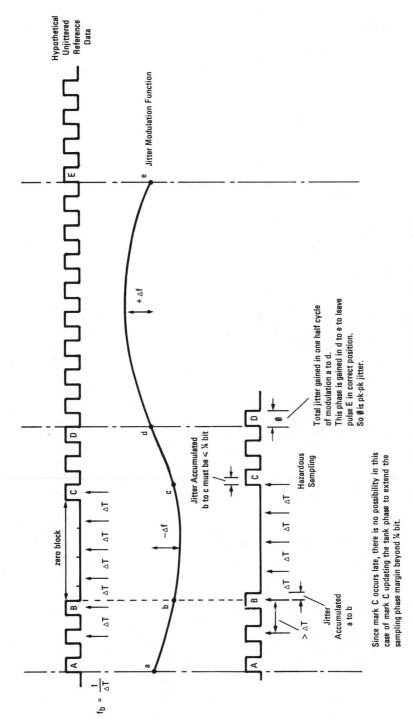

Figure 6.5 Zero block in $-\Delta f$ half of modulation period.

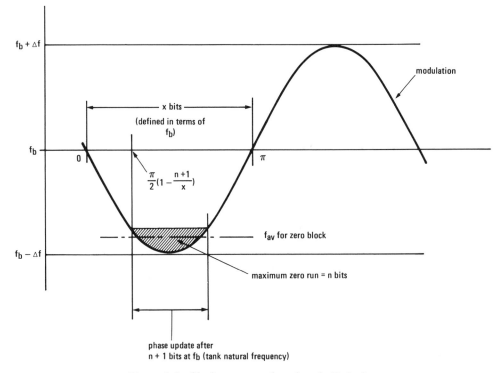

Figure 6.6 Clock recovery time domain limitation.

modulating waveform. By inspection of Fig. 6.6, we conclude that the average frequency of data during a zero block is

$$f_{av} = f_b - \Delta f \cdot 2 \int_{(\pi/2)\cdot[1-(n+1)/x]}^{\pi/2} \frac{\sin\theta \, d\theta}{(n+1)/x} \tag{6.7}$$

$$= f_b - \Delta f \cdot \frac{x}{(n+1)\pi} \cdot 2 \int_{(\pi/2)\cdot[1-(n+1)/x]}^{\pi/2} \sin\theta \, d\theta$$

Now, allowable $\frac{1}{4}$-bit phase error is accumulated at the difference frequency $(f_b - f_{av})$ and is equal to $(f_b - f_{av}) \cdot$ time for $n + 1$ bits at f_b. Thus we have

$$\frac{1}{4} = (f_b - f_{av}) \cdot \frac{n+1}{f_b} \tag{6.8}$$

From (6.7) and (6.8),

$$\frac{1}{4} = \frac{2\,\Delta fx}{\pi f_b} \cos\frac{\pi}{2}\left(1 - \frac{n+1}{x}\right) \tag{6.9}$$

The average frequency over x bits (i.e., over the negative half-cycle of the modulation) is

$$f_b - \Delta f \cdot \frac{1}{\pi} \int_0^1 \sin\theta \, d\theta = f_b - \frac{2}{\pi}\Delta f \tag{6.10}$$

The jitter (in periods of f_b) over x bits is

$$J_{pp} = \frac{2}{\pi} \Delta f \cdot \frac{x}{f_b} \tag{6.11}$$

Substituting the value for Δf obtained from (6.9), we obtain the jitter tolerance expression

$$J_{pp} = \frac{1}{4} \cdot \frac{1}{\cos(\pi/2)[1 - (n + 1)/x]} \qquad \text{where } x = \frac{f_b}{2f_j} \tag{6.12}$$

From (6.9) we find the corresponding frequency deviation.

$$\Delta f = \frac{\pi}{8} \cdot \frac{f_b}{x \cos(\pi/2)[1 - (n + 1)/x]} \qquad \text{where } x = \frac{f_b}{2f_j} \tag{6.13}$$

From (6.12) and (6.13),

$$\Delta f = \frac{\pi}{2x} \cdot f_b \cdot J_{pp} \tag{6.14}$$

Thus

$$\Delta f = \pi f_j J_{pp} \tag{6.15}$$

Equations (6.12) and (6.13) are valid for $(n + 1)/x < 1$.

If $(n + 1)/x > 1$, zeros can always fill a complete half-period of f_j. Consequently, only $\frac{1}{4}$ bit may be accumulated during this time.

Thus

$$J_{pp} = \frac{1}{4} \qquad \text{for } \frac{n + 1}{x} \geqslant 1 \tag{6.16}$$

$$\Delta f = \frac{\pi}{4} f_j \qquad \text{for } \frac{n + 1}{x} \geqslant 1 \tag{6.17}$$

Equations (6.12) and (6.16) are plotted in Fig. 6.7 as a function of maximum zero-block length (n) in the data. The graph is for RZ data. For Non-Return to Zero (NRZ) data (centrally sampled), the tolerance is doubled.

As already illustrated in Fig. 6.4, a further increase in **tolerance** is possible by the unorthodox approach of introducing a static phase error, since it is possible to accept more than $\frac{1}{4}$ bit deviation in the positive direction where a mark can

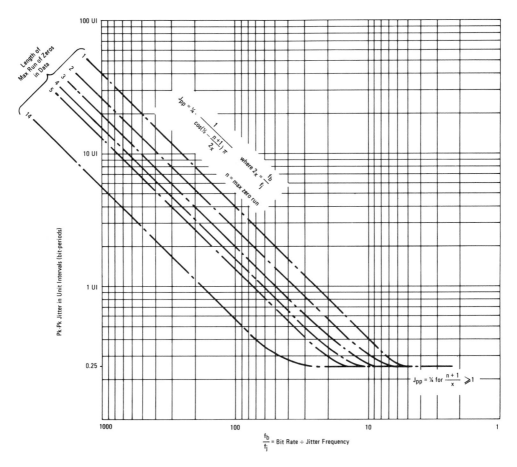

Figure 6.7   Fundamental time domain jitter-tolerance limits.

update its own sampling edge, whereas it is impossible to avoid sampling an extra zero in the negative half cycle of the modulation when $\frac{1}{4}$ bit is exceeded.

6.2.4 Limitation of Tolerance in Narrow-Band Clock Recovery Systems

Example 6.1

Consider an example using B3ZS coding. In this case, $n = 2$. Suppose we are interested in jitter to 1% of the bit rate. Let $f_b = 44.7$ Mb/s. Then $f_j = 447$ kHz. From Fig. 6.7, the fundamental limit for tolerance is 2.7 unit intervals (bits) peak to peak.

From equation (6.15), we have

$$\Delta f = \pi f_j J_{\text{pp}} = \pi \times 447 \times 10^3 \times 2.7$$

$$= 3.8 \text{ MHz}$$

Based on Carson's rule [Bell Labs, 6.4], the bandwidth occupied by this phase modulation is approximately

$$2(\Delta f + f_j) = 2(3.8 + 0.447) = 8.5 \text{ MHz}$$

Even accepting a 3-dB bandwidth implies a recovery tank Q of not greater than $44.7/8.5 \approx 5$. A higher Q than this will reduce the jitter tolerance correspondingly. By the definition of Q, a tank circuit can be considered as a low-pass filter to phase modulation with a cutoff $f_c = f_b/2Q$. The truncation of the Bessel spectrum of the phase-modulated data signal by the tank reduces its ability to track the phase of the data faithfully.

Values of Q of 40 to 50 and higher are typically employed in regenerators using similar codes. The effect of the clock recovery tank circuit may be considered to be that of a low-pass filter acting on the jitter modulation. This leads to the concept of a jitter-transfer function relating jitter loss or gain between input and output of a regenerator as a function of jitter-modulation frequency. This is illustrated in Fig. 6.8.

The loss of tolerance is traded to gain a narrower jitter-transfer function. Two opposed approaches are possible:

1. Use low Q recovery circuits with high tolerance and, consequently, wide jitter-transfer functions. Remove the accumulated jitter in one lump in an elastic store at the end of the transmission path.

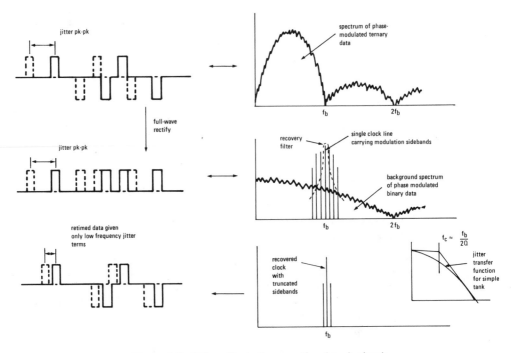

Figure 6.8   Filter effect of conventional tank circuit.

2. Alternatively, use a policy of high Q, low tolerance, and narrow transfer function for regenerators and so control the jitter accumulation along the link.

For conventional tank recovery, the tolerance mask has a similar form to that derived for the time-domain limit, with a -6-dB/octave fall to an asymptote of $\frac{1}{2}$ bit peak to peak. The asymptote derives from the data pulse width. The corner frequency of intersection of the -6-dB/octave slope and the $\frac{1}{2}$-bit peak-to-peak asymptote relates to the tank circuit roll-off. The narrower the tank, the lower is the jitter corner frequency and consequent position of the -6-dB/octave line. ■

Example 6.2

Assume

$$f_b = 1544 \text{ kHz}$$

$$f_j = 1\% \text{ of } f_b = 15 \text{ kHz}$$

Assume an AMI code with 14 zero limit; thus $n = \frac{1}{4}$. From Fig. 6.7, the fundamental limit of tolerance is approximately 0.57 bits peak to peak.

From equation (6.15),

$$\Delta f = \pi \times 15 \times 0.57 = 26.9 \text{ kHz}$$

Carson's rule gives the total bandwidth occupied as

$$2(26.9 + 15) \approx 84 \text{ kHz}$$

This time a maximum Q (allowing a 3-dB discrepancy) would be

$$\frac{154}{84} \approx 18.4$$

which is closer to conventional Qs.

It is interesting to note the large gain in potential tolerance obtained by using high-density codes such as B3ZS. ■

6.2.5 Jitter Contribution from Regenerators

Jitter contributions arise from the following:

1. Mistuning of the tank or oscillator circuit, which then runs during the zero block at a rate different from the nominal bit rate, thus accumulating phase error even in the presence of an incoming jitter-free data stream. This will add jitter to the marks following the zero block unless instantaneous updating is possible (which it is not for conventional bandwidth–limited recovery circuits).

2. Imperfect equalization will result in intersymbol interference (ISI) and subsequent pattern-dependent phase updating of the recovery tank.

3. Relative skew between positive and negative output marks can contribute high-frequency jitter terms, which may become significant in low-Q recovery circuits.

6.3 JITTER IN ASYNCHRONOUS MULTIPLEX-DEMULTIPLEX (MULDEX) SYSTEMS

6.3.1 Basic Principles of Digital Multiplexing

Time division multiplexing of several digital signals to produce a higher-speed stream can be accomplished by a selector switch that takes a pulse from each incoming line in turn and applies it to the higher-speed line. The receiving end will do the inverse of separating the higher-speed pulse stream into its component parts and thus recover the several lower-speed digital signals. The main problems involved are the **synchronization** of the several pulse streams so that they can be properly interleaved and the **framing** of the high-speed signal so that the component parts can be identified at the receiver end. Both of these operations require elastic stores, which constitute important parts of a multiplexer.

Information pulses arriving at the multiplexer must await their turn to be applied to the higher-speed transmission systems. Due to delay variations and frequency variations of the incoming lines plus the framing and synchronization operation of the multiplexer terminal, this wait is variable in time. It is this **muldex waiting time** that introduces the second major source of **timing jitter** into a composite digital transmission path.

Although there are many performance trade-offs in the design of digital multiplexer systems [Bell Telephone Laboratories, 6.4], such as reframe time, channel capacity, and stuffing margin, the practical significance of this timing jitter is best understood by considering a specific system like a second-order digital multiplexer, as shown in the block-diagram format of Fig. 6.9.

Looking more closely at the multiplexer operation (Fig. 6.10), each input tributary (digital stream) is *written* into a store under the control of a timing waveform derived from the input digital stream by a timing recovery circuit.

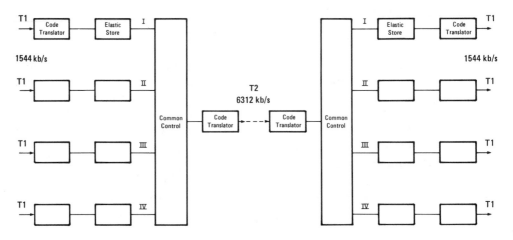

Figure 6.9 M12 digital multiplexer block diagram.

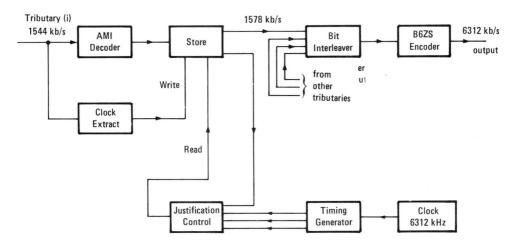

Figure 6.10   M12 digital multiplexer (positive justification).

The data is *read* out of the store at a higher rate under control of a timing waveform derived from the multiplex clock that is common to all tributaries.

The *read clock* is inhibited at the digit time slots when control information such as the frame alignment signal and **justification (stuffing)** control digits has to be inserted in the frame. It is also inhibited when the justifiable digit time slot is to carry a redundant digit.

In **pulse stuffing,** the elastic stores are usually of four or more cells and perform a key roll in the muldex operation. Consider an elastic store that operates like a commutator (Fig. 6.11). Segments of the commutator are connected to the storage cells. One rotating contact writes into the cells; another reads out of the cells. The angular velocities of the contacts correspond to the frequencies of the write and read clocks. Because the read clock is faster than the write clock, reading overtakes writing, and a block of digits will be repeated if reading is nondestructive. In this situation, the elastic store is said to have **spilled.** Pulse stuffing controls this spilling to allow the eventual recovery of the correct sequence of digits.

Two conditions are necessary for success:

1. The read clock must be faster than the write clock.
2. The insertion of extra digits must be done at prearranged times to permit proper removal.

Condition 1 is satisfied by assignment of nominal frequencies and allowed frequency tolerances. Condition 2 is satisfied by periodically monitoring the delay between writing and reading and a signal format, which allows for extra digits to be inserted at prearranged times, as shown for a M12 muldex in Fig. 6.12.

With pulse stuffing, even when the incoming tributary stream to the muldex is uniformly spaced, the outgoing stream is not. It will have occasional interruptions

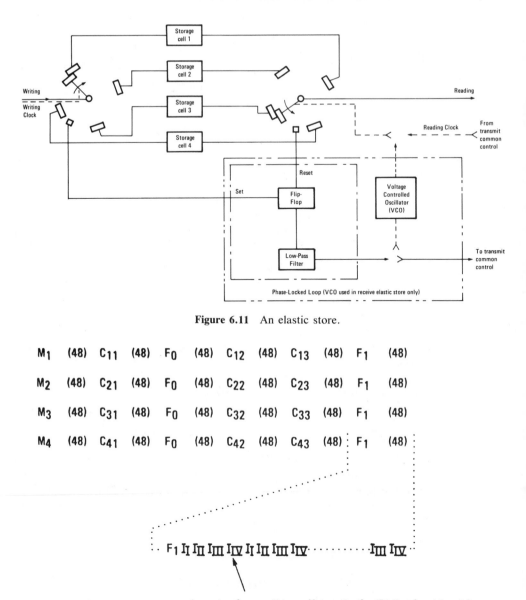

Figure 6.11   An elastic store.

M_1	(48)	C_{11}	(48)	F_0	(48)	C_{12}	(48)	C_{13}	(48)	F_1	(48)
M_2	(48)	C_{21}	(48)	F_0	(48)	C_{22}	(48)	C_{23}	(48)	F_1	(48)
M_3	(48)	C_{31}	(48)	F_0	(48)	C_{32}	(48)	C_{33}	(48)	F_1	(48)
M_4	(48)	C_{41}	(48)	F_0	(48)	C_{42}	(48)	C_{43}	(48)	F_1	(48)

$$F_1 \; I_I \; I_{II} \; I_{III} \; I_{IV} \; I_I \; I_{II} \; I_{III} \; I_{IV} \cdots\cdots\cdots I_{III} \; I_{IV}$$

Location for possible stuffing pulse for T1 line for this subframe

Figure 6.12   M12 multiplexer format.

where the information pulses are interleaved with the inserted pulses (framing, justification control, and stuffed digits). Thus, even after successful removal of the inserted pulses, the received information digits have jitter that must be smoothed before the digits can be processed further. **Smoothing** of this **jitter** is the function of the receiving elastic store and its phase-locked loop. The phase-locked loop (PLL) is there to give the best running estimate of the average input frequency and use it as the demultiplexer read clock. Although the demultiplexer PLL can be made into an extremely narrow band, it is finite and therefore cannot remove jitter frequencies that lie within its bandwidth.

6.3.2 Waiting-Time Jitter

In practice, the jitter frequencies that the demultiplexer PLL sees from the line system regenerators and from the systematic insertion and removal of framing and stuffing control pulses are relatively high in frequency (typically kilohertz) and can be readily smoothed. However, there is a more subtle jitter caused by the fact that pulse stuffing takes place only in certain allowed time slots (Fig. 6.12). This process can be demonstrated by a few simple examples:

1. If the rate of the phase advance of the read clock relative to the write clock is such that stuffing occurs at an integer number of stuffing opportunities, then the stuffing ratio will have a numerator equal to 1, and there will be no waiting-time jitter. The only jitter present will be the stuffing jitter, which will be a regular sawtooth with amplitude equal to one unit interval (see Fig. 6.13).

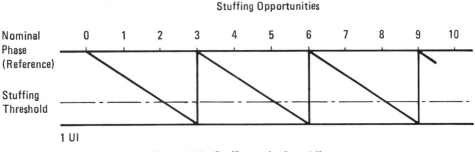

Figure 6.13   Stuffing ratio $S = 1/3$.

Next let us consider the simplest cases in which waiting-time jitter occurs. We will refer to these as the *first structure* of waiting-time jitter.

In each case, the waiting-time jitter appears as a slow sawtooth displacement of the regular stuffing-jitter sawtooth. If the stuffing ratio S is inverted and expressed as a decimal number, this represents the time position of the first crossing of the stuffing threshold relative to the reference phase (expressed in stuffing opportunity periods).

2. If the fractional part of $1/S$ is greater than 0 and less than or equal to 0.5, then the waiting-time jitter sawtooth will have a slope of opposite sense to the stuffing jitter, and the maximum phase excursion will take place at the first stuffed bit following nominal phase (see Fig. 6.14).

3. If the fractional part of $1/S$ is greater than 0.5 but less than or equal to 1,

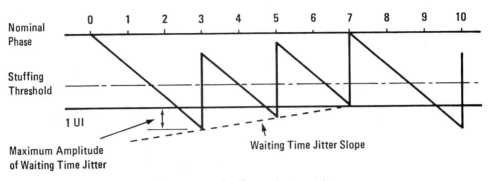

Figure 6.14   Stuffing ratio $S = 3/7$.

the waiting-time jitter sawtooth will have the same sense as the stuffing jitter, and the maximum phase excursion will take place at the stuffed bit immediately prior to the stuffed bit at nominal phase (see Fig. 6.15).

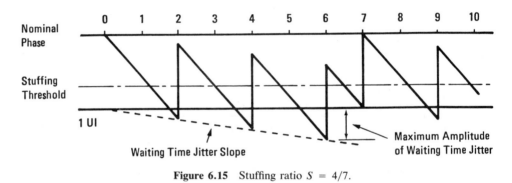

Figure 6.15   Stuffing ratio $S = 4/7$.

If the stuffing is not a simple integer ratio (which is more than likely), then the waiting-time jitter will have the pattern shown in Figs. 6.14 and 6.15, except that a small error will be accumulated by the end of each pattern. After several repetitions of the first structure, the error will build up to a point where a stuff occurs earlier than its regular position. The pattern may repeat at this point (which gives us a second structure of waiting-time jitter), or another small error may still exist, which will build up over the repeats of the second structure to reveal a third structure, and so on. These structures will clearly exhibit lower and lower repetition frequencies. In general, for stuffing ratios S that are not simple ratios, the total peak-to-peak amplitude of all the structures approaches S unit intervals, and this is the maximum amplitude of waiting-time jitter:

$$\text{Waiting-time jitter amplitude} = \frac{n-1}{m} \text{ unit intervals peak to peak for } S = \frac{n}{m}$$

Figure 6.16 shows waiting-time jitter amplitude plotted against stuffing ratio. The abrupt discontinuities correspond to simple ratios, which have few structures.

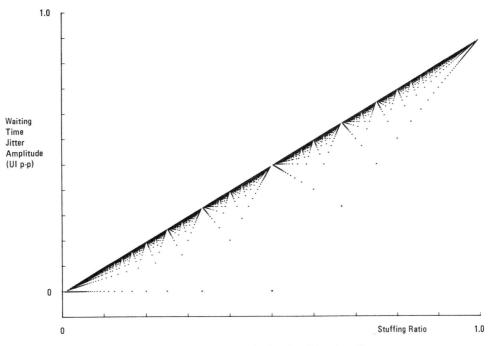

Figure 6.16 Maximum amplitude of waiting-time jitter.

In practice, the low frequency component of waiting-time jitter tends to accumulate linearly with the number of tandem multiplexers. This will not cause any difficulty in digital transmission because low-frequency jitter is preserved by the stuffing operations, PLLs, and the line regenerators. However, this low-frequency jitter is passed on to the decoded baseband analog signal, but as long as the frequencies of the residual jitter are held low enough, negligible degradation of the signal results.

6.4 PHILOSOPHY OF JITTER SPECIFICATIONS FOR COMPOSITE DIGITAL SYSTEMS

Initially looking at bit error studies, reference connections form a useful basis because digital errors accumulate over the whole connection and are more a function of circuit length than a function of the transmission bit rate of the circuit [CCITT, 6.5].

This is because the hypothetical reference connections for telephony and an overall acceptable error rate lead to design error rates for line links, digital muldexes, and digital exchanges.

Concerning models for jitter studies, the following aspects are essential:

1. Jitter is a characteristic related to a digital signal of a defined bit rate.

2. Jitter does not necessarily accumulate in a predetermined manner but can be reduced by proper equipment (TDM exchanges and demultiplexers work as *jitter reducers*).

3. If jitter at the input of an equipment exceeds a certain upper limit, bit errors occur; to avoid serious impairments caused by jitter, the probability of exceeding this upper limit should be very low in all practical equipment configurations.

4. The length of the model is of minor importance compared to its structure.

Hypothetical reference digital paths and the existence of upper limits for permissible jitter lead to reference paths with a worst-case structure. The aim of jitter studies for hypothetical reference paths is to permit real systems to be built by the application of simple network planning without the need for measurement on the system as a whole. The number of line sections and muldex sections in the hypothetical reference digital path and the bounds of the upper limits of the permissible jitter must be specified by the telecommunications network controlling body. The upper limits of permissible output jitter and input jitter tolerance should be specified for each type of element (digital line section, muldex, exchange). This ensures that any digital output port can be connected to a corresponding digital input port with negligible probability of occurrence of unacceptable jitter-induced digital impairments.

6.5 JITTER MEASUREMENTS

The types of measurements frequently required are shown in Fig. 6.17 and consist of the following:

> Maximum peak-to-peak jitter measurement at the system output: in the absence of input jitter (intrinsic jitter)
>
> Tributary peak-to-peak output jitter measurements: in the absence of input jitter (intrinsic jitter)
>
> Tolerance to input jitter measurements: a lower limit of sinusoidal jitter tolerance which, when applied to a recommended pseudorandom binary sequence simulating live traffic, should produce no errors
>
> Jitter transfer function measurements: ratio of system output jitter to applied input jitter (again a sinusoidal phase modulation is used)

6.5.1 Intrinsic Jitter Measurements

The statistics of timing jitter depend upon the transmitted signal, pseudorandom binary sequence (PRBS) ($2^{15} - 1$ or $2^{20} - 1$) or live traffic. The type of signal is important as it directly affects the correction factor, which has to be applied when measurements are done with PRBS stimulus as opposed to live traffic.

For the practical measurement of maximum peak-to-peak jitter at the output

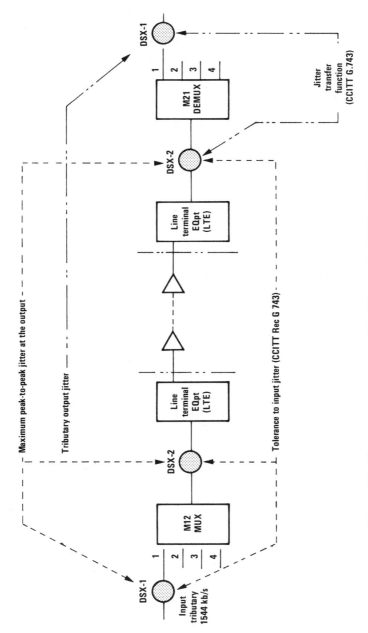

Figure 6.17 Jitter measurements on a digital line section, MUX or DEMUX.

and tributary peak-to-peak output jitter, each measured at the DSX interconnection and on PRBS test patterns, knowledge of similar measurements must be obtained with live traffic as the transmitted signal. This allows a safety margin to be built into the output-jitter specifications, thus allowing the measurements to be made on PRBS test patterns but still ensuring a negligible probability of jitter-induced transmission errors when live traffic is carried.

In digital transmission systems, binary errors due to jitter are caused by peak excursions exceeding the tolerance of the particular equipment. It is, therefore, important to measure the *maximum peak-to-peak* jitter amplitude rather than the *instantaneous peak-to-peak* jitter amplitude. Hence, we should specify the maximum peak-to-peak output jitter at both multiplex and demultiplex (tributary output).

The maximum peak-to-peak jitter amplitudes must have a safety margin on the input-jitter tolerance mask. This means that, for example, the tributary peak-to-peak output jitter (from a demultiplexer) must be less than the input-jitter tolerance of the following equipment.

6.6 *JITTER TOLERANCE MEASUREMENTS*

The upper limit of sinusoidal jitter tolerance of a narrow-band clock recovery circuit is shown in Fig. 6.18. Because of its narrow bandwidth, such a clock recovery circuit requires many signal pulses to update its phase, and, as a consequence, its phase at the start of a zero block will tend to be the average phase of the jittering data stream. This will effectively result in an available eye width of $\pm\frac{1}{4}$ bit, or $\frac{1}{2}$ bit peak-to-peak.

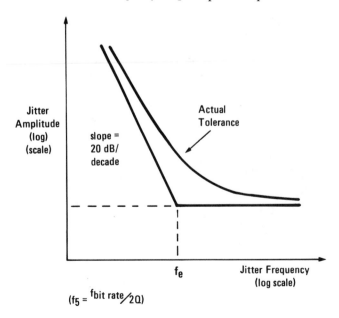

$$\left(f_5 = \frac{f_{bit\ rate}}{2Q}\right)$$

Figure 6.18 Regenerative repeater.

   Jitter tolerance is a measurement of the point at which the equipment starts to introduce errors due to the applied jitter amplitude. This point gives an upper limit of maximum tolerable jitter, that is, the system crash point.

   Referring to Fig. 6.18, it should be noted that the horizontal lines of the tolerance plot are lines of constant eye-width, and 20-dB/decade sloping lines are lines of constant bandwidth. The plot consists of an eye-width constraint at high frequencies and a bandwidth constraint at low frequencies. The eye-width represents a low level of tolerance, and it is therefore clear that it is the bandwidth that is the dominant tolerance effect, as it controls the frequency at which the eye-width becomes the limitation. When the tolerance of a composite system (Figs. 6.19 and 6.20) is examined, it is the item with the narrowest bandwidth that dominates. Thus a jitter reducer is dominated by its narrow-band phase-locked loop, and a muldex is dominated by the demux phase-locked loop. In order to alleviate the bandwidth effect, both the reducer and the demux have buffer storage included in their construction, and this item effectively increases the eye-width at the sampling point. This provides an increase in tolerance from the point of intersection of the bandwidth and eye-width asymptotes (f_a in Figs. 6.19 and 6.20) up to the frequency where the eye-width intersects with the next bandwidth asymptote (f_b in Figs. 6.19 and 6.20).

   In the jitter reducer this is the input timing-recovery bandwidth, and in the muldex system it is the effective bandwidth of the justification (stuffing) process. Since the justification bandwidth is relatively narrow (a few kilohertz), this would

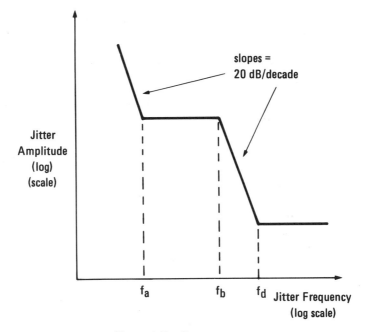

Figure 6.19 Jitter reducer.

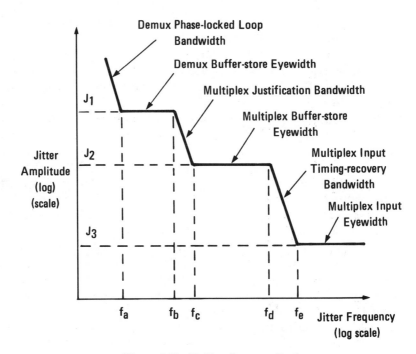

Figure 6.20 Muldex jitter amplitude.

dominate the jitter tolerance of the system were it not for the inclusion of the multiplex buffer store. The store increases the effective eye-width for jitter-modulation frequencies up to the point of intersection with the multiplex input timing recovery bandwidth (f_d in Fig. 6.20).

These plots lead to the definition of a performance standard for the minimum value of tolerance to input jitter that can be applied to all systems. This is *the lower limit of maximum tolerable input jitter,* and when applied in conjunction with the output-jitter specification, ensures compatibility.

The lower limit of maximum tolerable jitter (**jitter mask**) is tested by applying sinusoidal jitter to a PRBS-simulating traffic signal. The sinusoidal jitter used for test purposes does not simulate the noiselike jitter that is generated within the network, but it does provide information regarding the peak-to-peak alignment jitter tolerance (A_2), timing-recovery-circuit bandwidth (f_5), and buffer storage capacity (A_2 or A_1).

The positioning of this lower limit of maximum tolerable input-jitter mask is shown in Fig. 6.21. This lower limit permits simple correlation between measured output jitter and input-jitter tolerance. The maximum permissible output jitter is measured in two frequency bands:

1. Peak (or peak-to-peak) measurements in the band from f_1 to f_2 in Fig. 6.21 are conducted using a band-pass filter with corner frequencies at f_1 and f_4.

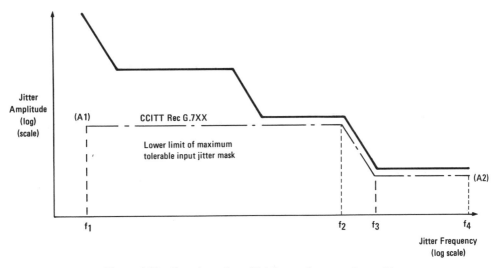

Figure 6.21   Complete sinusoidal jitter tolerance of a muldex.

2. Peak (or peak-to-peak) measurements in the band from f_2 to f_4 are conducted using a band-pass filter with corner frequencies at f_3 and f_4. Since the tolerance mask has 20-dB/decade slope and the filter has an equal and opposite slope from f_2 to f_3, the overall response is flat from f_2 to f_4.

In each case, the high-pass part of the filter acts to remove the lower-frequency jitter components which, conceivably being of larger amplitude, could swamp the measurement.

A compromise is necessary when establishing the jitter amplitude (A_1). This compromise is necessary because pattern-dependent jitter from digital line sections can increase without limit as the number of regenerators is increased, whereas at low frequencies, equipment such as muldexes has an upper limit to the amplitude of jitter that can be accommodated, which is determined by the size of buffer stores. Peak-to-peak jitter amplitudes in excess of several unit intervals are possible from digital line sections having more than 100 regenerators.

The value of A_1 should, therefore, represent a reasonable compromise between the lengths of digital line sections that are likely to occur, the size of buffer stores in systems, and the additional cost of providing jitter reducers where these are shown to be necessary. (For the high-rate digital line sections, however, jitter reducers can be incorporated in line terminal equipment with little or no cost penalty).

Having provided specifications for the control of jitter within component parts of a digital line section, it remains only to specify the jitter-transfer function of the section so that they may be cascaded in a controlled manner.

This jitter-transfer function will, in practice, reflect the characteristic of the demultiplexer PLL timing reconstruction circuit and therefore takes the form of Fig. 6.22.

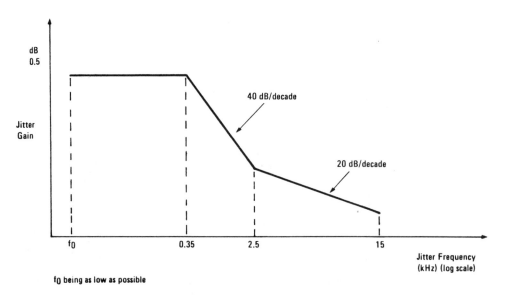

fₒ being as low as possible

Figure 6.22   Jitter transfer characteristic (M_{21}).

6.6.1 Brief Summary of Test-Equipment Requirements

The following test equipment and requirements are essential:

1. (a) A jitter generator capable of providing sinusoidal phase modulation onto a PRBS test pattern in order that
     (i) The lower limit of maximum tolerable input jitter can be established easily.
     (ii) The upper limit of maximum tolerable input jitter can be examined—that is, the safety margin.
     (iii) A known input-jitter amplitude can be supplied for jitter-transfer-function measurements.
   (b) The jitter frequency and jitter amplitude ranges should be wide enough to satisfy (a).
2. A jitter-measuring circuit capable of
   (a) Recovering a jittered clock signal from the incoming jittered data *without* inadvertently filtering any of the jitter components that should be passed to the jitter demodulator and the measurement circuit.
   (b) providing a measurement of the instantaneous peak-to-peak jitter amplitude to perform a measurement of jitter-transfer function on sinusoidal jitter stimulus.
   (c) providing a variable threshold so that the number of occasions that peak-to-peak jitter amplitudes exceed this threshold can be counted.
   (d) providing a measurement of the maximum peak-to-peak jitter amplitude

over a specified gating interval to conform to the specification of maximum permissible output jitter for digital equipment.

(e) providing the filtering to specify measurement (d) fully.

3. All tests should be performed at standard interconnection interfaces.

This chapter has presented the fundamental limits of jitter performance and related these to jitter in a digital network. The current approach to specification and control highlighted the need for practical measurements on the accumulation of jitter within networks, taking account of the inherent jitter reducing properties of some equipment (demultiplexers, digital exchanges). All these factors must be taken into account when devising design and installation rules for a digital network, and this is a matter of study nationally and internationally.

The nature of jitter in a digital network may be compared with that of noise in an analog network. Although both are impairments that propagate through the network, the great advantage is that jitter can be reduced to a practical level to ensure that the evolving Integrated Digital Services Network (ISDN) is built on a foundation of well specified digital elements.

6.7 ACKNOWLEDGMENTS

Thanks to Malcolm Rix (Hewlett-Packard QTD) for verification of Fig. 6.16 and for offering a useful insight into the jitter mechanism.

REFERENCES

[6.1] CCITT Rec. G.104. *Yellow Book,* Vol. 3, Fascicle III.1, Recommendation G.104, pp. 28–30.

[6.2] Byrne, C. J., B. J. Karafin, and D. B. Robinson. "Systematic jitter in a chain of digital regenerators," *Bell System Technical Journal,* Vol. 42, No. 6, November, 1963, pp. 2679–2714.

[6.3] Duttweiler, D. L. "Waiting time jitter," *Bell System Technical Journal,* Vol. 51, No. 1, January, 1972, pp. 165–207.

[6.4] Bell Telephone Laboratories. *Transmission Systems for Communications,* pp. 608–625.

[6.5] CCITT Rec. G701-G941. *Yellow Book,* Vol. 3.3, "Digital Networks," Transmission Systems and Multiplexing Equipment Recommendations G701-G941.

[6.6] CCITT Rec. 171. *Yellow Book,* Vol. 4, Fascicle IV.4, Recommendation 171, pp. 98–105.

7

DIGITAL RADIO MEASUREMENT TECHNIQUES

GEOFF WATERS

Hewlett-Packard Ltd.

7.1 INTRODUCTION

In both analog and digital microwave radio systems, it is essential that demodulated baseband signals have an adequate signal-to-noise ratio, S/N, and that the system is reliable, that is, has high availability. Availability is affected by fading of the received signal and equipment failure or may be reduced by interference on one or more hops.

In an analog system, a reduction in **received signal level** (RSL) results in a proportional increase in noise. The system is designed to meet the noise performance requirement. Space diversity can provide some protection against fading, whereas frequency diversity provides protection against both fading and equipment failure.

The digital system is quite different. The final judgment of the performance quality is the number of bits received in error in a particular time period—that is, how large the probability of error or the bit error rate (BER) is. The error rate and noise in a voice channel are independent of the RSL until a low carrier-to-noise ratio (CNR) exists where the BER changes rapidly with small changes in RSL followed by loss of synchronization. For example, a BER of 10^{-4} at a CNR of 20 dB, for instance, may define the minimum acceptable performance and the minimum RSL under faded conditions. For signal levels a few decibels above this, the error rate can be considered to be negligible and the system to be error-free most of the time.

A second important difference between analog and digital microwave systems is the manner in which they respond to the effects of multipath fading, which produces in-band amplitude and phase dispersion. It can be estimated that the outage of a 45-Mb/s–8-PSK system (without diversity or adaptive equalization) is 60 times greater than that of a 1200-channel frequency division multiplexed (FDM) system [CCIR, 7.1] under similar path conditions and bandwidth. In addition, the susceptibility of analog systems to in-band distortion decreases with the number of voice channels carried. Because multipath activity is common at night and in the early morning, the effect on system availability during the critical busiest hours is reduced. A digital system is fully loaded continuously. Therefore, it is essential to test a digital system under both simulated flat-faded and selectively faded conditions.

The following are the more important factors that affect the performance of digital microwave systems:

1. AM to PM conversion in power amplifiers
2. Compression and expansion in nonlinear amplifiers
3. Excessive intersymbol-interference (ISI)
4. Nonorthogonal modulators
5. Timing errors
6. Carrier phase offset and jitter
7. Excessive thermal noise
8. Selective fading due to multipath activity
9. Cochannel and adjacent-channel interference due to reduction in discrimination as a result of multipath activity

The first six factors are related to hardware faults or deterioration and may result in an increase in error rate at unfaded signal levels and a reduction in the fade margin for minimum acceptable performance. The principles of BER measurement and subsequent error analysis are fully covered in Chapter 5. Jitter tolerance measurements are likewise covered in Chapter 6. In addition, many of the techniques described in Chapters 8 and 9 find important application in the testing of digital radio systems.

This chapter examines the application of the constellation display (phase state display or cluster oscilloscope) to the diagnosis of below-standard error margins and the generation of BER versus CNR and carrier-to-interference ratio, C/I, plots using additive white noise and interference. The effectiveness of the countermeasures against multipath fading is increasingly judged by measuring the radio signature with a simplified three-path model. This is discussed, along with the use of the signature to predict the outage of a link. This chapter does not present RF spectrum occupancy measurements or RF power or frequency measurements. These are more than adequately covered in the literature and in Hewlett-Packard application notes.

7.2 WHITE NOISE

Before studying methods of establishing a required CNR, we examine some of the fundamental properties of the additive noise.

Random noise has a uniform power distribution over a wide range of frequencies (up to greater than 1000 GHz), beyond which the power distribution decreases. Because white light also covers a broad band of frequencies, such noise is known as **white noise**. Relevant examples are thermal noise in a resistor and shot noise in transistors. Thermal noise of resistors has a Gaussian distribution and is thus referred to as Gaussian white noise.

7.2.1 The Gaussian (Normal) Distribution

The Gaussian probability density function (pdf) is the function most frequently used for the description of noise and random signal sources. The Gaussian pdf represents with high accuracy the noise sources of a number of thermal noise generators, including the front-end noise of radio receivers.

If the average value (dc component) of the noise source is zero, the Gaussian pdf is given by

$$p(v) = \frac{1}{\sigma\sqrt{2\pi}} \cdot e^{-v^2/2\sigma^2} \tag{7.1}$$

The area obtained by multiplication of the pdf by an infinitesimal width dv represents the probability that the signal (or noise) has a value in the interval v to $v + dv$. The probability that the value of a signal or noise sample is less than a predetermined numerical value is known as the **cumulative probability distribution function** (CPDF). The CPDF represents the probability that the signal $v(t)$ has a value $V < x$, where x has a specified value.

The CPDF of the Gaussian noise (average value zero) is given by

$$F(v) = p(v \leqslant v) = \frac{1}{\sqrt{2\pi}\,\sigma} \int_{-\infty}^{v} e^{-u^2/2\sigma^2}\, du$$

where u is a dummy variable of integration.

The variance σ^2 would be the ac power dissipated by that noise voltage in a 1-Ω load. It equals the difference between the mean-square value and the square of the mean voltage when the latter is not zero. The square root of the variance is σ, which is the root-mean-square (rms) value of the ac component.

The normalized Gaussian pdf of a noise source is obtained if it is assumed that the average value of the variable is zero and that it has a 1-V rms voltage ($\sigma = 1$ V rms). This unit normal Gaussian pdf is given by

$$p(v) = \frac{1}{\sqrt{2\pi}} \cdot e^{-v^2/2} \tag{7.2}$$

and the corresponding CPDF is

$$F(v) = p(V \leqslant v) = \frac{1}{\sqrt{2\pi}} \cdot \int_{-\infty}^{v} e^{-u^2/2}\, du \tag{7.3}$$

Figure 7.1 is reproduced from Chapter 2 of Feher's book [Feher, 7.2] to remind the reader of the application of the pdf and CPDF. The reader is referred to the same chapter to obtain a further insight into the application of statistical methods in digital transmission systems analysis. The CPDF shows that the cumulative probability that a noise sample is less than 2σ is 0.977. The probability that a noise sample exceeds the value of 4σ is less than 10^{-4}. In other words, the likelihood that a measured value exceeds four times its rms value is only 0.01%. Although this is small, it is still finite and is the major contributor to the error-generating mechanisms in analysis of digital transmission systems.

7.2.2 The Rayleigh pdf

The propagation of radio signals through fading media can be described by the Rayleigh pdf. In addition, the envelope of a narrow-band Gaussian process is Rayleigh distributed. The front end of a radio receiver is considered to be narrow-band if the receiver bandwidth is small compared to the carrier frequency. The Rayleigh pdf is defined by

$$p(v) = \begin{cases} \dfrac{v}{\delta} \cdot e^{-v^2/2\sigma^2} & 0 \leq v \leq \infty \\ 0 & -\infty \leq v \leq 0 \end{cases} \tag{7.4}$$

where σ^2 is the variance of a Gaussian process with envelope v. In a digital radio we are interested in the two-dimensional noise distributed around each state in the phase plane. The noise can be characterized in two ways. A three-dimensional picture is given by the product of two orthogonal Gaussian distributions with the same standard deviation. Alternatively, with polar coordinates centered on the undeviated position of the state, the radial distribution of the noise is described by the Rayleigh distribution.

The variance for the Rayleigh distribution is $2\sigma^2$ and is the same as the sum of two orthogonal Gaussian distributions, each of standard deviation σ. This latter distribution is a special case of the *bivariate normal probability function.*

7.2.3 Finite Crest Factor Noise

Theoretically, an ideal Gaussian noise process has infinitely high peaks; in other words, there is a finite chance that a peak as high as 7σ (or even higher) will occur, even though this probability is only 10^{-12}.

The **crest factor** (c) of a waveform is the ratio of the peak voltage to the rms voltage. For example, the crest factor of a sine wave is $\sqrt{2}$, or 3 dB. Noise with a 15-dB crest factor is limited in its voltage excursions to peaks 5.6 times the rms value.

Most studies of the effect of noise on communication systems assume Gaussian or Rayleigh noise of infinite crest factor. This section considers what happens in practice with realizable noise sources, amplifiers, and power meters. Because all these involve amplification and the output of any amplifier must be

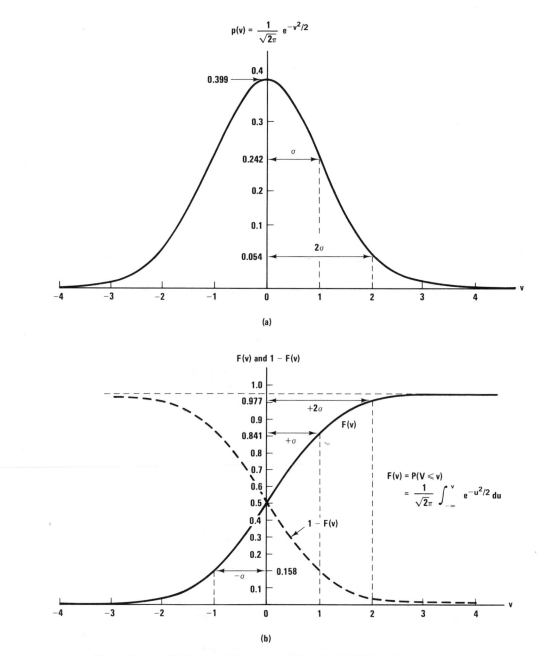

Figure 7.1 (a) Unit normal (Gaussian) PDF and (b) CPDF. (From [Feher, 7.2] with permission from Prentice-Hall, Inc.).

limited by its power supplies, the maximum noise voltage is constrained. We will restrict the analysis to hard or abrupt limiting.

7.2.4 The Limiting CNR Above Which No Errors Are Produced by Noise of a Finite Crest Factor

Gaussian noise with an infinite crest factor will always produce a small, but finite BER, no matter how large the CNR is. However, when the crest factor is finite, reducing the noise below a definite threshold (or, equivalently, raising the CNR above a definite limit) makes the BER actually zero. This is illustrated by means of examples.

PSK

In Fig. 7.2, OX and OY are orthogonal noise-free coordinates of the carrier in an M-phase system, and OA is a decision threshold midway between them. In the case illustrated (QPSK), angle $AOX = 45°$. In general this angle $= \pi/M$. S is the instantaneous position of the noise-free carrier and v is an instantaneous noise voltage that adds vectorially to S.

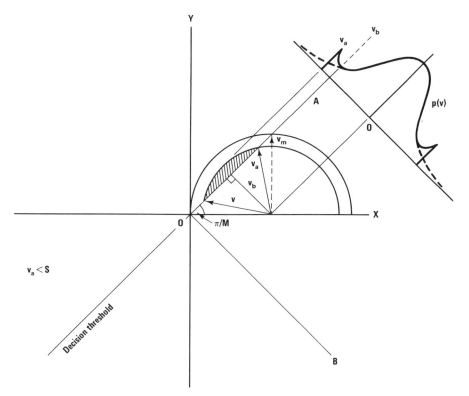

Figure 7.2 Error mechanism in an M-ary PSK system with additive finite crest factor noise.

For v to cause an error, it must be at least of length v_b. Consequently, in this limiting case it must also be perpendicular to OA. Let c = crest factor of noise expressed as a voltage ratio

Then

$$v_b = c\sigma_m$$

where σ_m = minimum rms value of finite crest factor noise that is required to cause one or more errors.

From Fig. 7.2 we have

$$\frac{v_b}{S} = \sin\frac{\pi}{M}$$

where S = peak signal

Thus

$$\frac{S}{\sigma_m} = \frac{c}{\sin\dfrac{\pi}{M}} \tag{7.5}$$

The CNR in a PSK system is given by

$$\frac{C}{N} = 10\log\frac{S^2}{2\sigma^2}\ \text{dB}$$

where σ = rms noise voltage

Therefore, the CNR above which no further errors can occur due to noise of finite crest factor is given by

$$\frac{C}{N} = 20\log\left(\frac{S}{\sigma_m}\right)\cdot\frac{1}{\sqrt{2}} \tag{7.6}$$

Example 7.1

For 8-PSK, $M = 8$ and for noise of 15 dB, the crest factor c is

$$c = \text{antilog}\ \tfrac{15}{20} = 5.62$$

From equation (7.5), $S/\sigma_m = 5.62/(\sin\pi/8) = 14.69$

Therefore, from equation (7.6),

$$\frac{C}{N} = 20\log\frac{14\cdot69}{\sqrt{2}} = 20.3\ \text{dB} \qquad\blacksquare$$

In an 8-PSK system with additive white Gaussian noise of a 15-dB crest factor, no further errors are caused by the noise above a CNR of 20.3 dB. In other words, with such a "finite crest factor" additive noise source, we would measure a BER of $10^{-\infty}$ at CNRs greater than 20.3 dB. An ideal infinite crest factor noise source would lead to a symbol error rate of 2.12×10^{-8} at this CNR.

16 QAM

Figure 7.3 shows the signal space diagram for 16-state Quadrature-Amplitude-Modulation (16 QAM). The minimum distance between any two signal points

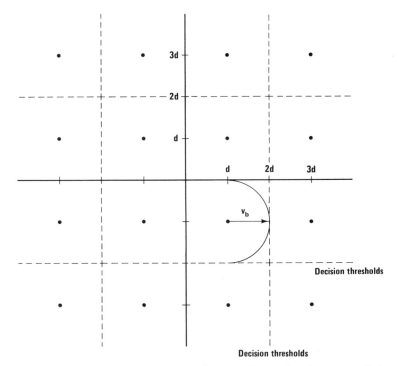

Figure 7.3 Ideal 16-QAM constellation diagram. Measured 16-QAM constellation diagrams are illustrated in Fig. 7.22.

is $2d$. Assuming rectangular transitions between states, the average power in the constellation is given by

$$\frac{(18d^2 + 10d^2 + 10d^2 + 2d^2)}{4(\sqrt{2})^2} = 5d^2$$

The $(\sqrt{2})^2$ term corresponds to the crest factor of the sinusoidal carrier. The average carrier-power-to-noise-power ratio for 16 QAM is

$$\frac{C}{N} \text{ dB} = 10 \log \frac{5d^2}{\sigma^2} \qquad (7.7)$$

Here, d is the distance between any noise-free signal position in the constellation and the nearest decision threshold level.

It is a simple matter to calculate the CNR above which no further errors occur for noise of finite crest factor. For an error to occur, the noise vector v_b must be of minimum length d.

$$v_b = c\sigma_m$$

where σ_m again is the minimum rms value of the finite crest factor noise which can cause an error. Also,

$$\sigma_m = \frac{v_b}{c} = \frac{d}{c}$$

From equation (7.7)

$$\frac{C}{N} = 10 \log 5c^2 \text{ dB}$$

Example 7.2

For a 16-QAM system with injected noise of a 10-dB crest factor, the CNR above which no further errors are attributable directly to the noise is

$$\frac{C}{N} = (10 \log 5)(\text{antilog } \tfrac{10}{20})^2$$

$$= 17 \text{ dB} \qquad \blacksquare$$

From the solution of this example we conclude that, assuming that we have an ideal modem and a 10-dB crest factor Gaussian noise source, then we have a BER of $10^{-\infty}$ (i.e., no errors). On the other hand, we note that with an ideal (infinite crest factor) Gaussian source, the BER performance of the same modem would be 10^{-3} if CNR = 17 dB. Thus the low crest factor noise source would give totally false optimistic measured results for the real system performance.

64 QAM

In an M-ary QAM system, the average power of the constellation is given by

$$P = \frac{d^2(4N^2 - 1)}{3}$$

$$\text{where } N = \frac{\sqrt{M}}{2}$$

For 64 QAM, $P = 21d^2$ and

$$\text{CNR} = 10 \log \frac{21d^2}{\sigma^2}$$

The distance to the decision threshold is still equal to d, and the CNR above which no errors occur is now

$$\text{CNR} = 10 \log 21c^2$$

These CNR limits are tabulated for PSK and QAM for different crest factors in Table 7.1.

TABLE 7.1 CNR (dB) Upper Bound for Zero Errors Due to Finite Crest Factor Noise

	Noise crest factor				
Modulation scheme	10.0	12.0	15.0	20.0	dB
4 PSK	10.0	12.0	15.0	20.0	dB
8 PSK	15.3	17.3	20.3	25.3	dB
16 PSK	21.2	23.2	26.2	31.2	dB
16 QAM	17.0	19.0	22.0	27.0	dB
64 QAM	23.2	25.2	28.2	33.2	dB

7.2.5 Finite Crest Factor Gaussian Noise

The cumulative probability distribution function $F(v)$ for Gaussian noise with finite crest factor c, standard deviation σ, and zero mean is

$$F(v) = \begin{cases} \int_{-\infty}^{v} 1/\sqrt{2\pi} \cdot \sigma \cdot e^{-u^2/2\sigma^2} \, du & -c\sigma < v < c\sigma \\ 0 & v \leq -c\sigma \\ 1 & v \geq c\sigma \end{cases} \tag{7.8}$$

This is shown in Fig. 7.4(b), and the corresponding pdf, $p(v)$, is shown in Fig. 7.4(a). The complementary cumulative distribution function $Q(x)$ is of more

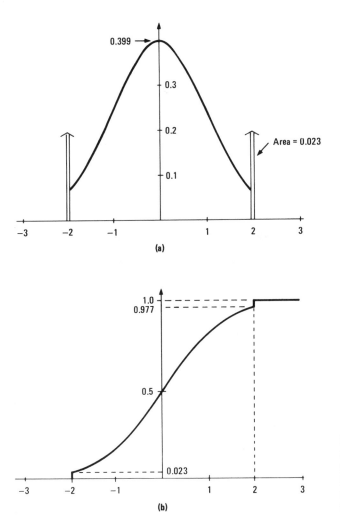

Figure 7.4 Unit normal Gaussian PDF and CPDF for finite crest factor noise. (a) Unit Gaussian PDF with finite crest factor $c = 2$. (b) Unit Gaussian cumulative probability distribution function with finite crest factor $c = 2$.

interest as it can be directly used for calculations of error probability. It is

$$Q(v) = 1 - F(v) = \begin{cases} \int_v^\infty \dfrac{1}{\sqrt{2\pi}} \cdot \sigma \cdot e^{-u^2/2\sigma^2} \, du & -c\sigma < v < c\sigma \\ 0 & v \geq c\sigma \\ 1 & v \leq -c\sigma \end{cases} \tag{7.9}$$

Note that this is identical to the infinite crest factor case for all values of v between the limiting values. This integral cannot be evaluated explicitly, but a suitable polynomial approximation (Abramowitz, 7.3) normalized to unit standard deviation is

$$Q(v) = \frac{1}{\sqrt{2\pi}} \cdot e^{-v^2/2} \, (b_1 t + b_2 t^2 + b_3 t^3 + b_4 t^4 + b_5 t^5) \qquad 0 \leq v < c$$

$$t = \frac{1}{(1 + pv)}$$

$$p = 0.2316419 \qquad b_1 = 0.31938153 \qquad b_2 = -0.356563782 \tag{7.10}$$

$$b_3 = 1.781477937 \qquad b_4 = -1.821255978$$

$$b_5 = 1.330274429$$

The error in this approximation is 1% at $v = 8$ ($Q(v) = 1 \times 10^{-15}$) and increases to 10% at $v = 26$ ($Q(v) = 1 \times 10^{-148}$).

Because of the limited precision inherent in all calculators and computers, it is very important to use $Q(v)$ directly rather than find it indirectly from $1 - F(v)$.

7.2.6 Effect of Truncation on Standard Deviation

Limiting restricts the excursions of a waveform and must, therefore, reduce the total power in that waveform. The limiting operation is a good first-order approximation of saturated (hard-limited) amplifier operation. The variance of $F(v)$ is then

$$E(v)^2 = \int_{-c\sigma}^{c\sigma} v^2 p(v) \, dv + 2 \cdot Q(c\sigma) \cdot (c\sigma)^2 \tag{7.11}$$

Numerical integration of this gives the results of Table 7.2. Note that the error is still only 0.02 dB when the crest factor has been limited to 3, or approximately 10 dB.

7.2.7 Finite Crest Factor Rayleigh Noise

The truncated Rayleigh distribution is shown in Fig. 7.5. The area of the impulse at $c\sigma$ is found from

$$q(c\sigma) = \int_{c\sigma}^\infty \frac{r}{\sigma^2} \cdot e^{-r^2/2\sigma^2} \, dr = e^{-c^2/2} \tag{7.12}$$

TABLE 7.2 Error (dB) in
Noise Power Due to Truncation
at Finite Crest Factor of
Gaussian Noise

Crest factor		Error dB
Ratio	dB	
6	15.56	-1.7×10^{-8}
5	13.98	-4.8×10^{-6}
4	12.04	-5.2×10^{-4}
3	9.54	-2.2×10^{-2}
2.5	7.96	-9.9×10^{-2}
2.0	6.02	-3.6×10^{-1}
1.4	2.92	-1.3
1.3	2.28	-1.6
1.2	1.58	-2.0
1.1	0.83	-2.4
1.0	0.00	-2.9

On the phase plane this Rayleigh distribution and the associated angular function can be viewed as a three-dimensional Gaussian hump contained by a circular wall of radius $c\sigma$ and volume $q(c\sigma)$ [Proakis, 7.19]. This is a reasonable representation of realizable noise sources contained in test equipment such as carrier-noise test sets.

Of real interest is the effect of truncation on the demodulated baseband of, for example, a digital radio. The pdf of this is the addition of two parts; the

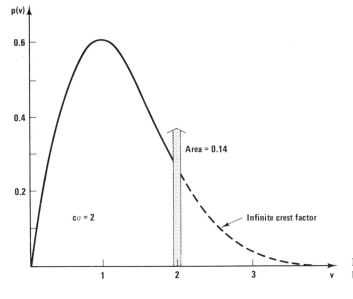

Figure 7.5 Truncated Rayleigh probability density function.

first is due to the truncated Gaussian hump (p_1) and the second is due to the surrounding wall (p_2). Writing P_1 using Cartesian coordinates X and Y gives

$$p_1(x, y) = \frac{1}{2\pi\,\sigma^2} \cdot e^{-(x^2 + y^2)/2\sigma^2} \qquad 0 \leq x^2 + y^2 < (c\sigma)^2$$

where $v^2 = x^2 + y^2$, $dx,\ dy = v\ dv\ d\theta$

After demodulation along, for example, the Y-axis, the pdf of the baseband signal is given by:

$$p_1(y) = \frac{1}{\sqrt{2\pi}\cdot\sigma} \cdot e^{-y^2/2\cdot\sigma^2} \int_{-(c^2\sigma^2 - y^2)}^{+(c^2\sigma^2 - y^2)} \frac{1}{\sqrt{2\pi}\sigma} \cdot e^{-x^2/2\sigma^2}\ dx$$

$$= \frac{1}{\sqrt{2\pi}\sigma} \cdot e^{-y^2/2\sigma^2} \left\{ 1 - 2\,Q\left[c^2 - \left(\frac{y}{\sigma}\right)^2 \right]^{1/2} \right\}$$

where $Q(x)$ is the complementary cumulative Gaussian function already described (equation (7.9)).

The contribution from p_2 after demodulation is found from the CPDF $P_2(y)$:

$$p_2(y) = 0.5 + \frac{1}{\pi} \sin^{-1}\left(\frac{y}{c}\right)$$

Differentiation of this with respect to y gives the pdf:

$$p_2(y) = e^{-c^2/2} \cdot \frac{1}{\{\pi \cdot [(c\sigma)^2 - y^2]^{1/2}\}}$$

This is a similar procedure to deriving the pdf of a sinusoid [Proakis, 7.19].

Adding these two contributions and normalizing to unit sigma, the pdf for demodulated truncated Rayleigh noise is

$$p(v) = \frac{1}{\sqrt{2\pi}} \cdot e^{-v^2/2} [1 - 2Q(c^2 - v^2)^{1/2}] + e^{-c^2/2} \cdot \frac{1}{[\pi \cdot (c^2 - v^2)^{1/2}]}$$

This can be readily evaluated for any c and is plotted in Fig. 7.6. When analyzing system performance, what is really required is the equivalent of $Q(v)$ for truncated Rayleigh noise:

$$Q_r(v) = \int_v^c \frac{1}{\sqrt{2\pi}} \cdot e^{-u^2/2} [1 - 2Q(c^2 - v^2)^{1/2}]\ du + e^{-c^2/2} \cos^{-1}\frac{(v/c)}{\pi} \qquad (7.13)$$

This can be integrated numerically for any given c (Table 7.3) and then used in predictions of system performance.

7.2.8 Theoretical PSK Error Rate with Finite Crest Factor Noise

Referring again to Fig. 7.2, consider the case when v is greater than the limiting case v_b and v_a is the maximum possible noise excursion due to clipping. Let v_a

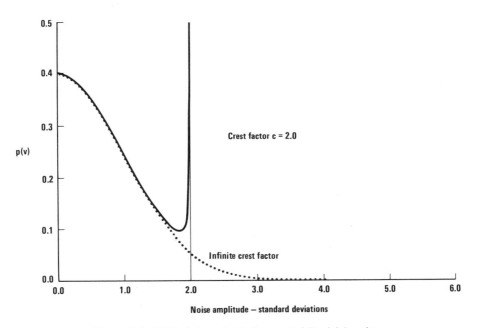

Figure 7.6 PDF of demodulated truncated Rayleigh noise.

$< S$, the peak signal. For an error to occur, $v_b \leqslant v \leqslant v_a$ and $|\theta|$ must be less than $\cos^{-1}(v_b/v)$. Since the magnitude of v and the phase angle of v are statistically independent, we may write the probability of error $p_e(v)$ for voltage v to $v +$ dv as

$$p_e(v) = p(v) \cdot dv \times p[|\theta| < \cos^{-1}(v_b/v)]$$

$$= v(v)\, dv \times \frac{1}{\pi} \cos^{-1}(v_b/v)$$

where $(1/\pi)\cos^{-1}(v_b/v)$ is the probability of the phase angle of v lying within a sector 2θ. So the total error probability for the shaded segment of Fig. 7.2, including the impulse at clipping level, is

$$p_e = \int_{v_b}^{v_a} p(v) \cdot \frac{1}{\pi} \cos^{-1}(v_b/v)\, dv + \int_{v_a}^{\infty} p(v) \cdot \frac{1}{\pi} \cos^{-1}(v_b/v_a)\, dv$$

where $p(v)$ is the pdf of the additive noise. Remembering that the envelope of a narrowband Gaussian process is Rayleigh distributed, then for Gaussian noise of rms value σ,

$$p(v) = p\left(\frac{v}{\sigma}\right) = \frac{v}{\sigma^2} \cdot e^{-v^2/2\sigma^2} \qquad (7.14)$$

where P is the normalized density function.

Referring to Fig. 7.2, a similar shaded area is intersected by OB, and

TABLE 7.3 Complementary Cumulative Distribution
Function $P\ (v > Y_\sigma)$ for Truncated and Ideal
Demodulated Rayleigh Noise

Y	Ideal Rayleigh	Truncated Rayleigh: crest factor =		
		10 dB	12 dB	15 dB
0.0	5.00E-001			
0.1	4.60E-001			
0.2	4.21E-001			
0.3	3.82E-001			
0.4	3.45E-001			
0.5	3.09E-001			
0.6	2.74E-001			
0.7	2.42E-001		Results in this region	
0.8	2.12E-001		agree to ideal case to	
0.9	1.84E-001		more than three places	
1.0	1.59E-001			
1.1	1.36E-001			
1.2	1.15E-001	1.15E-001		
1.3	9.68E-002	9.67E-002		
1.4	8.08E-002	8.07E-002		
1.5	6.68E-002	6.67E-002		
1.6	5.48E-002	5.47E-002		
1.7	4.46E-002	4.45E-002		
1.8	3.59E-002	3.58E-002		
1.9	2.87E-002	2.86E-002		
2.0	2.28E-002	2.26E-002		
2.1	1.79E-002	1.77E-002		
2.2	1.39E-002	1.37E-002		
2.3	1.07E-002	1.06E-002		
2.4	8.20E-003	8.01E-003		
2.5	6.21E-003	6.01E-003		
2.6	4.66E-003	4.44E-003	4.66E-003	
2.7	3.47E-003	3.22E-003	3.46E-003	
2.8	2.56E-003	2.28E-003	2.55E-003	
2.9	1.87E-003	1.55E-003	1.86E-003	
3.0	1.35E-003	9.78E-004	1.34E-003	
3.1	9.68E-004	4.86E-004	9.60E-004	
3.2	6.87E-004	0.0	6.79E-004	
3.3	4.83E-004		4.75E-004	
3.4	3.37E-004		3.28E-004	
3.5	2.33E-004		2.22E-004	
3.6	1.59E-004		1.48E-004	
3.7	1.08E-004		9.50E-005	
3.8	7.24E-005		5.73E-005	
3.9	4.81E-005		2.90E-005	
4.0	3.17E-005		0.0	
4.1	2.07E-005			
4.2	1.34E-005			
4.3	8.55E-006			
4.4	5.42E-006			

4.5	3.40E-006			
4.6	2.11E-006			
4.7	1.30E-006			1.30E-006
4.8	7.94E-007			7.91E-007
4.9	4.80E-007			4.77E-007
5.0	2.87E-007			2.84E-007
5.1	1.70E-007			1.67E-007
5.2	9.98E-008			9.69E-008
5.3	5.80E-008			5.48E-008
5.4	3.34E-008			2.98E-008
5.5	1.90E-008			1.47E-008
5.6	1.07E-008			4.30E-009
5.7	6.01E-009			0.0

therefore the total symbol error probability for $v_a < S$ is

$$p_{(E)} = \frac{2}{\pi} \int_{S \cdot \sin \pi/M}^{v_a} p\left(\frac{v}{\sigma}\right) \cdot \cos^{-1} \frac{(S \cdot \sin \pi/M)}{v} \, dv$$

$$+ \frac{2}{\pi} \int_{v_a}^{\infty} p\left(\frac{v}{\sigma}\right) \cdot \cos^{-1} \frac{(S \cdot \sin \pi/M)}{v_a} \, dv$$

where $p(v/\sigma)$ is given by equation (7.14) and $v_b = S \cdot \sin \pi/M$.
Similarly, for $v_a > S$, the total symbol error probability is

$$p_{(E)} = \frac{2}{\pi} \int_{S \sin \pi/M}^{S} p\left(\frac{v}{\sigma}\right) \cos^{-1} \frac{(S \sin \pi/M)}{v} \, dv$$

$$+ \frac{1}{\pi} \int_{S}^{v_a} p\left(\frac{v}{\sigma}\right) \left[\frac{\pi}{2} - \frac{\pi}{M} + \cos^{-1} \frac{(S \sin \pi/M)}{v} \right] dv$$

$$+ \frac{1}{\pi} \int_{v_a}^{\infty} p\left(\frac{v}{\sigma}\right) \left(\frac{\pi}{2} - \frac{\pi}{M} + \cos^{-1} \frac{(S \sin \pi/M)}{v_a} \right) dv$$

where v_a = clipping level of noise in volts
  S = peak signal level in volts
  M = number of phase states in PSK modulation
  σ = rms value of noise

The probability of symbol error rate has been computed for 4, 8, and 16 PSK and for 10-, 12-, 15-, and 20-dB crest factors. The results are shown in Fig. 7.7. The effect of the crest factor on the error rate can clearly be seen together with the bound in CNR discussed in Section 7.2.3, where the error rate falls to zero.

To simulate accurately a flat fade over the range of CNR that produces error rates of 10^{-3} to 10^{-8}, the additive white Gaussian noise should have a crest factor of at least 15 dB. If this is not the case, the BER versus CNR plot for the radio under test may significantly deviate from the theoretical curve because of this fact alone, and different noise sources will produce different deviations. In addition, slight differences may occur when comparing the measured

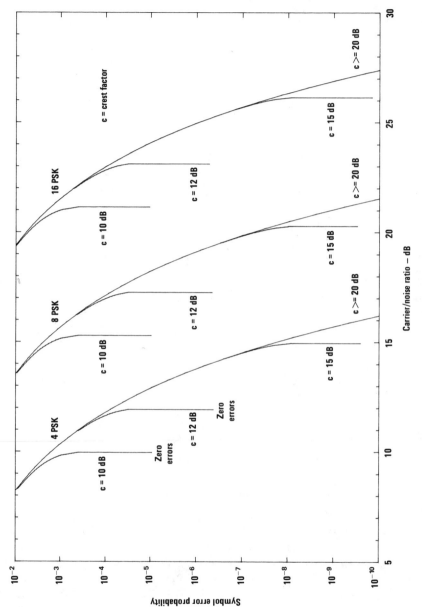

Figure 7.7 PSK symbol error probability vs. CNR for finite crest factor noise.

plot obtained by fading the radio with an RF attenuator (where the IF amplifier may define the noise crest factor) and by the injection of noise after the IF amplifier (at nominal RF RSL). Both methods define the CNR and are discussed in Section 7.5.1.

Note that the results for crest factors larger than 20 dB are for all practical purposes indistinguishable from ideal Gaussian noise over the ranges of BER plotted.

7.2.9 QAM

The theoretical error rate for 16 QAM with finite crest factor additive white Gaussian noise (AWGN) is derived in Section 7.6.1 in the course of deriving the theoretical error rate of 16 QAM in the presence of sinusoidal interference.

7.3 NOISE BANDWIDTH OF RADIO RECEIVERS

The concept of noise bandwidth, associated with the ideal rectangular passband characteristic, provides a convenient means by which digital microwave systems may be compared and receiver thresholds may be calculated; further, it provides a means to establish a desired carrier-to-noise-density ratio in the course of demodulator testing.

7.3.1 Calibration of Noise Bandwidth

Let the power gain versus frequency characteristic of the filter be as shown in Fig. 7.8. The figure also shows the equivalent noise filter of bandwidth B_e and gain g_1.

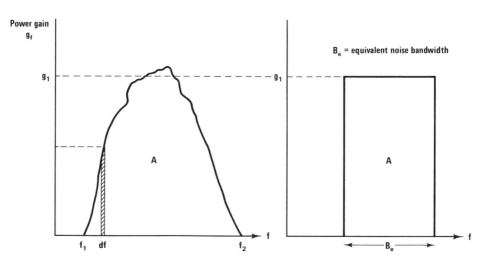

Figure 7.8 Actual filter power response and equivalent rectangular response.

The equivalent filter is said to have a noise bandwidth B_e, where

$$\int_0^\infty g_f \, df = A = g_1 B_e$$

In practice, the power gain versus frequency characteristic is integrated between frequencies f_1 and f_2 to give area A and a suitable value of g_1 selected so that B_e may be established from

$$B_e = \frac{A}{g_1} = \frac{\int_0^\infty g_f \, df}{g_1} \tag{7.15}$$

Presuming that the characteristic of the filter has been accurately integrated, the only significant error then lies in the choice of the reference gain g_1. The integration of the filter response can be carried out in at least three ways:

1. By plotting the power gain (ratio) response of the filter onto squared paper and counting the squares under the curve!

2. By using a network analyzer such as the Hewlett-Packard 8505A under computer control to characterize the filter across the required frequency range. The computer can subtract the instrument and connecting cable "back-to-back" response before converting the filter response to power ratios and numerically integrating.

3. A similar method uses a HP-IB programmable signal generator and power meter such as the HP8660 or HP8662 and HP436 or HP438 in place of the network analyzer. Again, the back-to-back response of the test equipment can be subtracted from the filter response before integration.

All these methods are time consuming, and the first one is not very accurate. The Hewlett-Packard 3708A Noise and Interference Test Set (which is discussed in more detail in Section 7.5.6) contains a wideband noise source and a power meter and can measure filter noise bandwidth directly in a few seconds. In order to do this, the 3708A must know the filter reference gain, and it can measure this at the (IF) carrier frequency. When the reference gain must be chosen in some other way, the chosen insertion loss can be entered via the keyboard before the noise bandwidth is measured. During CNR testing, the noise density obtained at a frequency separate from the (IF) carrier frequency is dependent upon the value selected for g_1 in the evaluation of B_e. In FM/FDM systems, this results in an error in baseband signal-to-noise ratios in the appropriate telephone slots, and this error can be minimized by making g_1 equal t, the arithmetic mean (in decibels) of the filter loss over the required passband. The procedure used to calculate the mean value is dependent upon the filter characteristics and the specific application.

The measurement of the noise bandwidth of digital radio IF filters differs from the FM/FDM situation, and it is customary to measure the reference gain at the center of the IF carrier frequency band—that is, usually 70 or 140 MHz.

7.3.2 Noise Bandwidth and 3-dB Bandwidth

Many radio engineers presume that the noise bandwidth of a filter is the same as its 3-dB bandwidth. Although this is approximately true for high-order filters, it is not true for many filter types of lower order used in the IF of digital radio systems. To illustrate this, consider the Butterworth filter. The power response is given by

$$|H(\omega)|^2 = \frac{1}{1 + x^{2n}} \tag{7.16}$$

Noise bandwidth is

$$B_e = \frac{1}{2\pi} \frac{\int_0^\infty |H(\omega)|^2}{|H(\omega_0)|^2} \, d\omega$$

Normalizing the transfer function at dc for low-pass (or center frequency for bandpass) reduces equation (7.16) to

$$B_e = \frac{1}{2\pi} \int_0^\infty |H(\omega)|^2 \, d\omega$$

Thus the noise bandwidth of a Butterworth filter is

$$B_e = \frac{1}{2\pi} \int_0^\infty \frac{1}{1 + \omega^{2n}} \, d\omega$$

The result for $n = 1$ to 5 is tabulated in Table 7.4, where it can be seen that the noise bandwidth is larger than the 3-dB bandwidth by from 57% (2 dB) for $n = 1$ to 1.66% (0.007 dB) for $n = 5$.

It is interesting to compare the difference between the noise bandwidth and the 3-dB bandwidth of the raised-cosine and the root raised-cosine Nyquist channel for both pulse and impulse transmission. The square root of the raised-cosine characteristic is studied because of the practice of partitioning the overall filtering function between transmitter and receiver. The amplitude characteristics

TABLE 7.4 Noise Bandwidth (B_e) and 3-dB Bandwidth (B_{3dB}) for Butterworth Filters

n	Be (Hz)	B_{3dB}	$\dfrac{B_e}{B_{3dB}}$
1	0.25	0.159	57
2	0.177	0.159	11.2
3	0.167	0.159	4.7
4	0.163	0.159	2.6
5	0.162	0.159	1.66

and functions for the Nyquist channel for both impulse and pulse transmission can be found in Chapter 1 of this book.

Table 7.5(a) shows the difference between the noise bandwidth and the 3-dB bandwidth for $\alpha = 0$ to 1 for the Nyquist channel for *pulse* transmission. For $\alpha = 0.1$, the difference is 31.7%, or 1.2 dB. Similarly, Table 7.5(b) shows the difference for the square root of the Nyquist channel for pulse transmission. For $\alpha = 0.1$, the difference is 14.1%, or 0.57 dB.

TABLE 7.5(a)　Noise Bandwidth (B_e) and 3-dB Bandwidth (B_{3dB}) of Raised-Cosine Nyquist Channel for *Pulse* Transmission

x	B_e	B_{3dB}	$\dfrac{B_e}{B_{3dB}}$ (%)
0	1.39	1.0	139
0.1	1.33	1.01	131.7
0.2	1.28	1.01	126.7
0.3	1.24	1.02	121.6
0.4	1.20	1.03	116.5
0.5	1.17	1.045	112.0
0.6	1.15	1.06	108.5
0.7	1.13	1.08	104.6
0.8	1.12	1.10	101.8
0.9	1.16	1.12	103.6
1.0	1.13	1.16	97.4

TABLE 7.5(b)　Noise Bandwidth (B_e) and 3-dB Bandwidth (B_{3dB}) of the *Square Root* of the Raised-Cosine Nyquist Channel for *Pulse* Transmission

x	B_e	B_{3dB}	$\dfrac{B_e}{B_{3dB}}$ (%)
0	1.17	1.0	117
0.1	1.17	1.025	114.1
0.2	1.17	1.05	111.4
0.3	1.18	1.08	109.3
0.4	1.19	1.12	106.3
0.5	1.21	1.15	105.2
0.6	1.23	1.20	102.5
0.7	1.25	1.25	100.0
0.8	1.28	1.31	97.7
0.9	1.32	1.38	95.7
1.0	1.39	1.48	93.9

The noise bandwidth and the 3-dB bandwidth of the square root of the Nyquist channel for *impulse* transmission are both equal to the Nyquist bandwidth. This can be verified by inspection of the amplitude characteristics of the channel (Chapter 1).

In CNR testing of digital radios, it is fairly common to reference CNRs to the output of the main IF amplifier (i.e., after the IF filter). The CNR at the detector input is higher by the ratio of the system noise bandwidths at the respective points, assuming that the modulated carrier was filtered at the transmitter, and the receiver filter has an impact only on noise. A correction factor is normally used, and great care should be taken in the measurement of the IF filter bandwidth, together with any assumptions about it being equal to the 3-dB bandwidth.

7.4 C/N, C/N₀, Eb/N₀, AND SOME USEFUL CONVERSIONS

The ratios of carrier power to noise power (C/N) and carrier to noise density (C/N_0) are well known and apply to analog or digital microwave systems.

The ratio of energy per bit to noise density (E_b/N_O) is not so obvious and requires further explanation. In the case of digital systems, a convenient standard for comparing different modulation systems is provided by the concept of average signal energy per transmitted bit (E_b) in the presence of a given noise spectral density (N_O)(noise in 1-Hz normalized bandwidth). This concept results in a figure of merit for comparing modulation systems that emphasizes the minimization of the total transmitted energy required to convey a given amount of data.

The bit energy, E_b, is obtained by dividing the carrier power, C, by the bit rate, f_b.

$$E_b = \frac{C}{f_b} = C \cdot T_b$$

where T_b = bit duration

Therefore,

$$\frac{E_b}{N_O} = \frac{C}{N_O} \cdot \frac{1}{f_b} \tag{7.17}$$

Since E_b has the unit joules and N_O is power per hertz, which is (joules/second) × seconds, E_b/N_O is dimensionless.

Expressed in decibels,

$$\underset{\text{(dB)}}{\frac{E_b}{N_O}} = \underset{\text{(dB-Hz)}}{\frac{C}{N_O}} - \underset{\text{(dB-Hz)}}{10 \log f_b}$$

If the receiver noise bandwidth is B_e (hertz), and the total noise power measured at the receiver output is N, then

$$\underset{\text{(W/Hz)}}{N_O} = \frac{N \ (\text{W})}{B_e \ (\text{Hz})}$$

Substitution in equation (7.17) gives

$$\frac{E_b}{N_O} = \frac{C}{N} \cdot \frac{B_e}{f_b}$$

or, again in decibels,

$$\underset{\text{(dB)}}{\frac{E_b}{N_O}} = \underset{\text{(dB)}}{\frac{C}{N}} - \underset{\text{(dB)}}{10 \log \frac{f_b}{B_e}}$$

If the noise bandwidth equals the bit rate, then

$$\frac{E_b}{N_O} = \frac{C}{N}$$

The ratio C/N_O relates the carrier power to the noise power in a 1-Hz bandwidth.

To convert C/N_O to C/N, which is the ratio of carrier power to noise power in a defined bandwidth, the following equation applies:

$$\frac{C}{N} = \frac{C}{N_O} - 10 \log B_e$$

(7.18)

$$(dB) \quad (dB\text{-}Hz) \quad (dB\text{-}Hz)$$

Finally, in terms of received signal level, front-end noise figure, and bit rate,

$$\frac{E_b}{N_O} = (RSL) - (10 \log f_b + 10 \log kT + 30 + F)$$

(7.19)

$$(dB) \quad (dBm) \qquad\qquad (dBm)$$

where RSL = received signal level in dBm

f_b = bit rate in b/s

$k = 1.38 \times 10^{-23}$ J/K (per degree Kelvin)

$T = 290$ K or 17°C

$kT = -174$ dBm/Hz at 17°C

30 = conversion factor from dBW to dBm

F = noise figure in dB

The ratio E_b/N_O provides a useful figure of merit for comparing the efficiency of digital modulation systems; it is the minimum input signal power required per bit of information per second in the presence of a given noise power spectral density of N_O watts per hertz for a given error rate.

7.5 FLAT-FADE SIMULATION AND BER VERSUS C/N MEASUREMENTS

BER measurements of a digital microwave system at nominal or unfaded RSLs are time consuming and provide limited evaluation of system performance. Under these conditions, multihop digital microwave systems are effectively error free for most of the time; the residual BER may be less than 10^{-13} and the mean time between errors of the order of 500 h. It is important to characterize the system at the minimum RF input levels encountered during hostile propagation conditions. Manufacturers and operating companies frequently specify the RSL or C/N necessary to produce a BER of 10^{-6} and 10^{-3}. Regular checks are made to ensure that the radio is performing acceptably. In addition, a plot of BER versus C/N down to the receiver threshold allows comparisons to be made against theoretical curves.

7.5.1 Establishing a C/N or C/N_O Ratio

Frequently in the laboratory or factory, a variable RF attenuator is inserted into the input waveguide to attenuate the RSL. The thermal noise from the receiver front end defines the C/N_O ratio.

The value C/N_O can be calculated simply from a knowledge of the RSL and the front-end noise figure.

$$\text{Noise density } N_O = kT + (F - 1)kT_{Ot} \qquad (7.20a)$$

$$= kTF \text{ (W-s or W/Hz) at 290 K}$$

where T_O = 290 K
$\quad k$ = Boltzmann's constant = 1.38×10^{-23} J/K
$\quad T$ = temperature in Kelvin = $T°C + 273 = \frac{5}{9}T°F + 255$
$\quad F$ = noise figure (ratio)

Converting equation (7.20a) to milliwatts per hertz and expressing noise density in terms of decibels per hertz gives:

$$N_O = -174 + F$$

where F is now the noise figure in decibels (dBm/Hz).

The value of C/N_O at the receiver input is, therefore,

$$\frac{C}{N_O} = \frac{C}{kTF}$$

$$= \text{RSL} + 174 - F \qquad \text{(dB-Hz)}$$

Note that

$$10 \log kT = -203.97 \text{ dBW/Hz} = -174 \text{ dBm/Hz} \qquad \text{at } T = 290 \text{ K or } 17°C$$
$$(7.20b)$$

Example 7.3

If the receiver noise figure is 6 dB and the RSL is -70 dBm for a 10^{-6} BER, then:

$$\frac{C}{N_O} = -70 + 174 - 6 = 98 \qquad \text{(dB-Hz)}$$

This ratio remains unchanged throughout the receiver RF, IF, and predetection circuits because signal and noise are amplified equally (assuming the gain is constant over the range of signal and signal plus noise—i.e., there is no amplitude compression). The unfaded C/N_O would be about 40 dB higher, or 138 dB-Hz, and the error rate would revert to its residual value of less than 10^{-10} or even a much smaller value. It can be seen that a 40-dB RF attenuator can be used to vary the C/N_O from (in this case) 98 to 138 dB-Hz, enabling a BER curve to be measured down to the receiver threshold.

The ratio C/N in the receiver varies at points throughout the RF/IF chain and is dependent on a knowledge of the appropriate noise bandwidths.

$$\frac{C}{N} = \frac{C}{N_O} - 10 \log B_e$$
$$(7.21)$$

$$\text{[dB]} \quad \text{[dB-Hz]} \quad \text{[dB-Hz]}$$

where B_e is the noise bandwidth of the receiver in hertz. In terms of received signal level,

$$\frac{C}{N} = \text{RSL} - (10 \log kTB_e + 30 + F)$$

(7.22)

$$\text{[dB]} \quad \text{[dBm]} \quad \text{[dBW]} \quad \text{[dB]}$$

∎

Example 7.4

If the RSL is -70 dBm, $F = 6$ dB and

$$B_e = 30 \text{ MHz} \quad (30 \times 10^6 \text{ Hz})$$

$$\frac{C}{N} = -70 - (-129.2 + 30 + 6)$$

$$= 23.2 \text{ dB}$$

If B_e in this case represents the IF filter noise bandwidth, then the C/N given by either equation (7.21) or (7.22) will agree with that measured with a power meter at the IF amplifier output by measuring (under manual gain control) the carrier and noise powers separately. ∎

The BER performance of the radio can be compared to the theoretical curves by using the symbol rate bandwidth for B_e. Once this method is calibrated, the CNR can be related to the RSL for the purposes of flat-fade simulation. Note, however, that the method relies on a knowledge of the noise bandwidth and noise figure for every radio under test. The tolerance on the latter is unacceptably wide, remembering that BER can change by more than one order of magnitude for a CNR change of less than 0.5 dB.

The RF attenuator method is also time consuming and inconvenient. In the field, variations in RSL make it very difficult or impossible to fade the RF signal and make accurate BER measurements. This is particularly true when working near receiver threshold levels. A small scintillation fade can cause loss of synchronization. Equation (7.19) allows the calculation of the available system E_b/N_O from RSL, bit rate, and noise figure. The advantage of characterizing a digital radio in terms of E_b/N_O or C/N_O is that the noise bandwidth of the radio under test need not be known. Specialized test equipment is available, such as the Hewlett-Packard 3708, which can automatically define E_b/N_O at points throughout the IF stage without the need for an RF attenuator.

A second method of establishing a C/N, E_b/N_O, or C/N_O ratio is to inject relatively high level noise into the receiver demodulator or IF at normal received signal level. Many manufacturers of digital radio design their own carrier-to-noise test sets for demodulator and system testing. The block diagram of such a set is shown in Fig. 7.9. The carrier source is either the receiver IF signal, which is interrupted and has noise added to it, or a test modulator for use when a received signal is not available.

The wideband noise spectral density of the noise generator is filtered to the desired bandwidth and supplied via an attenuator-amplifier combination to

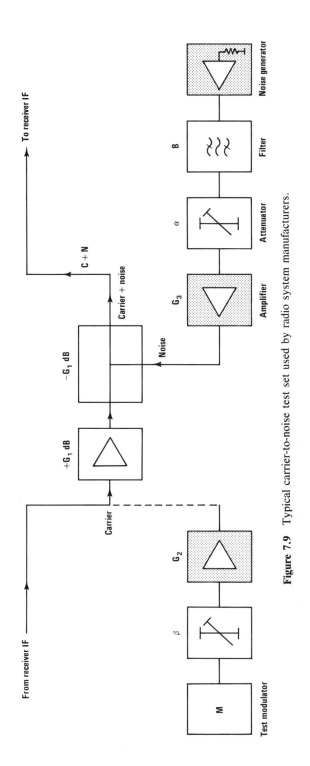

Figure 7.9 Typical carrier-to-noise test set used by radio system manufacturers.

a combining network, where it is combined with the carrier. In practice, a CNR is established in the noise bandwidth of the filter B. This corresponds to a specific value of C/N_O. Presuming that the test modulator is used, the procedure is as follows:

1. The noise attenuator α is set to maximum attenuation.
2. A power meter is connected to the $C + N$ output port of the combining network.
3. The carrier attenuator, B, is adjusted to a value of β_1, which produces the desired carrier power, C.
4. The carrier attenuator is set to maximum attenuation and the noise attenuator is set to α_1, which produces a noise power equal to the carrier power C in Step 3.
5. The carrier attenuator is readjusted to a value β_1, and the noise attenuator is adjusted to a value $\alpha_1 + C/N$ (dB), where C/N is the desired CNR.

The procedure with the IF signal as a carrier is similar except that the noise power is made equal to the IF signal level (say $+1$ dBm) to establish a 0-dB CNR. The noise is measured by turning the carrier off.

This test arrangement is attractive because the power meter is required to test only when two powers are equal and does not require absolute accuracy. The CNR error is then primarily dependent on attenuator accuracy. Other sources of error are discussed in the next section.

7.5.2 Sources of Error in Setting C/N

Other sources of error in setting C/N are as follows:

1. Incorrect measurement of the noise bandwidth of the noise band defining filter B.
2. Equivalent noise bandwidth of the out-of-band response of the filter B.
3. Attenuator errors.
4. Noise from amplifiers in the carrier path or after the noise-setting attenuator (G_1, G_2, and G_3 in Fig. 7.9).
5. Temperature variation: a change in thermal noise source temperature of 3°C will change the noise density by 0.05 dB (see equation (7.20a)).
6. Residual radio noise.

To put the subject of errors into perspective, it should be remembered that an order-of-magnitude change in error rate can occur for less than a 0.5-dB change in C/N. One should, therefore, aim for a C/N setting accuracy of typically 0.1 to 0.25 dB.

In this section of the chapter, we will examine errors due to 2, 4, and 6 in more detail.

7.5.3 The Equivalent Noise Bandwidth of the Out-of-Band Response of a Filter

It is possible for the stop-band response of IF filters—for example, 70- or 140-MHz filters—to rise with increasing frequency. The magnitude of this out-of-band response at hundreds of megahertz can approach the magnitude of the pass-band response.

When noise from a typical broadband noise source is passed through the filter and the resulting output power measured on a broadband power meter, there is an additional contribution from the equivalent noise bandwidth of the out-of-band response.

The **error caused by the out-of-band response** is derived as follows:

1. Let the filter have zero loss.
2. Let the noise density at the filter input be N_O milliwatts per hertz.
3. Let the noise bandwidth of the filter be B_e hertz, and the noise bandwidth of the out of band response be B_O hertz, as defined in Fig. 7.10.
4. Let the power ratio between the heights of the rectangular filters B_e and B_O be m (less than or equal to 1.0).

The power at the filter output N is given by

$$N = N_O (B_e + mB_O) \text{ mW}$$
$$= B_e N_O + mB_O N_O$$

The ideal filter of noise bandwidth B_e hertz would give an output power of $B_e N_O$ milliwatts.

The error due to the out of band response is

$$mB_O N_O \quad (\text{mW})$$

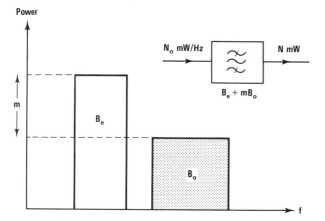

Figure 7.10 Equivalent noise bandwidth of a filter out of band response.

The power ratio of the error is

$$\frac{N_O B_e + N_O m B_O}{N_O B_e} = 1 + \frac{m B_O}{B_e}$$

Expressed in decibels, the error becomes

$$\text{error (dB)} = 10 \log\left(\frac{1 + m B_O}{B_e}\right) \qquad (7.23)$$

For less than a 0.1-dB error, $1 + m B_O/B_e = 1.023$.

A practical upper frequency limit must be defined in which B_O is measured, and 600 MHz would seem to be a sensible choice, considering the high f_T transistors used as amplifiers in noise sources. If at some frequency in this range, the out-of-band response is only 20 dB below the passband, then $m = 0.01$ and $B_O < 2.33 B_e$ for less than a 0.1-dB error. Results for 0.1- and 0.05-dB errors, with values of m corresponding to a maximum spurious response 10, 20, and 30 dB below the passband, are tabulated in Table 7.6.

TABLE 7.6 Error Caused by Equivalent Noise Bandwidth of a Filter Out-of-Band Response

Error (dB)	m (dB)	B_O/B_E
0.1	-30	23.3
0.1	-20	2.33
0.1	-10	0.23
0.05	-30	11.6
0.05	-20	1.16
0.05	-10	0.116

7.5.4 Error in Noise Density Output Due to Finite Noise Figure of an Amplifier of White Noise

The design of noise sources and methods of injecting noise onto a carrier are not discussed here. However, there are valid reasons for including amplifiers in positions such as those in Fig. 7.9. The thermal noise from these amplifiers adds either to the carrier or noise paths appropriately and causes an error in the ratio C/N_O. While this error is significant only at low levels of injected noise, it does place a lower limit on the magnitude of the noise density that can be injected for a given error. This limit becomes more important with the development of spectrally efficient digital radio. The error due to amplifier noise is derived as follows. The intrinsic noise density produced by the amplifier referred to its input is

$$n_i = FkT \times 10^3 \qquad \text{(mW/Hz)}$$

where F is the noise figure ratio. If n_O is the noise density applied to the amplifier signal input, the error in decibels is

$$\text{Error} = 10 \log \frac{n_O + FkT10^3}{n_O} \quad \text{(dB)}$$

This is plotted in Fig. 7.11 for magnitudes of noise where the error becomes significant.

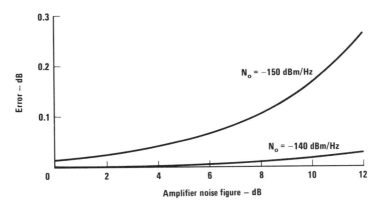

Figure 7.11   Error in noise density output due to finite noise figure of an amplifier.

7.5.5 Correction Factor for Residual Radio Noise

Injecting Gaussian noise into the IF stages of a radio receiver reduces the CNR as follows:

$$\frac{C}{N} = \frac{C}{N_r + N_i}$$

where C is the average carrier power, N_r is the residual radio noise power at the point of injection, and N_i is the injected noise power.
Expressing this in decibels gives

$$\frac{C}{N} = \frac{C}{N_i} - 10 \log\left(1 + \frac{N_O}{N_i}\right) \quad \text{(dB)} \tag{7.24}$$

Normally $N_i \gg N_r$ and equation (7.24) reduces to

$$\frac{C}{N} = \frac{C}{N_i} \quad \text{(dB)}$$

If this does not apply, then the correction factor is given by

$$10 \log\left(1 + \frac{N_r}{N_i}\right)$$

7.5.6 Noise and Interference Test Set: Principles and Operation

In Section 7.5.1, two methods of establishing a CNR were described. The second method described a noise injection into the receiver IF stages and avoided the need for an RF attenuator. The noise-injection method is developed even further in the HP3708A *Noise and Interference Test Set*. Noise of known spectral density is injected into the IF section of the receiver to establish the demanded carrier dependent ratio (C/N, C/N_O, or E_b/N_O). The receiver IF power is measured at the point of noise injection, and the noise density is adjusted automatically to maintain the required carrier-dependent ratio in the presence of received signal level variations. The instrument-response time is typically 10 ms, which enables BER testing to be carried out in the presence of rapid changes in RSL. Figure 7.12 shows the injected noise tracking sinusoidal variations in IF signal power at 50 dB/s and 10 dB/s. This feature also avoids loss of receiver synchronization due to RSL variations in the field when measuring BER near to receiver threshold CNRs.

The demanded carrier-dependent ratio is maintained at any point chosen for noise injection throughout the IF chain. In the C/N mode, the injected noise density is automatically adjusted to compensate for the receiver noise bandwidth, enabling C/N to be defined in, for example, the symbol rate bandwidth. Figure 7.13 shows a typical measurement setup.

Because the HP3708A contains a noise source, power meter (true rms) and microprocessor, it can be used to measure directly the noise bandwidth of a filter, avoiding lengthy numerical integration routines. The instrument covers all the common microwave radio IF frequencies from 10 to 140 MHz. The C/I facility is discussed in Section 7.6. A software support package is available, which allows computer control via the HP-IB interface bus of BER test set, the 3708A, and associated plotters and printers. The test setup and a list of suitable equipment are shown in Fig. 7.13. This enables automatic measurement of BER versus C/N (C/N_O, E_b/N_O) plots. Accuracy enhancement includes averaging a specified number of BER readings. Figure 7.14 shows such a plot for a 16-QAM radio averaged over five measurements with a 10-s gating period on the error receiver. The CNR is defined in the symbol-rate bandwidth. Figure 7.15 shows the system in use.

7.6 COCHANNEL AND ADJACENT CHANNEL INTERFERENCE

Cochannel and adjacent channel interference effects influence the design of radio hardware and are of interest to the frequency planner and regulatory agencies. Both can have a predominating influence on the BER performance of a radio. It is common to measure either the (C/N) threshold degradation that occurs in the presence of interference or the level of interference necessary to degrade, for example, the 10^{-6} BER threshold (at a specified C/N) to 10^{-5}. These mea-

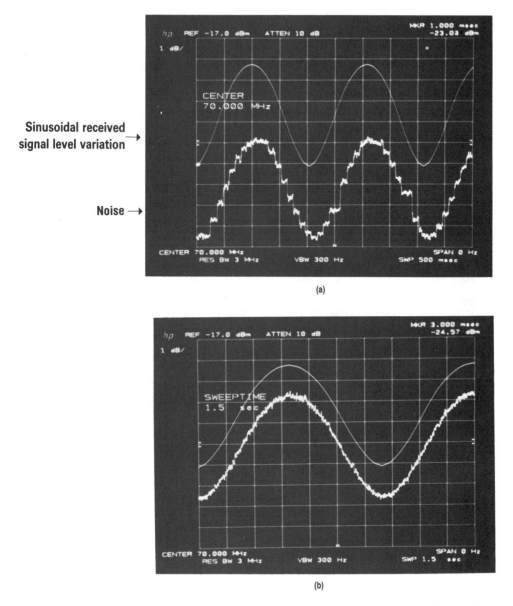

Figure 7.12 Injected noise from the HP3708 Noise and Interference Test Set tracing a sinusoidal change in carrier power to maintain a specified C/N. (a) 50 dB/s. Vertical scale: 1 dB/division; horizontal scale: 500 ms/division. (b) 5 dB/s. Vertical scale: 1 dB/division; horizontal scale: 1.5 s/division.

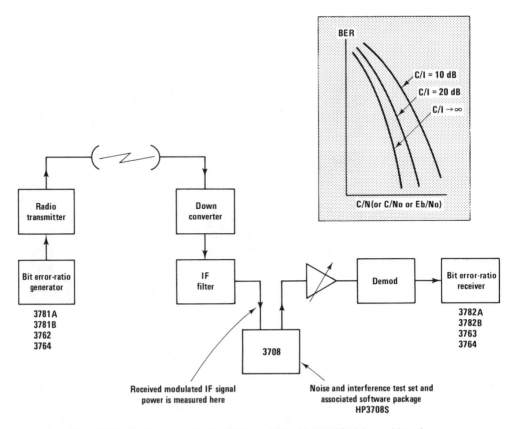

Figure 7.13   Equipment setup for BER vs. C/N with P3708S Noise and Interference Test Set and illustrative BER curve.

surements can then be plotted over cochannel and adjacent channel frequencies. Figure 7.16 shows a typical cochannel C/I plot for a 8-PSK radio. The degradation in threshold can be·clearly seen. The interference can be injected either at RF or IF frequencies.

7.6.1 Theoretical Error Rate in the Presence of Sinusoidal Cochannel Interference

Prabhu [Prabhu, 7.4, 7.5, and 7.6] has produced several in-depth papers on this subject, which unfortunately require an understanding of sophisticated mathematical techniques in order to thoroughly understand the analyses. Because of a growing interest in deterministic testing of 16- and 64-QAM radio systems with a sinusoidal interfering tone (rather than statistical testing with noise), we will examine the theoretical error rate for 16 QAM with a single sinusoidal interferer. This is used in Section 7.6.2 in the estimation of residual BER from a simple C/I measurement.

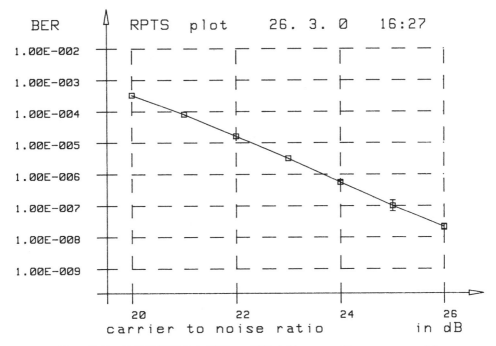

Figure 7.14   16-QAM 90 Mb/s BER vs. C/N plot (average of five measurements).

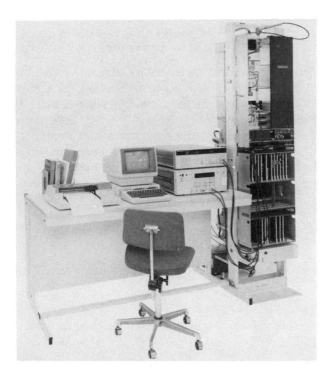

Figure 7.15   HP3708 and HP3764 under computer control, performing BER vs. C/N tests on a 16-QAM radio.

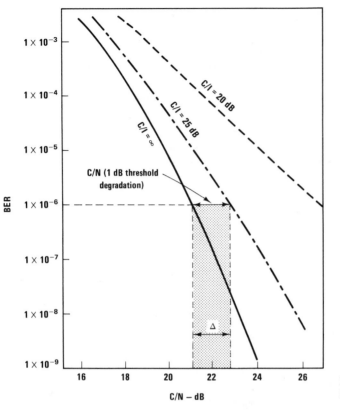

Figure 7.16 Typical 8-PSK cochannel interference plot.

Furthermore, we will illustrate how the complementary cumulative distribution function $Q(v)$ for truncated Rayleigh noise discussed in Section 7.2.6 may be used to calculate the effect of finite crest factor Gaussian noise on the BER versus C/N curves for specific values of C/I.

The derivations that follow assume that the reader does not regularly use sophisticated mathematical techniques and result in an equation that is very suitable for computation on desk-top or personal computers. Figure 7.3 shows the 16-QAM constellation. From Section 7.2.4, we remember that the average power of this system is $5d^2$. Thus, C/I is given by

$$\frac{C}{I} = 20 \log_{10} \frac{5d^2}{I_{rms^2}} \quad \text{(dB)}$$

$$= 20 \log_{10} \frac{d}{I_{pk}} + 10 \quad \text{(dB)} \tag{7.25}$$

where I_{pk} = peak interferer voltage

(Observe that $I_{pk} = d$, the distance from a nominal signal position to a decision threshold, when $C/I = 10$ dB.)

From equation (7.25)

$$I_{pk} = d.10^{[-(C/I)/20 + 0.5]}$$

From Section 7.2.4, we find that the CNR is

$$\frac{C}{N} = 10 \log_{10} \frac{5d^2}{\sigma^2} \qquad (7.26)$$

Hence

$$\sigma = \frac{d}{\sqrt{2}} \cdot 10^{[-(C/N)/20 + 0.5)}$$

(The CNR for the I and Q streams is unchanged by ideal coherent demodulation, and the noise variance is unchanged).

For either the I- or Q-stream, the symbol error rate, P_e, is given by

$$P_e = P[n(t) + i(t) > d]$$

Note. Here we use the term probability of error P_e instead of BER. The P_e term is more frequently used in theoretical references, whereas the practically equivalent BER term is used more in applied tests and reports.

The probability of the noise being greater than some value v is given by the complementary cumulative distribution function ($Q_G(v) = 1 - F(v)$, where $F(v)$ is given by equation (7.3) for Gaussian noise.

$$i(t) = I \cos \theta$$

where θ is uniformly distributed. We can, therefore, numerically integrate over $0 < \theta < \pi$. The error probability over this interval is the same as over 0 to 2π. At $\theta = 0$, there is a maximum reduction in effective noise margin to $d - I$. The contribution to error probability for an incremental angle $\delta\theta$ is given by

$$Q_q \frac{(d - I \cos \theta)}{\sigma} \frac{\delta\theta}{\pi}$$

(see Fig. 7.17). The total error probability is then

$$P = 2 \sum_{n=1}^{N} \frac{Q_q\left[d - I \cos \frac{\pi(2n - 1)}{2N}\right] \Big/ \sigma}{N}$$

where π/N is the incremental angle for the numerical integration. The multiplier 2 accounts for the second adjacent error threshold. The probability of error P applies to one of the four central states of the constellation, that is, the two middle levels in Q or I. The outer levels have only half the error probability of the inner ones. The overall error probability is $\frac{3}{4}$ that for a center state alone.

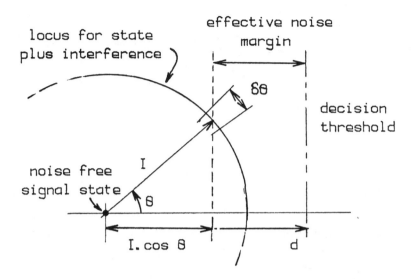

Figure 7.17   Effective noise margin for QAM with sinusoidal interference.

Hence

$$P_{eQ} = \frac{3}{2} \sum_{n=1}^{N} \frac{Q_q \left[d - I \cos \dfrac{\pi(2n - 1)}{2N} \right] \Big/ \sigma}{N} \tag{7.27}$$

P_{eQ} is the probability of I- or Q-channel symbol errors for 16 QAM with additive white Gaussian noise and one sinusoidal interferer. The derivation of a more general form of equation (7.27) applicable to other modulation schemes can be found in [Hewlett-Packard, 7.20].

It is now a simple matter to derive P_e for finite crest factor (truncated) noise and one sinusoidal interferer. We replace Q_a (v) in equation (7.27) by Q_r (v) for truncated Rayleigh noise discussed in Section 7.2.7 and given by equation (7.13).

The error probability for I- or Q-symbol errors is plotted in Fig. 7.18 for values of C/I of 13, 15 and 20 dB and with additive noise of finite crest factor equal to 10, 12 and 15 dB. Over the range plotted, the results for a 20-dB crest factor are indistinguishable from an infinite crest factor. Again, it can be seen that for measurement of threshold error rates (10^{-6} and 10^{-3}, for example) in the presence of interference, the noise injected to degrade the CNR should have a high crest factor. Similarly, if the fade is simulated by attenuating the received signal, the predetection and IF circuitry will determine the crest factor of the noise "hitting" the decision circuitry in the demodulator. In both these cases, unless the noise crest factor is at least 15 dB, the C/N plot in the presence of interference will deviate from the theoretical curve (infinite crest factor) for this reason alone.

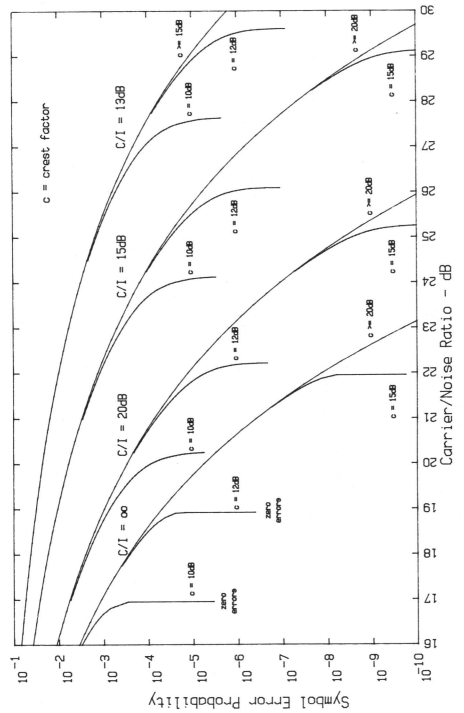

Figure 7.18 Probability of symbol error for a 16-QAM system in the presence of additive sinusoidal interference and noise of finite crest factor, plotted for $C/I = 13$, 15, 20 dB and ∞. The error probability can be seen to fall to zero at points dependent on the crest factor of the additive noise.

7.6.2 Useful Tests with a Sinusoidal Interferer

Threshold Degradation $\Delta C/N$

The measurement of threshold degradation is performed by many digital radio operating companies and post-telegraph and telephone administrations as part of regular out-of-service checks of a radio. Referring back to the cochannel interference plot for a typical 8-PSK radio in Fig. 7.16, the procedure is as follows:

1. An approximate C/N is established by using the relationship between RSL and C/N (equation 7.22) for the radio under test. In Fig. 7.16, this is a C/N value of 21 dB for a BER of 10^{-6}. (Either the transmitted power is reduced or the RSL is attenuated with an RF attenuator.)
2. The sinusoidal interferer is added—this is usually within the IF bandwidth at a frequency within a few megahertz of the carrier frequency. The level of the interferer is adjusted to establish the required C/I. In this example, $C/I = 25$ dB.
3. The CNR is readjusted until the error rate is once again 10^{-6}.
4. The change in C/N, $\Delta C/N$, is noted.
5. C/N is compared to the value that was measured during commissioning of the radio. A deviation from the original value indicates a reduction in error margin and initiates further investigation into possible causes.

The HP3708A *Noise and Interference Test Set* enables this measurement to be performed without RF attenuation and with improved accuracy by avoiding errors inherent in the RSL to C/N conversion.

Estimation of Residual BER from a Simple C/I Measurement

Channel imperfections are often evaluated by means of the well-known eye diagram. The percentage of eye opening may be defined on the conceptual phase plane, or constellation diagram, in Fig. 7.19. Measured constellation diagrams are presented in Section 7.7 (for instance, Fig. 7.22(a)). With a noise-free sinusoidal interferer, the eye opening is related to C/I in a 16-QAM system as follows. We rearrange equation (7.25) to give

$$\frac{d}{I_{pk}} = \text{antilog}\left[\frac{(C/I)_{dB} - 10}{20}\right] \tag{7.28}$$

The percentage eye opening (E.O.) is

$$\text{E.O.} = \left(1 - \frac{I_{pk}}{d}\right) \times 100\%$$

Example 7.5

If C/I is 30 dB, then from equation (7.28),

$$\frac{d}{I_{pk}} = 10$$

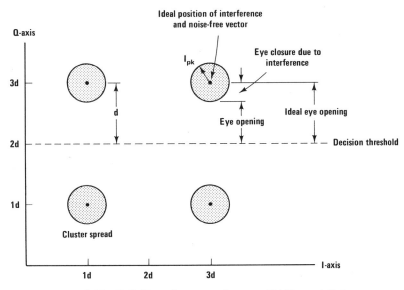

Figure 7.19 Definition of eye opening on a QAM constellation.

and
$$\text{E.O.} = (1 - 0.1) \times 100\% = 90\%$$

Therefore, the eye is 10% closed for this level of sinusoidal interference in an ideal 16-QAM system. The **error margin,** or the distance to the error threshold, of the radio under test can be inferred by a measurement of the level of noise-free sinusoidal interferer required to close the eye. ■

The error margin determines the background error rate, and a radio system must not only operate satisfactorily at threshold signal levels and in the presence of selective fading but also with a low residual error rate at normal signal level. This residual error rate ($<<10^{-10}$) is effectively impossible to measure because the mean time between errors can be from tens to hundreds of hours.

However, the value of C/I required to establish a specific error rate is dependent upon the amount of ISI and residual noise present around the phase states in the constellation. The measurement of this value of C/I can therefore be used to predict the **background,** or **dribble, error rate** for a modulated radio system. Figure 7.20(a) shows the symbol error rate derived from equation (7.27) for 16 QAM with additive noise of infinite crest factor and sinusoidal interference. We can use Fig. 7.20(a) to predict the dribble error rate of a 16-QAM system by making the assumption that residual noise plus ISI is like Gaussian noise. The procedure is as follows:

1. Establish a value of C/I corresponding to one of the theoretical curves of Fig. 7.20(a) that results in a BER measurable in minutes or seconds.
2. Convert the measured BER to a symbol error rate (taking into account any additional errors produced by the descrambler in the receiver).

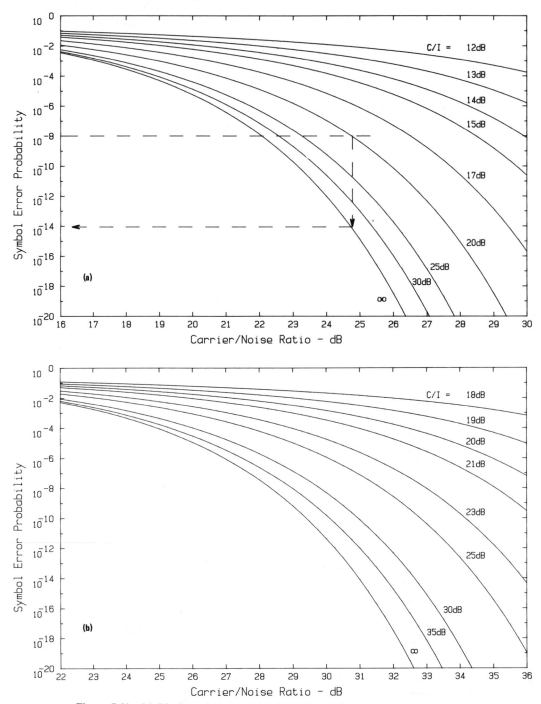

Figure 7.20 (a) Ideal symbol error rate for 16 QAM in the presence of Gaussian noise and sinusoidal interference. In this figure noise is specified in the double-sideband Nyquist bandwidth. For example, for a 90 Mb/s 16-QAM radio, that bandwidth would be 22.5 MHz. (b) Ideal symbol error rate for 64 QAM in the presence of Gaussian noise and sinusoidal interference. In this figure noise is specified in the double-sideband Nyquist bandwidth. For example, for a 135-Mb/s 64-QAM radio, that bandwidth would be 22.5 MHz.

3. Draw a horizontal line from this symbol rate to the chosen C/I curve.

4. Drop a vertical line from this point to the $C/I = \infty$ curve.

5. Draw a horizontal line from the point on the $C/I = \infty$ curve to intercept the symbol error-rate axes and read the residual symbol error rate of the radio system under test.

6. Convert this symbol error rate back to the residual BER.

As an example, Fig. 7.20(a) shows how a radio system that produces a symbol error rate of 10^{-8} at a C/I of 20 dB has a residual symbol error rate of 10^{-14} (i.e., in the absence of any additive noise and interference and at normal RSL). This graphical method is equally suitable for other modulation schemes, and suitable theoretical curves of symbol error rate versus C/N and C/I for M-ary PSK can be found in a Hewlett-Packard product note [Hewlett-Packard, 7.20]. Figure 7.20(b) is taken from this product note and allows the residual error rate of 64-QAM systems to be estimated graphically. The dribble error rate can be derived while computing the probability of error from equation (7.27), and this is plotted in Fig. 7.21 for an error rate of 10^{-8} and the equivalent residual

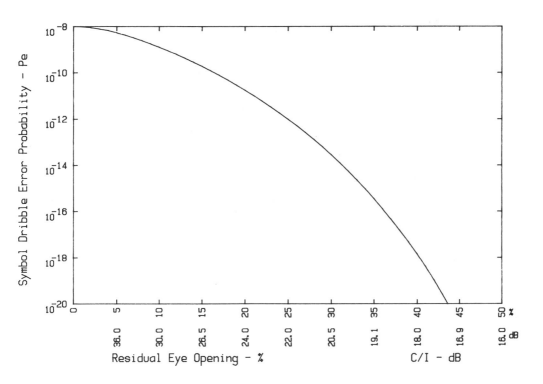

Figure 7.21 Dribble I or Q symbol error probability for Gaussian ISI/noise vs. C/I needed to raise error probability to 10^{-8} and the equivalent residual eye opening in the absence of interference.

eye opening or C/I ratio. One manufacturer of digital radios [Bates, 7.7] has stated that if the eye opening at a BER of 10^{-8} is greater than 20%, then one can guarantee a long-term background error rate of less than 10^{-10}. This would seem to agree with Fig. 7.21 and may mean that the assumptions made in deriving this figure can be ignored and the curve used with confidence to predict residual error rates. Furthermore, laboratory experiments at Hewlett-Packard with a 16-QAM radio system show that there is close correlation between this C/I technique and the traditional method of residual BER measurement [Hewlett-Packard, 7.20].

7.6.3 The HP3708A and C/I Testing

The noise and interference test set will inject an interfering signal into the receiver IF to establish a demanded C/I ratio. The carrier power is measured at the point of injection, and *the C/I is therefore maintained in the presence of received signal variations* and at any point through the IF stages of the receiver. (The interfering signal can either be an internal sinusoidal source of fixed frequency or an external sinusoidal or modulated signal.) In this mode of operation, the level of intrinsic noise injected along with the interferer is very low, and the C/I ratio established can be considered to be noise-free.

7.7 PHASE PLANE ANALYSIS AND THE CONSTELLATION DISPLAY

The last section showed that the eye opening as inferred from a C/I test is a direct measurement of error margin. Acceptable results from this test guarantee that many circuits are operating satisfactorily. However, the bane of the field maintenance engineer is the radio with dribbling bit errors from a nonspecific cause. If the dribble error rate, as predicted from the C/I test, is outside the specification limits, a lengthy process of readjustment and module replacement starts and continues until the fault is cleared. Phase plane analysis of a radio system with reduced error margin can guide the engineer to a specific cause and prevent replacement of good modules in a trial-and-error substitution exercise.

7.7.1 The Principle of the Constellation Display (or Cluster Scope)

The method of displaying eye diagrams on an oscilloscope is well known. A sampling oscilloscope can be used to display the eye opening at the sampling instant of the demodulator. If both I and Q decision circuit inputs are sampled at the same time and displayed in the X- and Y-directions on the sampling oscilloscope, the resulting display corresponds to the constellation of possible signal states (for example, 16 in the case of 16 QAM, as displayed in Fig. 7.22(a)). The tube face of the sampling oscilloscope is equivalent to the phase plane with

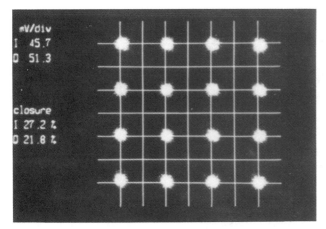

(a)

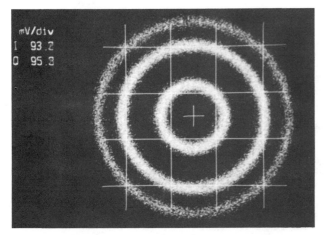

(b)

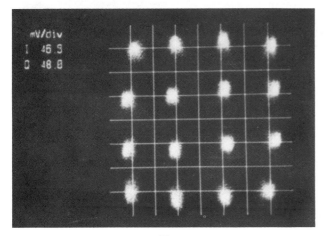

(c)

Figure 7.22 16-QAM constellation with geometric defects due to specific faults. (a) 16-QAM radio: Normal constellation plus eye-closure data. (b) Receiver out of lock. (c) 16-QAM radio: 3-dB transmitter power overdrive (AM-AM and AM-PM).

(continued)

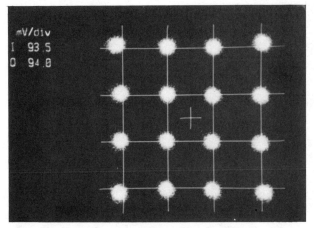

(d)

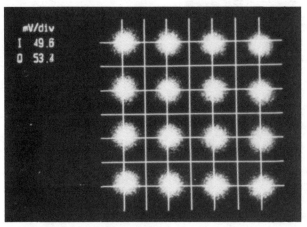

(e)

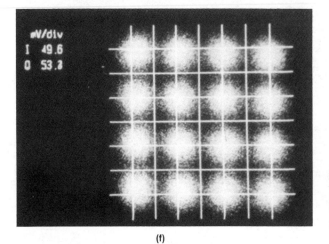

(f)

Figure 7.22 (continued) (d) Expansion due to underdrive of TWT. (e) 16-QAM radio: C/N ratio 20 dB (setup using 3708A). (f) 16-QAM radio: C/N ratio 15 dB (setup using 3708A).

(continued)

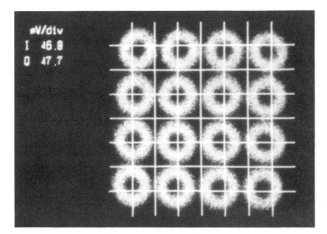

(g)

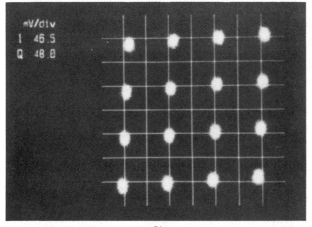

(h)

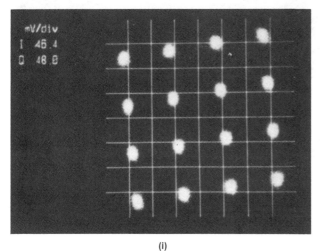

(i)

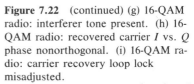

Figure 7.22 (continued) (g) 16-QAM radio: interferer tone present. (h) 16-QAM radio: recovered carrier I vs. Q phase nonorthogonal. (i) 16-QAM radio: carrier recovery loop lock misadjusted.

(continued)

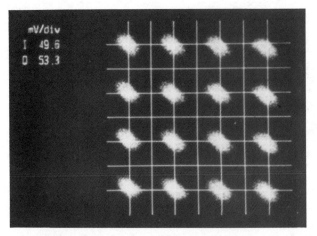

(j)

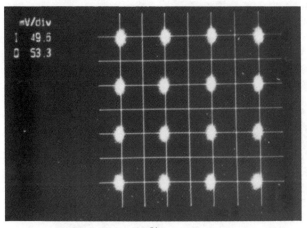

(k)

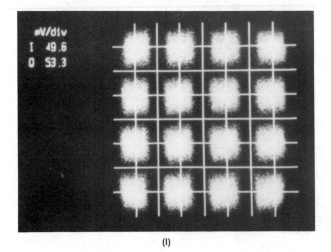

(l)

Figure 7.22 (continued) (j) 16-QAM radio: multipath fade (+6 dB slope, 50 MHz to 90 MHz). (k) 16-QAM radio: multipath fade (6 dB symmetrical notch, 70 MHz). (l) 16-QAM radio: multipath fade (10 dB symmetrical notch, 70 MHz).

226

amplitude increasing radially from the center and the phase separating the angle between these radials. The time axis comes perpendicularly out of the tube face.

Ideally, the 16-QAM constellation would consist of 16 small bright dots, but because of degradations in the radio system that produce ISI and thermal noise, the display actually consists of 16 small clusters. Table 7.7 [Bates, 7.7] breaks down the eye closure into constituent degradations.

TABLE 7.7   16-QAM Degradation Allocation at BER $= 10^{-8}$ (from [Bates, 7.7] with permission of the IEEE)

	16 QAM
Quadrative carrier offset amplitude imbalance	9%
Nyquist filters	15%
Timing recovery	6%
Carrier recovery	21%
Power amplifier	15%
Radio	2%
Total eye closure	68%
Resulting eye opening	32%

For QPSK, 16 QAM, and 64 QAM, the constellation should be square with equal spacings between phase positions. Deviations from the normal square constellation of small clusters form the basis for fault diagnosis, which can be carried out with the radio carrying traffic in service.

7.7.2 In-Service Fault Diagnosis with the Constellation Display

The engineers responsible for the development of a digital radio can easily produce a catalog of geometric defects of the ideal pattern (Fig. 7.22(a)) that results from faults identified during prototype testing and development. This may be done with either the HP3709A Constellation Display or, with less convenience, a sampling oscilloscope. The catalog can be comprehensive, and some defects are applicable only to certain types of radio; however, some common defects in the ideal 16-QAM pattern are shown in Fig. 7.22. Figure 7.22(b) shows the pattern that results when the receiver loses carrier lock. There are many causes of this, including excessive thermal noise or ISI, low received-signal level, inband amplitude slope, or lock-up problems in phase-locked microwave oscillators.

Figure 7.22(c) shows the effect of compression from amplifier overdrive. This compression may be in the receiver and could indicate a high RSL, or it could be in the transmitter, where the automatic level control and/or the predistorter may be incorrectly adjusted. The effect of AM-PM conversion is similar, except the innermost four states rotate in the opposite direction to the outer ones with respect to the ideal position. Figure 7.22(d) shows expansion (underdrive) due to a transmitter automatic level control (ALC) or predistorter fault. Figure 7.22(e) and (f) shows excessive thermal noise during a simulated flat fade at C/N ratios of 20 dB and 15 dB, respectively. Figure 7.22(g) shows the effect of sinusoidal interference at 70 MHz and $C/I = 14$ dB. Interference from an FM/FDM link into a digital link produces a similar effect, but if the interference is amplitude modulated, the center of the annulus fills in like the excess thermal noise case.

Figure 7.22(h) shows the effect of the recovered carrier I- versus Q-phase being nonorthogonal. Figure 7.22(i) shows the effect of misadjustment of the carrier recovery loop. Finally, Fig. 7.22(j), (k), and (l) shows the effect of simulated multipath propagation on the constellation.

7.7.3 The Hewlett-Packard 3709A Constellation Display

Although the phase plane can be displayed on a sampling oscilloscope, many controls have to be used and optimized simultaneously, and the result is not always satisfactory. Field maintenance engineers must accustom themselves with this procedure.

The HP3709A is, in effect, a sampling oscilloscope, simplified and optimized, to display either the eye diagram or the phase states of a digital radio with the minimum of adjustments. It is connected to I and Q monitor points ahead of the decision circuitry in the receiver demodulator. In addition, the calculated values for the mean separation between constellation states plus the average variance of the states are calculated, and the percentage of eye opening is displayed. As an aid to the identification of specific transmission impairments there is also available a range of internally generated constellation graticules for common modulation schemes.

7.8 DIGITAL RADIO AND MULTIPATH FADING

The susceptibility of digitally modulated signals to selective fading, which seemed to be a fundamental limitation to increased bit rates in the microwave common carrier bands, resulted in a great deal of experimental and theoretical investigation into microwave multipath propagation. Investigations on the multipath properties of line of sight links have been carried out since the 1950s, particularly at Bell Laboratories, in the course of designing reliable FM-FDM links. However, it was the investigation into high-capacity digital radio in the 1970s that saw a growth of interest in selective fading.

Atmospheric conditions often allow microwave propagation over two or more separate paths between two antennas in a line of sight link. This results in **multipath,** or selective, fading when the signals arrive out of phase. This fading was tolerated on microwave paths for many years because of the resistance of FM systems to the linear amplitude-distortion component that predominates in multipath-induced distortion [Hartmann, 7.8].

Following experiments during the summer of 1977 with a 78-Mb/s, 6-GHz 8-PSK radio system over a 26-mi link, Barnett [Barnett, 7.9] made the following conclusions.

1. The experimental results demonstrated that the hop missed short-term outage objectives by an order of magnitude without space diversity and was close to acceptable when diversity was activated.

2. The traditional average power fade depth was a poor indicator of BER performance. Average power fade depths corresponding to BER $> 1.2 \times 10^{-3}$ had a median of 28 dB with a 10 to 90% range of 14 dB.

3. Modest in-band amplitude dispersion, a linear 0.2 dB/MHz for example, caused the BER to exceed 1×10^{-3}. Amplitude dispersion was found to be a good indicator of BER performance.

The conventional method of calculating outages is based on the concept of flat fades and cannot be used for high-capacity digital radio systems. An increase in the flat-fade margin will reduce the thermal noise in an FM system; however, it cannot improve the performance of a digital radio system if the eye-diagram amplitude has already closed because of the effects of multipath propagation. An increase of transmitter power cannot be used to make the systems meet their outage requirements.

The use of complex modulation and higher data rates to achieve 4- to 5-bits/s/Hz signaling efficiency on digital transmission systems has resulted in systems that are 10 to 12 times more sensitive to multipath distortion than installed systems [Allen, 7.10]. These advances in transmission capacity have resulted in systems requiring advanced multipath countermeasures. For example, one administration uses correction techniques in space, frequency, and time domains (i.e., minimum dispersion combiner for space diversity, IF adaptive equalization (both under microprocessor control), and baseband transversal equalizers). In addition, the protection channel can be used in frequency diversity with hitless switchover equipment. Another administration has successfully developed an experimental cross-polarization interference canceller. The use of cochannel dual-polarization transmission can double the capacity of a digital radio route at little extra cost; however, during multipath fading, the cross-polarization discrimination may reduce to 0 dB, resulting in severe cross-polarization interference. The sophistication of these countermeasures illustrates the impact of multipath propagation on digital transmission. A significant proportion of the prolific output of papers on high-capacity radio concentrate on the measurement and characterization of the multipath fading channel and countermeasures to combat the effects of it.

7.8.1 Multipath Fading Models

The need to estimate the performance of digital radio during periods of selective fading has resulted in models to characterize fading channels statistically. The most widely accepted and used of these is the three-path model proposed by Rummler [Rummler, 7.11 and 7.12] in 1978. However, the simple two-path model will be examined first. Both of these models lead to "equivalent" frequency-dependent (nonflat) transmission medium characteristics. This equivalent model may be expressed in terms of amplitude and group delay characteristics similar to conventional filter characterization.

Two-Path Model

The simplest two-ray model of multipath propagation corresponds to the delayed ray being typically refracted by a layer of air above the direct path at night or in the early morning. Usually the direct ray is larger than the delayed ray (minimum-phase state), but under certain conditions the delayed ray can be larger. The channel is then said to be in a **nonminimum** phase state and the sign of the group delay variation reverses. Ramadan [Ramadan, 7.13] has shown that a two-ray model places a limit on the maximum amplitude slope that can exist across a 30-MHz channel. Amplitude slope can occur inside the channel bandwidth when the frequency of the notch is out of band. In addition, Rummler [Rummler, 7.12) found that the two-ray model cannot match an in-band maximum at an arbitrary fade level. Also, for 50% of the time during which the radio equipment was indicating errors, the channel could not be satisfactorily modeled with a two-path model. Thus the limit placed on amplitude slope by the two-path model is valid for only part of the time.

Simple Three-Path Model

Although the fading channel has been frequently modeled as a two-path medium, it was reported that a three-path model corresponded closer to the physical situation that occurred during selective fading [Lin, 7.14; Ruthroff, 7.15]. A three-path fade typically would reveal two paths of similar amplitude, transmitted over paths differing in length by a few half-wavelengths. These two components exhibit little relative delay. The resultant signal may still fade due to destructive interference between these two paths, which attenuates the signal over a wide bandwidth (with respect to the channel width) and produces little in-band amplitude dispersion. The weak signal provided via the third path of (relatively) long delay causes the selective fading in the channel.

Rummler considered a channel characterized by three paths of amplitude 1, a_1, and a_2 at the receiver. The second and third paths are delayed with respect to the first by τ_1 and τ_2 seconds, respectively, where $\tau_2 > \tau_1$. The simple three-path model is defined by requiring the delay between the first two paths to be small such that

$$(\omega_2 - \omega_1) \tau_1 \ll 1$$

where ω_2 and ω_1 correspond to the highest and lowest in-band frequencies. Rummler further simplified the three-path phasor diagram by designating the

amplitude of the vector sum of the first two paths by a; the angle of the sum
by $\phi = W_0\tau - \pi$, where τ is equal to τ_2, the delay difference in the channel;
and the amplitude of third ray by ab.

The complex voltage transfer function of the channel is given by

$$H(jw) = \mathrm{a}\,[1 - be \pm j\,(\omega - \omega_0)\,\tau_1] \tag{7.29}$$

If the third amplitude is greater than the sum of the first two, the assignments
of amplitudes a and ab are interchanged and the fade is nonminimum phase.
Figure 7.23(a) and (b) shows a minimum phase fade and a nonminimum phase
fade, respectively. Note that the amplitude characteristics are very similar, but
the sign of the group delay reverses.

The simple three-path fade cannot be used for a channel model because it
is possible for a family of simulated fades of widely varying parameters to be
almost indistinguishable from each other. Rummler was able to show that the
simple three-path fade overspecifies the channel transfer function if the delay is
less than $\frac{1}{6}B$ where B is the observation bandwidth. For a 30-MHz channel, this
critical delay is about 5.5 ns, and a unique set of parameters a, b, τ, and f_o
cannot be found for more than half the faded-channel conditions encountered.
In order to avoid this situation, one of the model parameters must be fixed.
Rummler was able to show that delay was the only parameter that could be
fixed to produce a reasonable model. Figure 7.24 shows the amplitude of the
channel transfer function of equation (7.29) for $\tau = 6.3$ ns. This value is frequently
used in the digital radio literature. When $\omega_0 = 0$, the fade minimum is at band
center and at frequencies spaced by $1/\tau$ from the center frequency. The selectivity
of the fade is determined by b, and changes in a correspond to flat fades across
the channel.

The validity of the model was confirmed by fitting measured propagation
data to the model derived from 25,000 swept-frequency scans representing 8400
s of fading activity. By analyzing the errors in fitting the observed characteristics
to the model, Rummler was able to show that the simple three-path fade model
is indistinguishable from a perfect model of a line-of-sight microwave radio
channel. The fixed-delay model is preferable to the three-path model, because
its three parameters can be uniquely determined from a channel scan. The delay
$\tau = 6.3$ ns was chosen to ensure that the spacing between notches of the model
function was a mathematically convenient multiple of the 1.1-MHz frequency
spacing at which the channel attenuation was measured during the 25,000 frequency
scans. The fixed-delay model transfer function can be thought of as the response
of a channel with a direct path of amplitude a and a second path of relative
amplitude b at a delay of 6.3 ns and with a phase of $\omega_0\tau + \pi$, independently
controllable at the center frequency of the channel. Furthermore, it is a mathematical convenience; no physical significance should be placed on the delay.
The model is based on fitting observed data to a mathematical function, and to
realize the independent control of phase at the center frequency requires synthesis
with three rays.

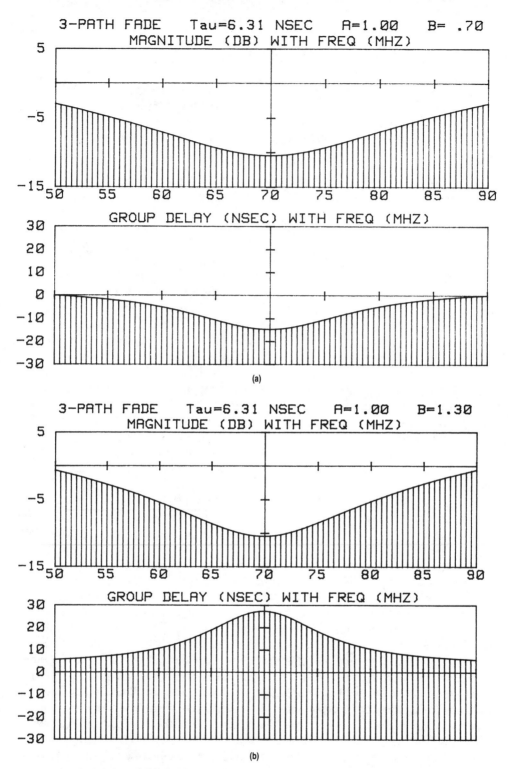

Figure 7.23 Minimum phase and nonminimum phase fade. (a) Minimum phase amplitude and group delay. (b) Nonminimum phase amplitude and group delay.

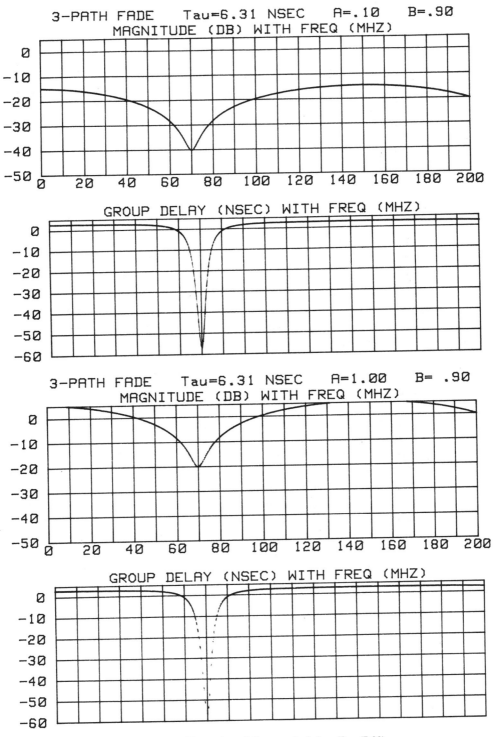

Figure 7.24 Examples of three-path fades: Eq. (7.29).

7.9 MODEL STATISTICS

The curve-fitting procedure resulted in 25,000 sets of values of a, b, and f_O. This enabled the joint frequency of occurrence of these parameters to be determined. These statistics allow the statistical generation of all the channel conditions that could occur on the 6-GHz hop. The important result of this work is that the determination in the laboratory of the parameter values that will cause unacceptable performance in a digital radio system allows the calculation of the time during fading activity that the error rate will equal or exceed this critical value.

The distribution of a, b, and f_O can be found in [Lundgren, 7.16].

7.10 DIGITAL RADIO SIGNATURE AND LINK OUTAGE CALCULATION

In 1978, Emshwiller [Emshwiller, 7.17] presented the system signature as a new method for predicting radio system outage. Analysis based on a fading model allowed the development of a system signature that provided a meaningful comparison between systems and a calculation of the contribution of selective fading to system outage. The *signature is a locus of fade-notch depths and frequencies that cause unacceptable performance.* Emshwiller showed that the fraction of time that the performance was unacceptable was related to the area under the signature. Figure 7.25 shows some typical signatures [CCIR, 7.1].

Lundgren and Rummler [Lundgren, 7.16] developed this methodology further using Rummler's simple three-path model. Although a detailed description of the method is outside the scope of this chapter, the procedure is as follows. The radio system is stressed in the laboratory using an IF multipath fade simulator, which provides a circuit realization of the simplified three-path model of equation (7.29). The IF simulator is set to specific shapes to determine the system response to multipath fading. Critical values of fade level A ($-20 \log a$) and notch depth B ($-20 \log (1 - b)$) are determined for a specified BER for each fade-notch frequency. This is done by plotting BER versus C/N (proportional to A) for each notch depth and frequency. The resulting curves allow critical contours of A and B for each notch frequency and BER chosen. The probability that A and B fall on the high BER side of a given critical contour is then calculated, leading to an estimate of the probability of all selective fades that produce unacceptable performance.

The procedure involved the measurement of 50 to 100 BER versus C/N curves. This can be speeded up by using the HP3708A *Noise and Interference Test Set* and chosen BER test set under computer control as already described. In addition, the insertion loss and selectivity loss of the IF multipath simulator will change for difference notch frequencies and depths. This alters the IF carrier power and requires the calculation of a compensation factor for each setting of notch depth and frequency and frequent readjustment of the noise source reference power.

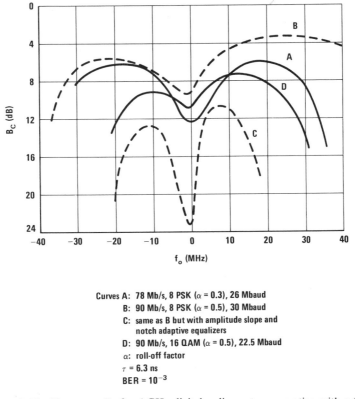

Curves A: 78 Mb/s, 8 PSK (α = 0.3), 26 Mbaud
     B: 90 Mb/s, 8 PSK (α = 0.5), 30 Mbaud
     C: same as B but with amplitude slope and
        notch adaptive equalizers
     D: 90 Mb/s, 16 QAM (α = 0.5), 22.5 Mbaud
     α: roll-off factor
     τ = 6.3 ns
     BER = 10^{-3}

Figure 7.25 Signatures B_C for 6-GHz digital radio system operating with a 30-MHz channel spacing. (From Report 784-1, CCIR XV Plenary Assembly, Geneva, 1982, with permission of the ITU.)

The noise that defines the CNR is injected after the fade simulator. The HP3708A can again simplify this procedure because the CNR is defined by measuring the IF carrier power at the point of noise injection; changes in the IF power are automatically compensated for.

Critical curves of A versus B (Fig. 7.26) are drawn for each notch frequency and for several values of BER, if required. In Fig. 7.26, the intercept with the A-axis is the flat-fade margin for a given BER. This is independent of the notch frequency. The B-axis intercept is the relative fade depth for a given notch frequency. For values of B to the right of this intercept, B_c, the critical value BER will not be obtained at any CNR for the given notch frequency.

The direct method of outage calculation calculates the probability of finding A and B outside all critical contours. Lundgren and Rummler [Lundgren, 7.16] note that outage due to the occurrence of A and B in region 1 of Fig. 7.26 is due to shape or selectivity, whereas outage in region 2 is due to the combined effects of low CNRs and selectivity. They found that for the radio system under study at a critical BER of 10^{-3}, most of the outage was caused by selectivity

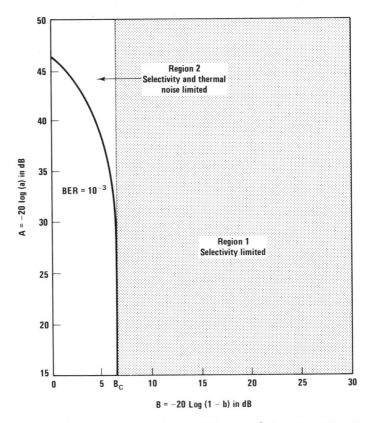

Figure 7.26 Critical curve of A vs. B for BER $= 10^{-3}$, $f_0 = 19.8$ MHz. (From [Lundgren, 7.16] with permission from AT&T Bell Laboratories Technical Journal.)

and that this depends on the relationship between B_c and the notch frequency. This is shown in Fig. 7.27 for a BER of 10^{-3}. The outage probability is the probability of finding B and f_O values above this curve. This curve corresponds to Emshwiller's signature.

From the work of Lundgren and Rummler, the relative outage due to frequency selection fading is given by

$$P_O \atop \text{rel} = \int e^{-B_c(f_O)/3 \cdot 8} \, df_O$$

and the signature is frequently used to estimate relative outages of radio systems. Generally a signature that is contained below a first signature will result in less outage than obtained for the first system.

Lundgren and Rummler describe the procedure to calculate actual rather than relative outages, and stress that outage calculated from sensitivity to in-band selectivity alone provides a quick estimate but is less accurate. For example,

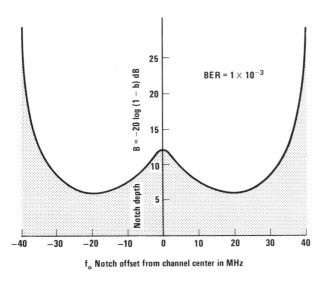

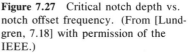

Figure 7.27 Critical notch depth vs. notch offset frequency. (From [Lundgren, 7.18] with permission of the IEEE.)

for the system studied at 10^{-3}, the full calculation of outage and the selectivity calculation agreed closely, but at 10^{-6} the results were 1860 and 2730 s, respectively.

The subjects of signature and outage calculation have not been discussed in depth, and to take advantage of the methodology, the quoted references must be studied in detail. They are included in this chapter as a further application of C/N testing.

7.11 ACKNOWLEDGMENTS

I would like to thank Ivan Young and Dr. Tom Crawford for their contribution to the sections on noise and Ian Matthews, Tony Lymer, and Daya Rasaratnam for stimulating discussions.

REFERENCES

[7.1] CCIR Rep. 784-1. "Effects of Propagation on the Design and Operation of Line-of-Sight Radio Relay Systems," CCIR Lth Plenary Assembly, Geneva, 1982.

[7.2] Feher, K. *Digital Communications: Microwave Applications*, Prentice-Hall, Englewood Cliffs, N.J., 1981.

[7.3] Abramowitz, M. and I. A. Stegun. *Handwork of Mathematical Functions*, Dover Publications, Inc., New York, 1965.

[7.4] Prabhu, V. K. "Error Rate Considerations for Coherent Phase-Shift Keyed Systems with Co-Channel Interference," *Bell System Technical Journal,* Vol. 48, No. 3, March, 1969, pp. 743–765.

[7.5] Prabhu, V. K. "The Detection Efficiency of 16-ary QAM," *Bell System Technical Journal*, Vol. 59, No. 4, April, 1980, pp. 639–656.

[7.6] Prabhu, V. K. "Co-Channel Interference Immunity of High Capacity QAM," *IEEE International Conference on Communications*, Denver, Colo., June 14–18, 1981, p. 68.8.

[7.7] Bates, C. P. and M. A. Skinner. "Impact of Technology on High-Capacity Digital Radio System," *IEEE International Conference on Communications*, Boston, Mass., June 19–22, 1983.

[7.8] Hartmann, P. and B. Bynum. "Adaptive Equalization for Digital Microwave Radio Systems," *IEEE International Conference on Communications*, Seattle, Wash., June 8–12, 1980, pp. 8.5.1–6.

[7.9] Barnett, W. T. "Multipath Fading Effects on Digital Radio," *IEEE Transactions Communications*, Vol. COM-27, No. 12, December, 1979.

[7.10] Allen, E. W. and J. A. Crossett. "6 GHz Digital Radio Propagation Experiments 1982/83," *IEEE International Conference on Communications*, Boston, Mass., June 19–22, 1983.

[7.11] Rummler, W. D. "A Multipath Channel Model for Line-of-Sight Digital Radio Systems," *IEEE International Conference on Communications*, June, 1978, pp. 47.5.1–4.

[7.12] Rummler, W. D. "A New Selective Fading Model: Application to Propagation Data," *Bell System Technical Journal*, Vol. 58, No. 5, May–June 1979, pp. 1037–1071.

[7.13] Rumadan, M. "Availability Prediction of 8 PSK Digital Microwave Systems During Multipath Propagation," *IEEE Transactions Communications*, Vol. COM-27, No. 12, December, 1979, pp. 1862–1869.

[7.14] Lin, S. H. "Statistical Behavior of a Fading Signal," *Bell System Technical Journal*, Vol. 50, No. 10, December, 1971, pp. 3211–3270.

[7.15] Ruthroff, C. L. "Multipath Fading on Line-of-Sight Microwave Radio Systems as a Function of Path Length and Frequency," *Bell System Technical Journal*, Vol. 50, No. 7, September, 1971, pp. 2375–2398.

[7.16] Lundgren, C. W. and W. D. Rummler, "Digital Radio Outage Due to Selective Fading—Observation vs Prediction from Laboratory Simulation," *Bell System Technical Journal*, Vol. 58, No. 5, May–June 1979, pp. 1073–1100.

[7.17] Emshwiller, M. "Characterization of the Performance of PSK Digital Radio Transmission in the Presence of Multipath Fading," *IEEE International Conference on Communications*, June, 1978, pp. 47.3.1–6.

[7.18] Lundgren, C. W. "A Methodology for Predicting Nondiversity Outage of High Capacity Digital Radio Systems," *IEEE International Conference on Communications*, Boston, Mass., June 10–14, 1979, pp. 32.3.1–32.3.6.

[7.19] Proakis, J. G. *Digital Communications*, McGraw-Hill, New York, 1983.

[7.20] Hewlett-Packard, 3708A Product Note 5953–5490.

8

FDM SYSTEM TESTING
AND NETWORK SURVEILLANCE

BOYD WILLIAMSON and ROBIN MYLES

Hewlett-Packard Ltd.

8.1 INTRODUCTION TO FDM SYSTEMS

Frequency division multiplexing (FDM) originated in the United States during World War I, with three voice circuits being carried on a pair of wires. The object of the exercise was, of course, to increase the capacity of a given amount of copper. From these early experiments, FDM has grown to today's huge networks, which carry thousands of circuits on a single coaxial tube or on a single radio carrier wave. The operating principle has remained unchanged.

8.1.1 Frequency Division Multiplexing

The FDM process begins with a number of voice channels. Each of these is translated to a different frequency so that the channels are stacked side by side in the frequency domain. With only a few channels to be multiplexed, it is practicable to shift each individual channel directly to its new frequency slot. As the number of channels increases, this direct modulation method becomes more difficult to implement.

Each channel is translated by mixing it with a carrier from a stable local oscillator and filtering the resultant output signal to extract the desired sideband. The design of the channel filter determines how closely the channels can be stacked, so a compromise is made between filter complexity and line capacity.

The cost of terminal equipment is balanced against that of line equipment. A high-capacity FDM trunk cable carries channels in a band from tens of kilohertz to tens of megahertz. Designing individual filters for each channel is now impossible, so to economize on filter costs, several stages of modulation are used, with intermediate frequencies chosen to suit available filter technologies.

With a few exceptions (e.g., some submarine cable systems), all current FDM systems translate 12 voice channels to form a **group.** Five of these groups are then stacked to form a **supergroup** of 60 channels. Supergroups are then further multiplexed until the desired complexity is achieved (see Fig. 8.1). This technique produces a pyramid structure of hardware, with thousands of channel modulators at the base, hundreds of group modulators at the next level, and so on. In order to minimize the overall cost, the system designer must minimize the cost of the most common element in the system. Reducing its cost tends to limit the performance of this element, and care must be taken to maintain a level of performance that will guarantee satisfactory overall system operation.

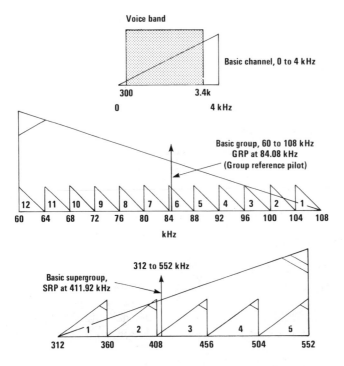

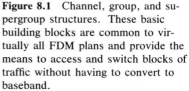

Figure 8.1 Channel, group, and supergroup structures. These basic building blocks are common to virtually all FDM plans and provide the means to access and switch blocks of traffic without having to convert to baseband.

Notes: 1. The triangle direction denotes ◿ erect or � inverted channels.

2. The corner bars denote number of stages of modulation, one for group, two for supergroup, etc.

8.1.2 Estimating System Performance

Rather than trying to estimate overall performance in individual cases, **hypothetical reference circuits** (HRC) can be used. The HRC represents a worst-case circuit over which some minimum standard of transmission must be maintained. In Europe, both the HRC and system performance are defined by the CCITT. The Bell system has its own definitions, which cater to the circuits of greater length that are found in the United States (see Fig. 8.2, the CCITT HRC).

HRCs assume that the link will split at various points to couple to other trunk circuits and link to local circuits. Links to the local networks must obviously

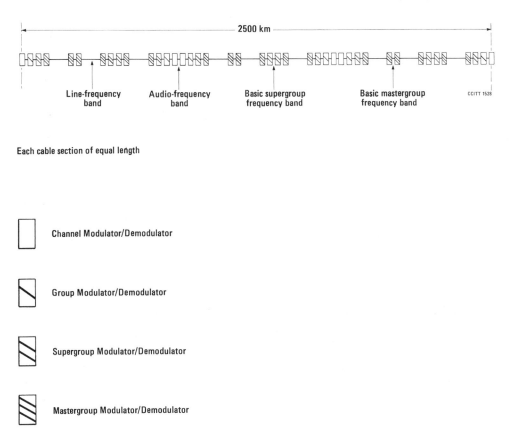

Noise over this link to be less than 10,000 $pwop$ (-50 dBm0p) during the busy hour.

Split as 2500 $pwop$ (-56 dBm0p) for terminal equipment and 7500 $pwop$ (-51 dBm0p) for the line.

Figure 8.2   CCITT Hypothetical Reference Circuit. The HRC represents a "typical" circuit in an FDM system. The performance of the various circuit elements is arranged to meet the desired link performance objectives with minimum overall cost.

be made at voiceband level, but it would be very costly if branching to other trunks could be done only at voiceband. Not only would this involve the cost of extra demodulators, but the performance of this equipment would degrade the overall system performance. By allowing access to the system at a level higher than voiceband, a poorer specification for the most common system element can be tolerated. This requirement for higher-level access means that systems from different manufacturers and installed in different countries must be common at levels higher than voiceband. This common-interface requirement is the reason for all systems having a common structure for groups and supergroups. At higher levels in the hierarchy, a variety of different standards have arisen for combinations of supergroups—for example, **hypergroups** of 15 supergroups (that is, 900 channels) or **mastergroups** and **supermastergroups** of 5 and 15 supergroups, respectively. These various structures are formed with spaces between each supergroup, between each hypergroup, and so on. These spaces make it possible to extract sections of a plan for branching purposes. The different structures simply reflect the needs of different administrations (see Fig. 8.3, the CCITT and Bell FDM plans).

8.1.3 Submarine Cables

These same arguments explain why submarine cables use both *nonstandard* FDM plans and also 3-kHz channel spacings. Access to the cable is possible only at either end, so branching considerations can be ignored. Instead, the cable cost dominates the system design. The cable is difficult to recover for repair, so the repeaters and the cable itself must be very reliable. This means that to minimize costs, the best possible use must be made of the available bandwidth. The extra cost of tighter channel filters and the better carrier balance required in the channel modulators is more than justified by the extra capacity achieved.

8.2 FACTORS THAT GOVERN SYSTEM PERFORMANCE

So far we have stated that individual elements will affect the overall system performance without defining either the element parameters or desired performance. It should be possible to communicate intelligibly over an FDM link. This requires some minimum bandwidth determined by the channel filter—usually 300 to 3400 Hz (submarine cable systems may use 200 to 3050 Hz) and also minimal impairment from noise, distortion or interference.

Interference may be caused by crosstalk from adjacent channels or signaling tones in the system. Distortion will result from overloading of amplifiers, mixers, and so on. Noise can be either thermal or intermodulation noise. With a large number of channels in a system, all behaving as independent signal sources, we can invoke the central limit theorem and treat the composite signal as Gaussian noise. Any nonlinear component in the system will produce intermodulation distortion and spread this noise to all channels.

8.2.1 *Loading Levels*

For a given FDM system, there will be an optimum loading level where the effects of intermodulation and thermal noise are balanced. In the interests of economy, this will be a fine balance, so it is important that the correct loading level is attained to meet the system performance objectives.

An individual voiceband will be shifted up and down in frequency and pass through many repeater amplifiers as it travels through the network. The gain of each of these stages must be accurately determined so that correct loading is achieved at all points in the system. However, even if the initial settings are perfect, the effects of time and temperature will soon cause changes in gain. Small gain errors will accumulate through the network, and the system will no longer be correctly loaded.

8.2.2 *Gain-Regulating Pilots*

To counter the effects of these gain errors, some kind of automatic gain control is needed. This control is exercised over groups rather than individual channels, and since the power in a group will change with its occupancy, a reference tone accompanies each group. The frequency and level of this group reference pilot is accurately defined at source, so it can be used to adjust the gain of modulators and amplifiers in the multiplex equipment. Similar pilots are added at each of the higher-order levels in the hierarchy.

The group and supergroup pilots are placed in the narrow guard bands between the channels. At higher levels, there are slots between each supergroup where the other pilots can be positioned without interfering with the traffic. The gain and frequency response of the line repeater amplifiers are similarly controlled by line-regulation pilots, which are positioned just above the highest traffic frequency. The need for accurate control of each repeater can be judged from the case of a trans-Atlantic submarine cable, which has an intrinsic loss of 20,000 dB at its highest operating frequency.

8.2.3 *Signaling Tones*

A communications system of any kind not only has to carry traffic but also signaling information, both in the *go* direction, so that the traffic reaches its correct destination, and in the *return* direction, to indicate fault conditions or number-engaged information, for instance. In the local network, a dc loop carries these signals. When a voice circuit passes over a frequency division multiplexed (FDM) link, the signaling information must travel with it. This signaling path can be either *in-band* or *out-of-band*. In both cases, the information is carried as a signaling tone. Some out-of-band systems use the channel *virtual carrier* as a signaling tone, the channel modulator is deliberately unbalanced, and the local oscillator signal passes through as a signaling tone. Other systems use a

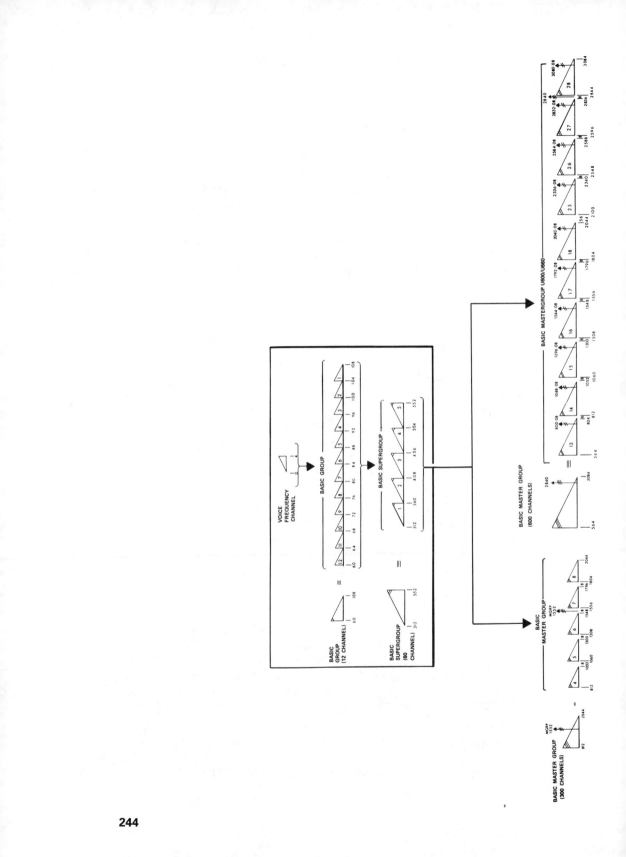

244

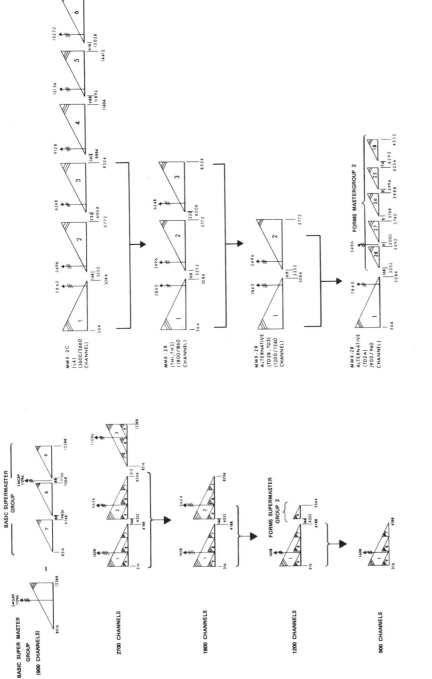

Figure 8.3 Typical CCITT and Bell FDM plans. A variety of other similar plans exist using many elements of these plans. They differ only in detail, not in overall architecture.

tone in the channel guard band of 3825 Hz. When in-band signaling is used, a tone near the top frequency of the voice band is used, chosen to minimize interference with the traffic.

8.2.4 Virtual Carriers

The traffic, reference pilots, and signaling tone make up the total wanted FDM signal. There is one more important signal that we have mentioned briefly, the virtual carrier.

When a channel is translated to its position in the basic group, the output of the channel modulator includes some local oscillator breakthrough. The local oscillator (LO) drive level is usually fairly high to achieve linear operating conditions in the modulator, so the resultant LO breakthrough is at a high level. At each subsequent stage of modulation, this tone will be retranslated along with the voiceband traffic, and new LO signals will also be added.

These tones are referred to as **virtual carriers,** as they are the carrier frequencies that would have translated, for example, a channel to its final place in the hierarchy if direct modulation had been employed. These signals serve no useful purpose (unless they are used as signaling tones), but they contribute to the system loading. They should, therefore, be suppressed as completely as possible.

A further problem is associated with the channel virtual carriers. The Group Reference Pilot (GRP) is only 60 Hz away, so it must be separated by a pilot filter before acting as an Automatic Gain Control (AGC) control signal. If the virtual carrier is not maintained at a sufficiently low level, it will interfere with the AGC action of the pilot and upset the system-loading level.

8.2.5 Signal Levels in a Correctly Loaded System

So that different systems can be linked together, the levels of the system pilots, carriers, signaling tones, and traffic-loading levels have all been set by international agreement. Countries that follow the CCITT recommendations use those in Table 8.1. How are these signals maintained in practice? The frequency of each carrier and pilot is carefully derived from a stable master oscillator in each station. (The master oscillator may be locked to a national standard.) The pilot levels are used as AGC signals at various points in the system, and if the level at any point violates preset thresholds, alarms are raised.

The virtual carriers are set up when the equipment is installed. Any further checks must be made with external measuring equipment—a **selective level measuring set** (SLMS). If the system is to give its best performance, then these various levels must be checked and adjusted periodically. The level of each signal is measured with an SLMS. With a 12- or 18-MHz system to check, this becomes a virtually impossible task with a conventional, manual SLMS, so level checking is done only to find faults rather than as part of a preventive maintenance policy.

TABLE 8.1   CCITT Loading Levels. With the Components of the Composite FDM Signals All at Their Correct Levels, the Best Overall System Performance Is Obtained

Signal	Level (dBmO)	CCITT recommendation
1. Mean power level per channel during the busy period (one direction, mean activity = 0.25)	−15 (32 μW)	G232
Consisting of (a) speech and echo, carrier leaks, and telegraphy	−16.6 (22 μW)	G223
plus (b) signaling tones	−20 (10 μW)	G223
2. Channel carrier leak on any one channel	< −26	G232
Sum of channel carrier leaks in one 12-channel group	< −20	G232
3. Group reference pilots		G232
84.080 kHz	−20	
84.140 kHz	−25	
104.080 kHz	−20	
4. In band-signaling times: short duration		G224
800 Hz	−1	
1200 Hz	−3	
1600 Hz	−4	
2000 Hz	−5	
2400 Hz	−6	
2800 Hz	−8	
3200 Hz	−8	
5. Band-signaling at 3825 Hz		Q.21
(a) Short duration	−5	
(b) Semicontinuous	−20	

8.3 NETWORK SURVEILLANCE

With the growth of telecommunications networks, increasing pressure is being placed on traditional maintenance methods. Trunk circuit testing on an individual basis is widely used; however, this approach requires that circuits are taken out of service with an ensuing loss of revenue and considerable inconvenience to the customer. It is, therefore, usually restricted to periods of low traffic activity and hence does not assess network performance under its most stringent operating conditions or on a continuous basis.

With FDM carrier systems still carrying the majority of long-haul traffic, ranging in capacity from a few voice channels to over 100,000 channels on a signal composite cable, this heavy concentration of traffic provides convenient monitoring points for network surveillance. Automatic monitoring can make a major contribution to the performance of a network by providing direct assessment of traffic and transmission-system performance on such long-haul routes. Operationally, automatic measuring systems can provide the dual benefit of continuous performance monitoring and fast, accurate fault location.

By making measurements of pilots, channels, carrier leaks, and so on, using an SLMS at various locations in a network, it is possible to monitor performance on a per-channel basis while simultaneously checking the operation of the associated multiplex and transmission equipment. System monitoring of this kind puts some special constraints on the design of an SLMS, as is described in some detail later on in this chapter.

With an SLMS and suitable equipment for test-point access and system control, reliable network monitoring can be provided to cover a wide range of functions, including

- System-fault control
- Level control
- Noise-performance monitoring
- Carrier-leak checks
- Detection of high-level traffic
- Location of spurious interfering tones
- Long-term assessment of system reliability

The monitoring of a number of points in a network independently (even on a continuous basis) does not, however, provide the capability to locate faults directly. It is, in effect, no more than a sophisticated alarm system that requires human communication between measurement positions to identify the location of a fault. What is needed is the collection of information from the monitoring sites at a central facility, allowing results to be correlated and minimizing loss of service by ensuring that repair action is focused at the location of a fault. Furthermore, it is usually possible to determine the extent of an out-of-service condition and take appropriate action to restore service by alternative routing. Continuous monitoring can also have a considerable impact on the overall network transmission performance. In general, nonvoice traffic (such as data and telegraphy) is less tolerant of poor circuit performance than conventional telephony, and the growth of this class of traffic, coupled with the increasing distance for the average connection, is putting greater emphasis on the performance of individual carrier systems. Whereas networks could tolerate individual systems that were not conforming to recommended noise-performance figures in the past, it is now increasingly important that every system perform to design specifications throughout its life. Continuous surveillance of the network at key locations can identify poor performance of an individual system before it steps far out of line from the rest of the network and affects service quality.

The most economical place to monitor a network is at the highest multiplex level—that is, the *line output or radio baseband*—as this maximizes the total amount of traffic monitored per test point. This permits direct measurement of the transmission-system performance for each link, as well as the performance of the complete transmit multiplex. To assess the performance of the receive (de-) multiplex, it is necessary to monitor at lower levels of the hierarchy. Although this is more costly, requiring many times more test points, monitoring at these levels can be particularly beneficial where circuits (e.g., groups or supergroups) are *through-routed,* as incorrect levels can have a major effect on the performance of the following multiplex. The parameters that have to be monitored at all levels of the multiplex are virtually identical.

8.3.1 Pilot-Level Monitoring

Level control is a complex problem, particularly when traffic is transmitted over a number of carrier systems and the transfer between systems takes place at different levels of the multiplex hierarchy. Inadequate level control is an increasing cause of poor noise performance as network routings become longer and traffic capacities increase. By monitoring pilot levels at several points in the network and correlating results at a central location, it is possible to determine the source of any significant level variations very quickly.

8.3.2 Noise-Performance Monitoring

In-service measurement of carrier-system noise performance is restricted to measurements in **intersupergroup slots** or on *quiet* channels allocated for such measurements. The first approach, however, has the disadvantage that the slot is "lost" whenever the multiplexed signal is demodulated to supergroup level. The second approach can permanently occupy many channels with the consequent loss of system capacity.

A surveillance system can be programmed to check the background noise level of each circuit while the system is in-service. This approach is still applicable in busy traffic periods, despite high channel-traffic occupancy, as each channel in the system will be unused for some time during the period; by looking for stable measurement results, it is possible to ensure that the measurement is indeed background noise.

8.3.3 Carrier-Leak Checks

Channel carrier leakage on newly installed equipment is usually less than -35 dBmO, a figure significantly better than the typical recommended figure of -26 dBmO. Unfortunately, carrier leaks on a network are rarely checked after installation, and therefore deterioration in performance is not detected. The reason for this delinquency in measurement of carrier leakage is very simple: It requires a large amount of effort. Carrier leakage can be introduced at any level of the multiplex hierarchy, from channel through to mastergroup, so on an 1800-channel system, 1983 virtual carrier measurements must be made, assuming that the carrier transmission is built up from channel level. Looking across a complete network and taking into account that at many stations where monitoring is performed, only supergroup and mastergroup carrier leaks need be measured, it is easy to understand why this parameter is so infrequently checked and why, when checked, leakage levels are often found to exceed the recommended limit. As a result, the carrier leakage makes a substantial contribution to system loading and, therefore, to intermodulation noise. It is a fact that stations with as many as 30% channel carriers in excess of -26 dBmO have been found using surveillance techniques.

8.3.4 High-Level Traffic Detection

The detection of high-level traffic is becoming increasingly important with the advent of data transmission and its continuous signal-level characteristics. For example, if a data signal specified to be at -10 dBmO is transmitted 6 dB high, it will have the equivalent system loading of 12 voice channels. This could result in a severe noise penalty at the channel demodulator for a fully loaded group. In practice, it is not uncommon for data-transmission levels to be far outside specified limits, up to 10 dB high.

Detecting such high-level data signals on carrier systems can be impossible using manual techniques, as the signals are often present only at certain times during the day. Monitoring routines used to detect high-level traffic can also detect *singing* circuits and high-level in-channel spurious signals.

8.3.5 Impairment Measurements

While all the previously mentioned parameters can be measured using an SLMS, it may be desirable—or even necessary—to measure voice-channel impairments such as phase jitter, gain hits, phase hits, dropouts (interruptions), and single- or three-level impulse noise. While some SLMSs may have a built-in limited-impairment measurement capability, a comprehensive capability is usually provided only by a dedicated data line analyzer. As it is necessary to measure impairments directly on the channel concerned, this can be achieved only by feeding the demodulated output from the SLMS to an ancillary data line analyzer. Figure 8.4 shows the measurement configuration and outlines the procedures required.

A useful addition to the measurement capability of an SLMS is the frequency measurement of a tone within the bandwidth of the tuned filter. This feature is particularly beneficial in the identification of interference sources. A comprehensive list of measurements required in an FDM surveillance system is shown in Table 8.2. Having identified the reasons for network surveillance and the measurements required, what can be achieved from analyzing the measurement results?

Specific Fault Conditions. Many faults, such as missing pilot tone and high-level carrier leak, can be *detected* with minimum analysis of the raw measurement data. *Locating* faults, however, requires comparison between the results of similar measurements taken at a number of locations. This, in turn, requires access to a data base containing circuit routing information so that a fault condition can be checked at all relevant locations. Isolation of a fault is then a logical

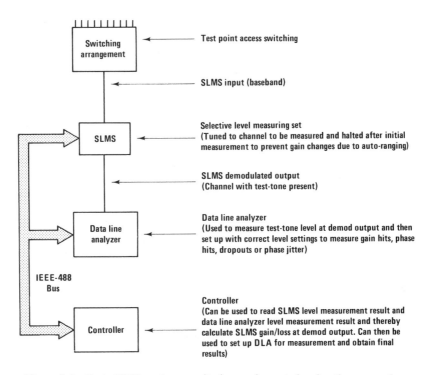

Figure 8.4 Basic FDM system monitoring equipment showing the connections between the controller, SLMS, data line analyzer and access switching arrangement.

TABLE 8.2 List of FDM Monitoring System Measurements Including Channel Impairment Measurements

Pilot level
Carrier leak (virtual carrier) level
Channel power level
Channel noise level
Channel-signaling tone level
Channel test-tone level
Group power level
Supergroup power level
Broadband power level
Intersupergroup noise level
Phase jitter
Phase hits
Gain hits
Dropouts (interruptions)
Impulse noise
Tone frequency

process, which may be handled from the central network-monitoring control station.

If the purpose of the network-monitoring system is to provide "immediate" indication of a fault condition, the central controller can provide audible or visible alarm indications. Note, however, that surveillance systems of this type are not ideally suited to the rapid detection of network faults, as is required for automatic changeover to standby facilities. This is because such factors as individual pilots are checked only on a sequential basis, and many minutes may elapse between these checks. It is normal, therefore, to provide a separate continual monitoring facility where auto-changeover is required.

Substandard Transmission Performance. Although the detection and location of major fault conditions is a capability provided by network-surveillance systems, it is in two other areas that the benefits of such systems are most significant. The first of these is the detection and location of transmission problems that are not "faults" in the sense of equipment failures, but are instead conditions that give rise to substandard transmission performance. Detection and location of such problems is frequently a time-consuming and often nearly impossible task on a manual basis; however, the facilities offered by a monitoring system make it considerably more straightforward.

There is, however, a fundamental difference in the approach to the problem using an automatic system. With the manual approach, it is normal to begin with the result of the problem—usually a report of a noisy channel, an interfering signal, low-level transmission, or similar problem. By tracing the channel routing and making measurements at various points in the network (both on the faulty channel and on the channels in the same group), it is usually possible for a skilled maintenance team to deduce the likely cause of a problem—for example, amplifier gain variation, unbalanced modulator, high-level user, or similar problems. A network-monitoring system can be used simply to facilitate this manual approach. Centralization of results, automatic searches for out-of-limits conditions, and the overall speed of a monitoring system make an enormous contribution here and can reduce the time taken to find the cause of this type of problem from days to less than an hour.

The second, and main, contribution of a monitoring system is when it is used to tackle the problem from the cause rather than the result. The capacity of a surveillance system to make a large number of measurements of many parameters on a continuous basis provides the capability of identifying the potential causes of reported faults immediately as they become apparent. Eliminating these problems at their sources has a major impact on the overall quality of transmission and the incidence of reported substandard transmission performance.

Trends in System Performance. The continuous measurement capability of a monitoring system can also be used to detect trends in system performance. These trends may be over short or long time scales and may be used to detect, in advance, the point at which a transmission system will become unserviceable.

For example, by monitoring pilot levels, it may be possible to detect a trend in amplifier gain variation before the system goes outside operating limits.

Other trends can identify below-standard performance in particular types or models of equipment across a network, with the objective of isolating and correcting potential fault conditions before the same problems occur on a wide scale.

8.4 INSTRUMENTATION HARDWARE

The previous sections of this chapter have introduced a variety of system measurements and some of the necessary hardware. In this section, we cover the instruments in a surveillance system in more detail, leaving the requirement of the control hardware and software for later discussion.

The instruments to be used are

1. SLMS
2. Access switches
3. Level generators
4. Telephone/data line analyzers

We shall discuss these in turn, beginning with the most important unit, the SLMS.

8.4.1 The Selective Level Measuring Set

The traditional manual SLMS was simply a wave analyzer, with a set of measurement filters chosen to suit the FDM signals. An instrument that can also be used by a computer has to be easier to use than a traditional set, which pleases the manual user. Let us examine why this is so.

The SLMS can be considered as a tunable bandpass filter, which selects the desired signal, and a level detector to measure the signal power. In practice, a frequency changer tunes to the desired signal, and the measurement is made at a fixed frequency. The instrument is, in fact, a *super-heterodyne radio receiver*.

The traditional SLMS bore a closer resemblance to a domestic radio than a complex measuring instrument. It was tuned by turning a knob and reading out the approximate frequency on a dial. The measured result was obtained from a combination of attenuator knob settings and meter readings. The meter usually had to be peaked to obtain an accurate reading, and there was always some doubt as to what was going on. Furthermore, tuning to a particular signal in the composite FDM spectrum is a bit like finding the proverbial needle in a haystack. The charts that map from FDM description to frequency are complicated and can be another source of error.

Although all of these functions can be carried out by a conscientious human operator, they are impossible for a computer, which can neither turn dials nor peak and read meters. A computer needs to have digital input and output, which implies a keyboard and digital displays on the front panel. Internally, the knob-tuned local oscillator is replaced by an accurate frequency synthesizer, and all gain and attenuation is controlled automatically. To dispense with meter peaking, accurate tuning from the synthesizer is combined with steep-sided flat-topped filters. As a further step, the SLMS's inevitable microprocessor holds a set of FDM plans to cover virtually every eventuality. Now the instrument is much easier to use, can be controlled by a computer, and includes some features that will improve its measurement accuracy.

An SLMS can be considered as having two roles in FDM system measurements. First, there is the manual SLMS used for troubleshooting and pinpointing a specific fault. Then there is the surveillance mode, where—under computer control—the complete network performance is monitored, looking for small degradations. In the former case, the operator will spend most of his or her time thinking about what to check, entering this information, and assessing the measured level. If the instrument takes a few seconds to autorange and has an automatic frequency correction (AFC) loop that must settle before delivering a result, no harm is done, and this time is insignificant in the total operation time.

In complete contrast, we have the monitoring operation. Here the computer does no thinking; it has been programmed with a set of measurements and knows what levels to expect at each point. It has a decision tree to follow at each stage and follows its program blindly. Having obtained a set of measurements, the computer can then analyze these at its (or rather our) leisure and compare present performance with past history, and so on. In this circumstance, the SLMS must measure as quickly as possible, because it may have to make thousands of measurements to extract a small amount of information.

Let us now examine the SLMS to see how accurate, fast measurements can be combined with ease of use, and how the instrument design is affected by the FDM system and the nature of the signals to be measured.

The SLMS block diagram has three main parts (see Fig. 8.5):

1. Receiver
2. Frequency synthesizer
3. Controller (microprocessor)

We begin by discussing the receiver.

Input Connections

In FDM systems, a number of impedances exist at various points in the system. At line, 75-Ω unbalanced cable is used, while channel connections are usually 600-Ω balanced, and intermediate points use 150-Ω (CCITT), 75-Ω (Japan), and (in North America) 124-Ω and 135-Ω cable, all balanced. Figure 8.6 shows the resultant tapping loss and return loss when 4.2 m of 75-Ω cable was used to connect a 75-Ω system to the SLMS high impedance input.

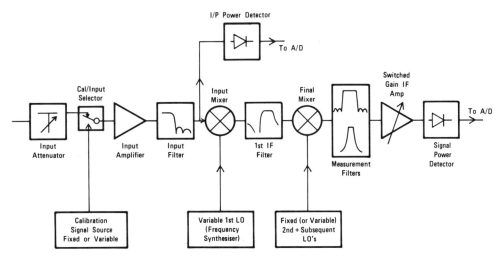

Figure 8.5 SLMS block diagram. All these elements are needed in any SLMS. Different technologies will reflect in different positions of some elements.

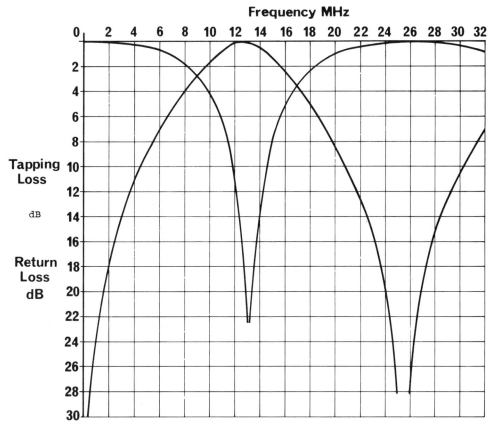

Figure 8.6 Tapping loss and return loss when bridging a 75-Ω system with 4.2 m of 75-Ω cable to the SLMS high impedance input.

Quite apart from the potentially disastrous effect on the system, the measured level results will also be affected by the mismatch, as shown in Fig. 8.7 for the same case as above. A diode ring mixer with series resistors and a square-wave local oscillator drive (Fig. 8.8) gives minimal distortion but has intrinsic conversion loss. This loss and filter insertion losses degrade the receiver noise floor, so an input amplifier is used to overcome these losses and hence restore the receiver noise figure.

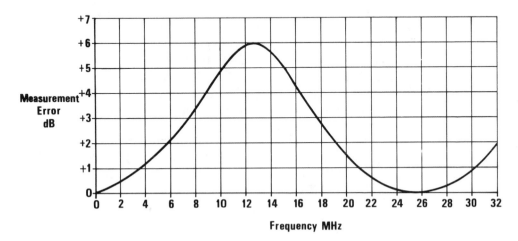

Figure 8.7   Measurement error caused by bridging a 75-Ω system with 4.2 m of 75-Ω cable to the SLMS high-impedance input.

The IF Strip

The IF strip is a series of mixers, filters, and amplifiers, which convert signals to frequencies where measurement filters can be realized. There are a number of technologies available for both the IF filters and the measurement filters—for example, LC, crystal, active, mechanical resonator, helical resonator—but, in general, narrower filters are best realized at lower frequencies.

Measurement Filters

An SLMS needs two standard measurement filters and optional special-purpose filters. The two standard filters are the *channel filter* and the *pilot filter*.

The Channel Filter

The **channel filter** is used to extract the complete voice band from any channel in the FDM system. The channel signal power, when carrying voice or data traffic can then be measured. Further measurements may be made through the channel filter either internally or, after demodulation to voice band, externally with other instruments. The channel filter must introduce minimum amplitude distortion in-band and have adequate rejection of interfering adjacent signals.

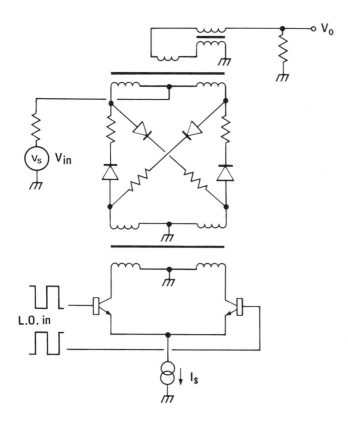

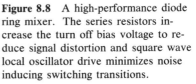

Figure 8.8 A high-performance diode ring mixer. The series resistors increase the turn off bias voltage to reduce signal distortion and square wave local oscillator drive minimizes noise inducing switching transitions.

The out-of-band signals are a mixture of pilots, signaling tones, virtual carriers, and adjacent channels, which may be carrying voice, data, or test signals. These signals are summarized in Fig. 8.9. Fig. 8.10 illustrates one realization of a filter to meet these rejection requirements. It is flat-topped to give an accurate measurement of traffic or white idle channel noise. If the noise is colored or contains tones, then a better estimation of the noise is obtained from a weighted filter, either C-message or psophometric. The stopband rejection of the pilot filter is determined by how closely tones are spaced in the FDM system. From Fig. 8.9, we find that the closest tones are the 84.080-kHz and 84.140-kHz group reference pilots.

The necessary passband width is determined by the frequency stability of the tones and of the receiver. Pilot frequencies are maintained to better than ± 5 Hz over a CCITT HRC, and receiver drift can account for a further ± 6 Hz. A passband that is flat across ± 11 Hz will, therefore, obviate the need for an automatic frequency-control scheme. Choosing about 40 dB of rejection of the adjacent pilot leads to the need for a fourth-order filter. Figure 8.11 shows the realized frequency response.

Textbook presentations of filter settling times—Fig. 8.12—indicate that higher-order filters have longer settling times, so it might be imagined that a

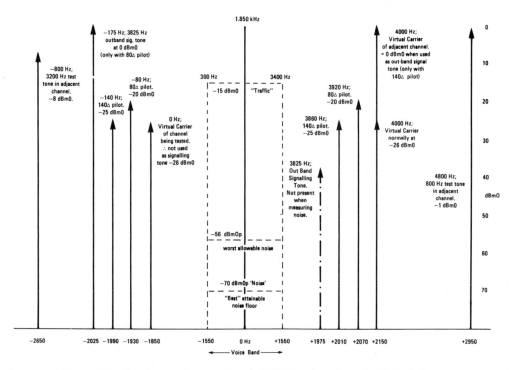

Figure 8.9   Signals round a worst-case CCITT voice channel. Not all these signals can be present simultaneously!

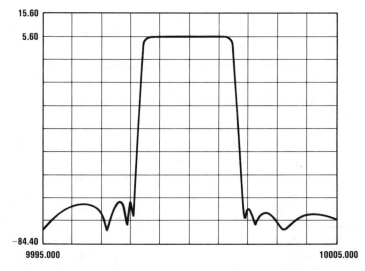

Figure 8.10   Measured SLMS channel filter response. The filter is "flat-topped" to measure all in band traffic signals but has very steep sides to reject even the virtual carrier.

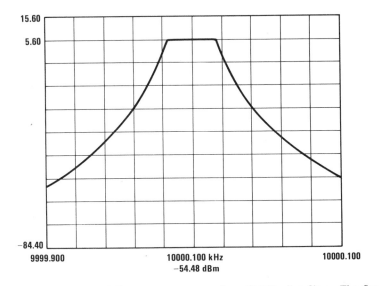

Figure 8.11 Measured frequency response of an SLMS pilot filter. The flat-topped response allows accurate level measurements even if the SLMS or pilot frequencies have drifted. About 40 dB of rejection is obtained at the adjacent interfering pilot frequency.

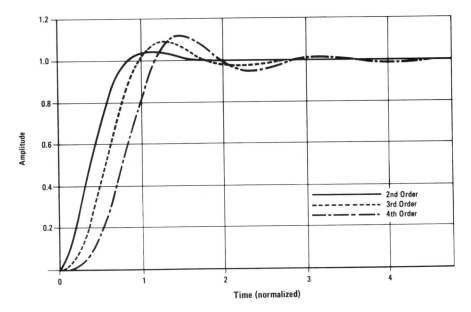

Figure 8.12 Filter settling times. These are all for 0.01-dB ripple Tchebychev filters, which are normalized to have the same 3-dB bandwidths. This is the normal filter handbook presentation.

lower-order, narrower filter, combined with an afc loop, would give faster measurements with the same accuracy.

However, the textbook curves assume that all filters have the same 3-dB bandwidths. If the curves are redrawn to show the case for equal 40-dB bandwidths, the situation changes as in Fig. 8.13 to show that the fourth-order filter now settles most quickly. So choosing a wider bandwidth, higher-order filter wins in measurement speed because the intrinsic settling time is faster, and no AFC loop is required.

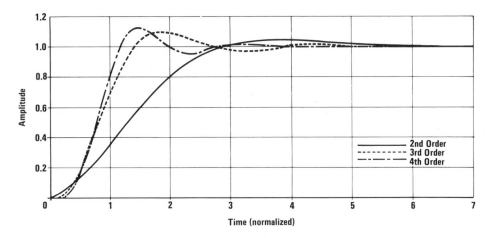

Figure 8.13 Filter settling times. These curves are also for 0.01-dB ripple Tchebychev filters, but now they have been normalized to have equal 40-dB bandwidths.

Figure 8.14 shows some measured time responses of our pilot filters. In the first case, a signal tuned to the center of the passband is switched on and off. The shape of the envelope agrees well with the theoretical response and has settled in about 200 ms. The second trace shows the effect of switching a signal on and off at the 40-dB point on the filter skirt. The settling time is still of the same order, but now there are massive overshoots on both transitions of the signal. These may cause saturation of subsequent stages of the receiver and can cause problems with autoranging if care is not taken in the detector design to ensure rapid recovery from transient overloads.

Optional Measurement Filters

As mentioned previously, the accurate measurement of idle channel noise requires an appropriately weighted filter. *Weighted filter shapes* are specified by CCITT and Bell (see Fig. 8.15). When compandors are used on the system to improve signal-to-noise ratio, it is necessary to apply a stimulating tone to the channel to activate the compandor and obtain the true noise floor. This tone must, of course, be removed before the noise measurement is made, so a **notch filter** is required (see Fig. 8.16).

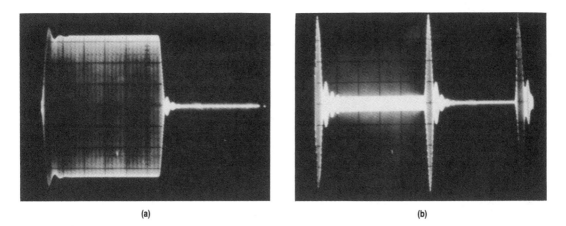

(a)   (b)

Figure 8.14 Measured SLMS pilot filter transient response. (a) The envelope in the first trace shows the same characteristic as the "textbook" case in Fig. 8.12. Time scale: 100 ms/division. (b) The second trace shows the response when an out-of-band signal is switched in level, massive overshoots occur on both the rising and falling edges.

Another useful filter is a **group filter.** The power in an individual group is a useful measurement per se, but the real advantage of a group filter is in speeding up the search for "hot tones" and "high talkers." A true high-order group filter that would accurately separate adjacent groups is not needed for this purpose; a simple design is adequate. An individual high channel will still sensibly affect the power in a complete group and so can be detected easily. A wider filter—for example, a supergroup filter—would not detect individual high channels. As mentioned previously, supergroups are the highest common denominator in all FDM plans. As such they are often the basic unit transferred across administrative boundaries, so a measure of the power in a complete supergroup is desirable. We can satisfy this requirement by using the group filter to measure the power in each group and summing the results. This method is sufficiently accurate to apportion responsibility for faults! Figure 8.17 shows the frequency response of a group filter realized as a crystal filter at about 50 MHz.

The synthesizer ensures accurate tuning of the SLMS and makes digital control of the LO possible from a microprocessor. The parameters of interest are frequency range, step size, phase noise, and spurious output signals. The frequency range of the synthesizer is determined by the desired tuning range of the SLMS and the first IF. The step size or resolution of the synthesizer can be determined by examining the FDM signals. They are all multiples of 10 Hz (with the exception of the 3825-Hz outband signaling tone), so a step size of 10 Hz would just be adequate, unless AFC was needed to cope with a narrow pilot filter.

Phase-noise sidebands on the local oscillator signal can impair the receiver's NPR specification or reduce the receiver selectivity, as can spurious sideband

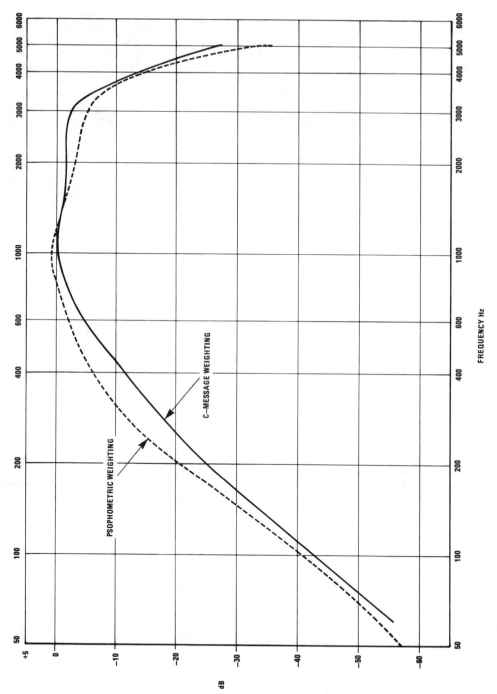

Figure 8.15 Psophometric (CCITT) and C-message (Bell) weighted filter frequency responses. These curves are designed to weight noise and interference as they annoy a human listener. This gives a better indication of the perceived noise on a telephone channel.

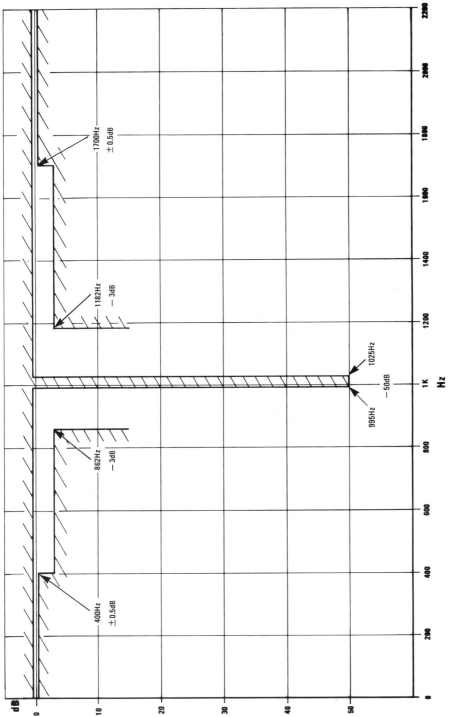

Figure 8.16 Notch filter frequency response. The notch filter is used to remove the inband tone on some idle channels—required by PCM companders, for example—to allow a noise measurement to be made.

263

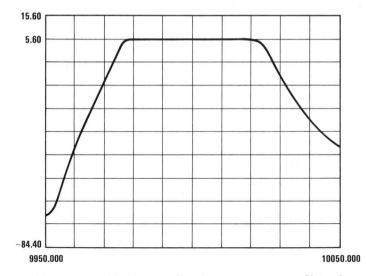

Figure 8.17 Measured SLMS group filter frequency response. Since channels are packed side by side through a *supergroup*, this filter cannot hope to extract a single group accurately. It does, however, give a quick way of measuring overall system loading and can rapidly find overloaded sections of a system.

signals. A conventional indirect synthesizer (Fig. 8.18) will need a number of *divide-by-N* and *summing* phase-locked loops (PLLs) to realize the desired step size with acceptable phase noise. Alternative techniques, such as *fractional-N* synthesis, can give equivalent or better performance with fewer loops, which

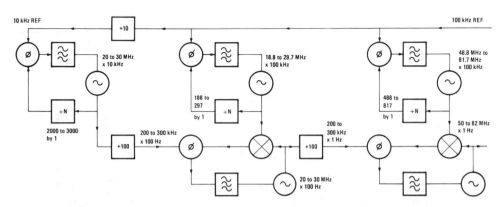

Figure 8.18 A "conventional" 50- to 80-MHz frequency synthesizer. This scheme requires three divide-by-*N* loops and two summing loops to obtain 1-Hz resolution. Each divide-by-*N* loop outputs only integer multiples of its input reference frequency. Noise problems preclude making the reference frequency and hence frequency resolution arbitrarily small.

not only reduces the overall cost and complexity of the synthesizer but also reduces the number of potential spurious sideband signals (Fig. 8.19).

Dynamic Range, Autoranging Attenuators, and Power Detectors

Specifying a filter shape alone does not determine the selectivity of an SLMS. The SLMS must also have sufficient dynamic range to prevent the wanted signal from being swamped by noise or distortion products. The crucial parameter in a SLMS is its **noise power ratio** (NPR), which is a very sensitive indicator of the receiver's performance under real traffic conditions.

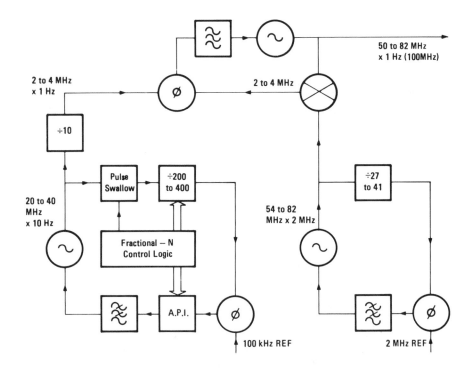

Figure 8.19 Fractional-N 50- to 80-MHz frequency synthesizer. Only two divide-by-N loops and one summing loop are required. One of the divide-by-N loops has extra circuitry to produce arbitrarily fine frequency resolution with a 100-kHz reference. Special control logic and extra analog phase interpolation circuits fill in the gaps between the conventional integer multiple outputs.

NPR measurements are made by using a broadband noise source, from which bands of noise can be removed with switchable bandstop filters. The receiver is tuned to the center of a noise slot, and the difference in measured noise level with notch in and out is the receiver's NPR.

At low input-signal levels, the slot becomes filled by thermal noise in the receiver. This effect reduces linearly as the input signal level is increased, so

NPR increases directly with input signal level. At higher signal levels, nonlinear effects in the receiver generate intermodulation noise products, which increase in amplitude faster than the applied input signal, so the NPR begins to degrade again. A plot of NPR against input power will produce a characteristic V-curve as in Fig. 8.20, with one level of signal where the effects of thermal noise and intermodulation noise balance and give the best receiver NPR.

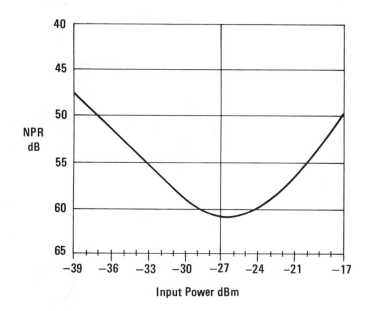

Figure 8.20 A typical NPR V-curve, or plot of S/N against input signal level. Thermal noise dominates the S/N ratio at low input signals so the curve initially (left to right) falls by 1 dB per 1-dB increase in input power. When thermal and intermodulation noise products are equal, the best S/N ratio is obtained at the bottom of the V. At higher input levels distortion mechanisms dominate and the curve steepens to 2 dB/dB, 3 dB/dB, and so on, as second-, third-, and higher-order effects become significant.

The SLMS input attenuator is used to adjust the signal level to the following circuitry to try to operate around the bottom of the V and maintain the best NPR spec. This gives rise to an overall curve that may resemble Fig. 8.21. Control of the input attenuator is derived from a broadband power detector. The measurement detector after the final selective filter will operate only over a finite signal range, so further gain or attenuation of the selected signal is needed to bring the level into the detector's operating window. Measurement speed is greatly influenced by how quickly this autoranging can be performed, and sophisticated algorithms can be employed to speed this process, Fig. 8.22.

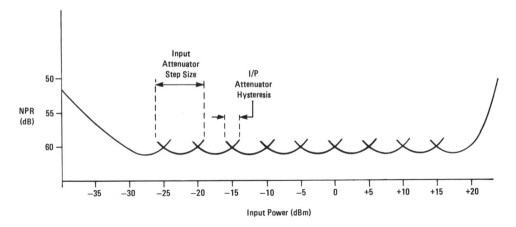

Figure 8.21 An SLMS NPR V-curve. This is a superposition of a number of curves as in Fig. 8.20. Below −30 dBm, thermal noise degrades the noise power ratio, and above +20 dBm distortion products dominate. Between these extremes, the receiver input attenuator adjusts the signal level to operate around the bottom of the curve.

Results Processing and Other Software Considerations

The processor in an instrument can contribute to the ease of use of the instrument and also to its accuracy. When operating under remote control, the processor can significantly increase the throughput on the system by reducing the information flow to and from the central computer. The first stage is to incorporate some results processing in the SLMS.

Programmable limits are set in the instrument, each measured result is compared to these limits, and only limit violations, either above or below the limits, are reported. When checking the channel virtual carriers, for example, only those above −30 dBmO, for instance, are of interest. With 3600 channels to scan, there can be a lot of redundant measurements. If results are being sent to a local printer, then only a record of relevant points is produced. Similarly, when reporting to a central computer, only the pertinent results from thousands of measurements are transmitted.

This technique can be extended when the same scan is repeated several times. If all pilot levels are being scanned overnight to look for drift, for example, any limit violations on the first pass are recorded, and thereafter only changes in the situation are noted. If, during the night, more pilot-level violations occur or some earlier violations are corrected, then these changes are reported. In the morning, there is a neat record of the changes in the system overnight on a local printer, or a central computer can have had to deal only with problems as they arose.

It is unlikely that the central computer will be sitting beside the SLMS, so a communication channel is needed beyond the instrument control bus. This can be a high-speed extended link over distances of a few hundred meters, but

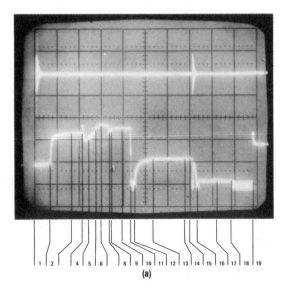

(a)

TIME SLOT	IF GAIN (dB)	DETECTOR RESOLUTION	EXPLANATION
1	85	0.1	Reset detector to remove overshoot and ringing
2	85	0.1	Detector is overloaded — try removing 5 dB of gain
3	80	0.1	Still overloaded — initiate fast ranging sequence
4	0	1.0	Reset having switched resolution and gain
5	0	1.0	Underloaded
6	60	1.0	Reset after 60 dB gain change
7	60	1.0	Overloaded
8	30	1.0	Reset after 30 dB gain change
9	30	1.0	Signal is now within detector range
10	30	1.0	Measure signal to 1.0 dB and calculate gain for 0.1 dB range
11	45	0.1	Reset after gain and resolution change
12	45	0.1	Allow detector to slew then check that calculated gain is correct
13	45	0.1	Integrate and a/d convert
14	45	0.1	Processor adds gain to a/d result and displays level
15	45	0.1	Instrument returns — to noise floor, reset detector
16	45	0.1	Detector underloaded — try adding 5 dB of gain
17	50	0.1	Still underloaded — initiate fast ranging sequence
18	0	1.0	Begin fast ranging sequence — underloaded
19	60	1.0	Still underloaded — will now try 90 dB etc

(b)

Figure 8.22 An SLMS autoranging scheme. A detector with variable dynamics and microcomputer control of gain stages is used to track signal changes over a 40-dB range. The SLMS is tuned from noise to a 40 dB higher signal, then background noise. The sequence begins as the instrument is tuned to the signal. Time scale is 50 ms/division. RMS detector output is logarithmic. Sophisticated techniques like this can squeeze extra performance from measurement circuits.

over longer distances, modems and the public telephone network are used. This places an immediate, severe limitation on the capacity of the data link. By returning only relevant data to the central computer, we can greatly increase the throughput of the system. The same concept works in the other direction. Compressing the data to the SLMS produces more savings.

The built-in FDM plans in the SLMS provide exactly this type of data compression. A complex series of measurements can be specified with a few front-panel keystrokes or a few bus commands from a remote controller. The alternative of sending every measurement frequency to the SLMS and then reading back every result for remote processing destroys the effectiveness of a remote surveillance system.

Cable Equalization

If a station has several test points linked through an access switch to the SLMS, it may be necessary to account for the frequency-dependent losses of the linking cables. If the cable runs are very long, then the losses at the top end of the band can cause measured results to violate the internal limits. Using wider limits is not very desirable; a better approach is to equalize the cable runs. If all the cables are made the same length from test points to access switch, then only one equalizer between the access switches and SLMS is needed. Alternatively, the cable characteristics can be stored in the SLMS, which now performs a software cable equalization.

Signal Statistics, Power Detectors, and Measurement Speed

There are three types of signals encountered commonly in FDM systems. These are tones, or sinusoids, noise, and speech.

Sine Waves

The simplest of these signals is the sine wave. Signaling tones, pilots, and virtual carriers all fall into this category. To measure the power in a sine wave, only the peak level is required. This can be captured within at most one-half cycle of the signal. RMS or average detection involves, by definition, averaging over some period of time, so it is an inherently slower process. Of course, this assumes that the signal-to-noise ratio is good and that the sine wave is free from distortion— that is, it contains only one frequency component.

Gaussian Noise

In contrast, noise contains many frequency components. It is often described as *white,* meaning that it contains equal energy at all frequencies. A peak detector is now useless, as it will overestimate the noise level by something in the region of 15 dB. This overestimation is the **peak-to-rms ratio,** or **crest factor,** of Gaussian noise. In theory, there is no upper limit on noise peaks; instead, the probability of exceeding a given level can be found from the well-known Gaussian probability distribution or, more usefully, the complementary error function erfc(x) (see Table 8.3).

TABLE 8.3 The Complementary Error Function (erfc(x))

Values of erfc(x) vs. x.*

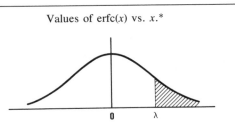

	0.00	0.01	0.02	0.03	0.04	0.05	0.06	0.07	0.08	0.09
0.0	.5000	.4960	.4920	.4880	.4840	.4801	.4761	.4721	.4681	.4641
0.1	.4602	.4562	.4522	.4483	.4443	.4404	.4364	.4325	.4286	.4247
0.2	.4207	.4168	.4129	.4090	.4052	.4013	.3974	.3936	.3897	.3859
0.3	.3821	.3783	.3745	.3707	.3669	.3632	.3594	.3557	.3520	.3483
0.4	.3446	.3409	.3372	.3336	.3300	.3264	.3228	.3192	.3156	.3121
0.5	.3085	.3050	.3015	.2981	.2946	.2912	.2877	.2843	.2810	.2776
0.6	.2743	.2709	.2676	.2643	.2611	.2578	.2546	.2514	.2483	.2451
0.7	.2420	.2389	.2358	.2327	.2296	.2266	.2236	.2206	.2177	.2148
0.8	.2119	.2090	.2061	.2033	.2005	.1977	.1949	.1922	.1894	.1867
0.9	.1841	.1814	.1788	.1762	.1736	.1711	.1685	.1660	.1635	.1611
1.0	.1587	.1562	.1539	.1515	.1492	.1469	.1446	.1423	.1401	.1379
1.1	.1357	.1335	.1314	.1292	.1271	.1251	.1230	.1210	.1190	.1176
1.2	.1151	.1131	.1112	.1093	.1075	.1056	.1038	.1020	.1003	.0985
1.3	.0968	.0951	.0934	.0918	.0901	.0885	.0869	.0853	.0838	.0823
1.4	.0808	.0793	.0778	.0764	.0749	.0735	.0721	.0708	.0694	.0681
1.5	.0668	.0655	.0643	.0630	.0618	.0606	.0594	.0582	.0571	.0559
1.6	.0548	.0537	.0526	.0516	.0505	.0495	.0485	.0475	.0465	.0455
1.7	.0446	.0436	.0427	.0418	.0409	.0401	.0392	.0384	.0375	.0367
1.8	.0359	.0351	.0344	.0336	.0329	.0322	.0314	.0307	.0301	.0294
1.9	.0287	.0281	.0274	.0268	.0262	.0256	.0250	.0244	.0239	.0233
2.0	.0228	.0222	.0217	.0212	.0207	.0202	.0197	.0192	.0188	.0183
2.1	.0179	.0174	.0170	.0166	.0162	.0158	.0154	.0150	.0146	.0143
2.2	.0139	.0136	.0132	.0129	.0125	.0122	.0119	.0116	.0113	.0110
2.3	.0107	.0104	.0102	.00990	.00964	.00939	.00914	.00889	.00866	.00842
2.4	.00820	.00798	.00776	.00755	.00734	.00714	.00695	.00676	.00657	.00639
2.5	.00621	.00604	.00587	.00570	.00554	.00539	.00523	.00508	.00494	.00480
2.6	.00466	.00453	.00440	.00427	.00415	.00402	.00391	.00379	.00368	.00357
2.7	.00347	.00336	.00326	.00317	.00307	.00298	.00289	.00280	.00272	.00264
2.8	.00256	.00248	.00240	.00233	.00226	.00219	.00212	.00205	.00199	.00193
2.9	.00187	.00181	.00175	.00169	.00164	.00159	.00154	.00149	.00144	.00139

Values of erfc(x) for large x.

x	10 log x	erfc(x)	x	10 log x	erfc(x)	x	10 log x	erfc(x)
3.00	4.77	1.35E-03	4.00	6.02	3.17E-05	5.00	6.99	2.87E-07
3.05	4.84	1.14E-03	4.05	6.07	2.56E-05	5.05	7.03	2.21E-07

x	$10 \log x$	erfc(x)	x	$10 \log x$	erfc(x)	x	$10 \log x$	erfc(x)
3.10	4.91	9.68E-04	4.10	6.13	2.07E-05	5.10	7.08	1.70E-07
3.15	4.98	8.16E-04	4.15	6.18	1.66E-05	5.15	7.12	1.30E-07
3.20	5.05	6.87E-04	4.20	6.23	1.33E-05	5.20	7.16	9.96E-08
3.25	5.12	5.77E-04	4.25	6.28	1.07E-05	5.25	7.20	7.61E-08
3.30	5.19	4.83E-04	4.30	6.33	8.54E-06	5.30	7.24	5.79E-08
3.35	5.25	4.04E-04	4.35	6.38	6.81E-06	5.35	7.28	4.40E-08
3.40	5.31	3.37E-04	4.40	6.43	5.41E-06	5.40	7.32	3.33E-08
3.45	5.38	2.80E-04	4.45	6.48	4.29E-06	5.45	7.36	2.52E-08
3.50	5.44	2.33E-04	4.50	6.53	3.40E-06	5.50	7.40	1.90E-08
3.55	5.50	1.93E-04	4.55	6.58	2.68E-06	5.55	7.44	1.43E-08
3.60	5.56	1.59E-04	4.60	6.63	2.11E-06	5.60	7.48	1.07E-08
3.65	5.62	1.31E-04	4.65	6.67	1.66E-06	5.65	7.52	8.03E-09
3.70	5.68	1.08E-04	4.70	6.72	1.30E-06	5.70	7.56	6.00E-09
3.75	5.74	8.84E-05	4.75	6.77	1.02E-06	5.75	7.60	4.47E-09
3.80	5.80	7.23E-05	4.80	6.81	7.93E-07	5.80	7.63	3.32E-09
3.85	5.85	5.91E-05	4.85	6.86	6.17E-07	5.85	7.67	2.46E-09
3.90	5.91	4.81E-05	4.90	6.90	4.79E-07	5.90	7.71	1.82E-09
3.95	5.97	3.91E-05	4.95	6.95	3.71E-07	5.95	7.75	1.34E-09

*From J. S. Bendat and A. G. Piersol, *Random Data: Analysis and Measurement Procedures, N.Y.*: Wiley-Interscience, 1971, and D. B. Owen, *Handbook of Statistical Tables*. Reading, Mass.: Addison-Wesley Pub. Co., 1962, both by permission; courtesy of the U.S. Energy Research and Development Administration.

Less well known is the chi-squared distribution, which can be used to find the equivalent probability distribution of the power or rms value of a Gaussian noise source. This distribution depends on the bandwidth of the noise and the detector integration time.

Appendix 8.1 shows how to use the chi-squared distribution for any general case, but for the moment we can use the following formulae from Schiesser [Schiesser, 8.17]: 90% of results will lie within a spread of $20/(\sqrt{(2BT)} - 1)$ dB and 98% of results will lie within $29/(\sqrt{(2BT)} - 1)$ dB, where B is the equivalent noise bandwidth of a measuring filter and T is the detector integration time. When measuring channel noise, we have a filter with 3.1-kHz noise bandwidth and integration times of either 30 or 300 ms, so we can expect our results to have 98% spreads of about 2 dB and 0.7 dB, respectively. Figure 8.23 shows the recorded results of 100 noise measurements. As can be seen, the results agree well with the theory.

Speech

The statistics of speech are not available in an analytical form. However, some deductions about speech statistics can be made based on simple observations. The actual sounds produced are similar to noise, but they are broken up by silences between phrases, words, and syllables. In conversation, there are longer pauses while listening occurs.

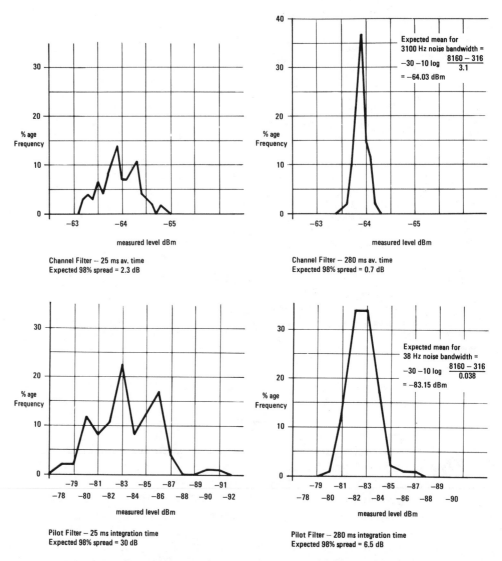

Figure 8.23 Spread in results when measuring noise through an SLMS channel filter. The improvement with extra averaging time is evident, as is the close agreement with predicted performance.

Speech signals are often analyzed by recording the rms signal integrated for $\frac{1}{8}$ s (this interval approximates the integration time of the human ear). This technique shows that the maximum power in any interval is about 18 dB above the long-term rms level. If the signal during the $\frac{1}{8}$-s interval is at all noiselike,

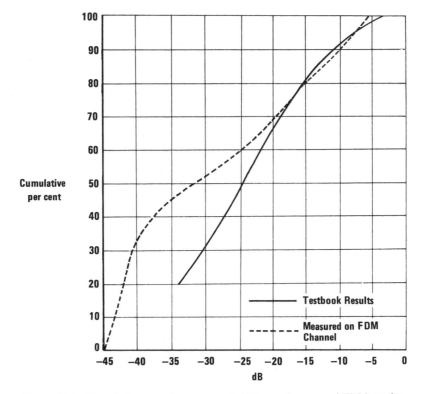

Figure 8.24 Cumulative speech power probability, and measured FDM results. In theory, an FDM channel will only carry one-half of a conversation, but in practice, hybrid echo and acoustic coupling will allow the other party in the call to breakthrough at a lower level. This gives rise to the "bulge" in the measured results on the FDM system.

then it will have a crest factor of perhaps 10 dB in this interval. So the overall crest factor of speech would approach 30 dB.

Measurements made on an FDM channel were plotted as a cumulative probability density function (pdf). The result, along with an example, are shown in Fig. 8.24. The curves are in reasonable agreement except at low power levels. The low-level bulge is caused by the presence of speech power from the other party in the conversation adding at a low level from acoustic coupling in the telephone handset and hybrid echo. This curve does not show the random nature of the signal as well as the original data in Fig. 8.25, where the problem that must be tackled by an autoranging routine can be seen. This diagram shows that to obtain an accurate measurement of traffic level in one channel, a very long integration time is needed. In practice, measuring groups or supergroups provides a sufficiently accurate estimate of traffic loading for link engineering purposes.

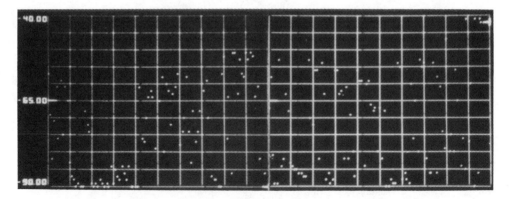

Figure 8.25 FDM speech measurements. These were made using a channel filter with 30-mS integration time, at a rate of one per 100 mS. The rapid changes in level can play havoc with autoranging schemes.

8.4.2 Access Switches

A single repeater station may handle tens of different trunk connections or only a few. To cope with these different situations, an expandable access-switch arrangement is required. We use 10 input switches, which can be cascaded to three levels providing 1000 inputs with 111 switches.

On an HP-IB control loop, only 15 device addresses are allowed, so individual control of each switch would limit the number of inputs available. To overcome this problem and reduce the overhead cost, a single switch controller is linked to the control bus. Individual switches are then controlled by simple, serial pulses from this unit. Since up to three switches can be cascaded between the equipment and the SLMS, the frequency responses of each switch must be rigorously controlled. Isolation must be specified, both between unselected inputs and the output to avoid measurement errors and also between any two inputs so that signals are not connected on the FDM system. Distortion and noise must be held at a low level and insertion loss should ideally be zero.

The pulse-control signals can drive more than one switch load, so a distribution switch can be connected in parallel with some or all of the access switches to provide a comprehensive stimulus and measurement system.

8.4.3 Level Generators

If a signal generator is wrongly connected to an FDM system which is carrying traffic, it can cause all kinds of trouble! Obviously, it must not interfere directly with traffic, nor should it upset any of the system pilot monitors, so its frequency must be precisely controlled. The same applies to its spectral purity, as any harmonics, spurious outputs, or even phase noise can cause problems. One obvious use of a level generator is to check the baseband flatness of a complete system. Tones can be injected in the intersupergroup slots with no interference to traffic, and the frequency response can be measured with an SLMS. The SLMS can often control the level and frequency of the generator to provide a tracking generator function and even allow FDM description tuning.

When changing from one ISG slot to another, the signal should be blanked rather than sweeping through the supergroup. This blanking should be *soft*, because a step change in amplitude produces sidebands on the carrier that can interfere temporarily with adjacent traffic. The same range of impedances and connector options needed in an SLMS are required so that signals can be injected at any point in the system from channel to line, and of course the level must be accurate across this range of frequencies from a few hundred hertz to tens of megahertz.

The design problems of an accurate level generator are similar to those in an SLMS. Both require a frequency synthesizer, accurate attenuator pads, low distortion amplifiers, and accurate power detectors. Apart from its use in traffic, a generator can be useful when fault-finding on an out-of-service link or for installation use. With the generator at one end of a link and the SLMS at the other, HP-IB control may not be possible. If the generator can step across a band of frequencies, then the SLMS can be set to step across the same band in an *open-loop track* mode. The two instruments will step in synchronization across the band.

8.4.4 Data Line and Telephone Line Analyzers

A communications link that is to carry data rather than speech needs some extra characterization. Humans are fairly insensitive to snaps, crackles, and pops, minor frequency shifts, group delay slope, phase jitter, or amplitude distortion. Data modems are not so forgiving, so telephone lines have to be monitored, either to fault-find on a troublesome circuit or to characterize a dedicated line of guaranteed performance.

Telephone line analyzers measure amplitude and delay distortion, whereas data line analyzers handle the other impairments. There is no hard-and-fast boundary, however, and some overlap between instruments will be found. These instruments are usually connected to an audio line, but they can also be connected to an SLMS audio output. This allows a link to be tested at each stage of modulation so that faulty equipment can be identified. The parameters of interest are as follows.

Amplitude or Attenuation Distortion. An ideal transmission medium would not alter the transmitted signal in any way, but of course practical channels will have amplitude ripples, slopes, and roll-off at the band edges. These effects are called **amplitude distortion.** CCITT recommends limits of less than 2.2 dB ripple from 600 Hz to 2.4 kHz and less than 9 dB roll-off from this band to the band edges at 300 Hz and 3400 Hz, all measured relative to the level at 800 Hz.

Group Delay Distortion. Sending data signals over a link that suffers from nonflat group delay will cause smearing of the received symbols and degrade the link performance. Typical voice channels have about 2 ms of relative delay distortion in band. To obtain compatibility between instruments from different manufacturers, CCITT specifies not only the channel characteristics but also the measurement technique (Recommendation 0.81). This method entails using a special test signal (see Fig. 8.26).

The receiver recovers the envelope of this signal and, by comparing its amplitude and phase during the test and reference periods, can measure amplitude and delay distortion simultaneously.

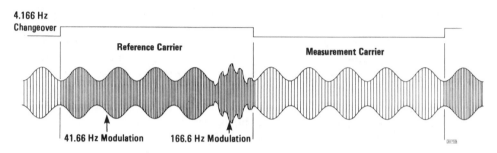

Figure 8.26 Telephone line analyzer test signal. By alternating a reference carrier and a variable frequency measurement carrier over a link, only the path under test is required to make measurements of amplitude and group delay distortion. Modulation on the reference carrier allows the receiver to lock in and distinguish the various portions of the signal.

Frequency Shift. The **frequency shift** across an HRC link should be less than 2 Hz if it is to carry data. This shift can be measured by transmitting a reference tone of known frequency and counting the frequency of the received tone. An alternative method is to transmit two tones, one of which is at exactly twice the frequency of the other. This can easily be done with a digital flip-flop. The actual frequency of the tones does not matter as long as they are both in-band. The two tones (F and $2F$) will be shifted by the same amount (f) across the link. In the receiver, the tones are separated and the frequency of the lower frequency tone is doubled to ($2F + 2f$). Subtracting this signal from the upper frequency tone at ($2F + f$) gives f directly without needing a precision frequency source.

Phase Jitter. **Phase jitter** is caused when a signal is modulated by other signals in the equipment. Common sources are power-line hum and ringing tones. The effect is measured by transmitting a clean test tone over a link and using a PLL in the receiver as a phase modulation detector (see Fig. 8.27). The phase modulation is measured as peak-to-peak phase deviation in the band 20 to 300 Hz (optionally from 4 to 300 Hz).

Impulse Noise. Steady-state noise on a telephone line can arise from thermal noise, intermodulation noise, or crosstalk from other circuits. **Impulse noise** can be created by natural causes—for example, lightning—but is more often artificially

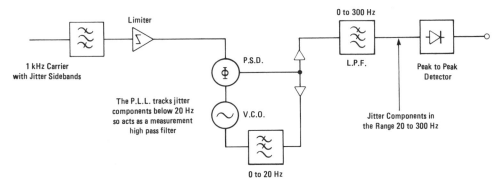

Figure 8.27   Phase jitter detector. The PLL locks onto the received 1-kHz test signal. The oscillator control signal contains any components at the jitter rates and these are filtered and peak detected.

created from switching signals, power-line transients, automobile ignition, dialing pulses, and so on. It appears as sporadic spikes of noise far above the background steady-state noise level. Instead of trying to measure the power in each spike, the occurrence of each spike above some threshold level is recorded, and a rate in terms of number of occurrences in a 5-, 10-, or 15-min period is derived.

Such measurement involves setting up three voltage comparators at the desired thresholds and counting pulses that exceed these thresholds for more than 50 ms. Multiple pulses in a 125-mS period (143 in Bell) are treated as being one impulse so that filter ringing effects do not cause false counts. Since the rate is expected to be low, the chances of missing a true impulse in this dead time are remote. A typical channel should have less than five impulses above −35 dBmO in a 5-min interval during the busiest hour of the day.

Gain Hits and Dropouts.   Another form of transient impairment occurs when the level of the received signal changes rapidly with time. This may be caused by impulse noise affecting an AGC loop or—more likely—by a dirty relay somewhere in the system. If the signal level momentarily increases, this is called a **gain hit**; a decrease in level is a **dropout**. These impairments are measured by observing the level of a transmitted tone and using threshold detectors and counters as for impulse noise. The gain hit thresholds are defined as +2, +4 and +6 dB, while a dropout is defined as being less than −12 dB for more than 4 mS.

Phase Hits.   The phase of a received signal will also be affected by transient path disturbances. The phase jitter circuit is used again, with its detected output being connected to threshold comparators and counters.

Simultaneous Event Recording

With a test tone connected to a channel, many of these impairments can be measured simultaneously. Simultaneous measurement is very desirable, as each test may last for several minutes. Phase jitter, phase hits, gain hits, and dropouts all require the tone as a reference, whereas weighted and impulse noise are measured after notching out the tone (a true noise reading will require the tone to be present on the channel to activate compandors).

Obviously, a single transient can be recorded as more than one impairment; for example, the sidebands produced by a gain hit may be interpreted as impulse noise. To obtain a truer record of disturbances, the following counting hierarchy has been established. When a dropout is detected, the counting of phase and gain hits is blocked for a short period or until the test tone recovers. If a gain hit or phase hit occurs, impulse-noise counting is similarly blocked.

8.5 SYSTEM ARCHITECTURE

8.5.1 Interfacing

Having identified and described the test equipment required to provide the measurement capability for FDM network surveillance, we must now focus our attention on the various types of system configuration and control hardware needed in such applications. Connection of an external device, or devices, to provide the controlling function in a monitoring system implies the ability to interconnect the equipment for the transfer of information to take place. This transfer of information is accomplished over some form of interface. Many interface types are in existence, but one that lends itself to most applications is the one defined in *IEEE Standard* 488-1978 [IEEE, 8.3].

This interface has become so universally accepted since its introduction that now only the simplest of test equipment does not have such an interface available. Apart from meeting the great majority of needs, its success stems from the virtual elimination of custom interface engineering, which is otherwise necessary. Through the use of this interface, integration of instruments into systems for telecommunications and other applications is now a relatively routine task.

The interface employs 16 signal lines with bidirectional information transfer, allowing communication between up to fifteen devices on one local contiguous bus. The mechanical, electrical, and functional elements of the interface are defined, leaving the operational aspects to be implemented as required within a particular instrument or device. A condensed description of this interface is provided in Appendix 8.2; however, if you wish to gain a more detailed understanding of the interface, then you should read the appropriate references given at the end of Appendix 8.2.

When the interface was devised, the designers initially tackled the immediate problems of local bench-top and rack-mounted systems, and no direct attempt was made to solve the problems of controlling remote instrumentation. Indeed, the interface requires that the total interconnecting cable length should not exceed 2 m per device or 20 m, whichever is less. While this interconnection limitation was not a significant issue at the time, the capability to provide remote operation is now a virtual necessity, with distributed systems becoming increasingly important in areas such as telecommunications. Where measuring stations require data to be transmitted over longer distance than can be handled directly by the instrumentation interface, the obvious carrier for this information is the conventional data circuit. Since this is a serial transmission medium, a conversion unit is required to serialize the parallel interface information for communication via data sets (modems) over the link (see Fig. 8.28). Such conversion units are more commonly known as **extenders,** as their function is to extend the interface bus (see Fig. 8.29).

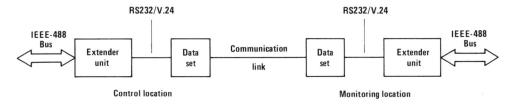

Figure 8.28   Interface extension technique showing how the IEEE-488 interface can be used to interconnect two remote sites over a standard telephone circuit using extender units and data sets.

The capability of a typical extender product includes the following:

- Interface extension over unlimited distances via a telephone channel and data sets
- Error detection and correction
- Dial-up operation: the ability to pass a telephone number to an autodialler via an appropriate interface, such as RS366/V25
- Multipoint operation: the selection of one of a number of remote sites connected to a single telephone channel
- The ability to detect (1) a breakdown in the communication link, (2) completion of an auto-dialed call, (3) failure to complete a call, and (4) completion of a message transmission

Corruption of data in transit between devices is not a problem that is usually encountered with benchtop- or rack-mounted systems, but it is of great importance in remote systems, particularly with dial-up modem connections, where data transmission errors due to transient phenomena are highly probable. Since even a single-bit error in a message transmitted over the interface is enough to cause a complete operational breakdown, such errors obviously cannot be tolerated. Long-distance extenders, therefore, must effectively eliminate the effects of communication errors.

The speed and type of data set to be used in a system will be determined to a large extent by the configuration of the system, the amount of information

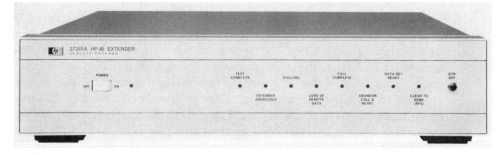

Figure 8.29 HP37201A long-distance extender used to convert the IEEE-488 interface into RS232C format.

to be transferred between the system equipment, the data rate of the equipment, and the quality and types of the available data circuits. For systems that are operating on a continuous basis, dedicated (leased) lines are ideal. However, where only periodic control of a measuring station is required, a dial-up line will suffice. The probability of data corruption with dial-up lines is higher than with dedicated lines, but dial-up does have the advantage that a channel failure can be surmounted by dialing up over another line.

8.5.2 Typical System Configurations

Irrespective of the type of monitoring and the equipment used, there are several basic system configurations that can be identified. These configurations are shown in Fig. 8.30(a)–(e). The simplest system is shown in configuration (a), where the controller is connected directly to the monitoring equipment via the instrumentation interface bus. Such a configuration could be used for an in-station monitoring system comprising up to 15 interface-compatible devices connected side by side. For in-station applications (b), where the system controller is located in a different area of a building than the monitoring equipment, two extender units can be incorporated in the system to provide extension up to a distance of 1000 m using coax or optical fibre as the extended transmission medium. Where the monitoring equipment is truly remote from the system controller, two extender units and data sets can be used to provide communication between the sites, as shown in configuration (c).

In configurations (a), (b), and (c), multiple monitoring stations can be connected individually to the system controller, either locally or remotely, providing simultaneous surveillance. Configuration (d) shows a system where multiple monitoring sites are connected to the controller by a single leased data circuit. Unlike the previous three configurations, only one site can be controlled directly at any time. This configuration is known as **multidrop** (or **multipoint**) and for such an application, the data sets must have switched-carrier capability controlled by the RS232C interface signal request to send (RTS). All the remote data set receive

(a) Side by side

(b) Intrafacility

(c) Point to point

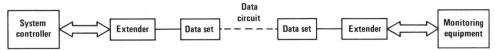

(d) Multidrop

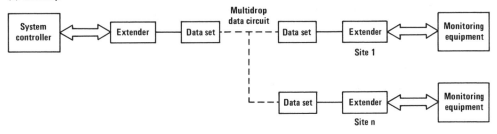

(e) Dial-up

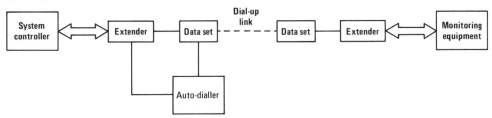

Figure 8.30 Basic system configurations for connecting monitoring and measuring equipment to a system controller.

ports are connected functionally in parallel with the local (controller-end) data set transmit port and all remote transmit ports are connected in parallel with the local receive port. All remote data sets receive continuously so that each extender can wait, listening for its own multidrop address (preassigned by switch selection in each remote extender). When the controller wishes to communicate with a remote site, it sends the multipoint address of the appropriate remote extender (for example, M3) to the local extender, which, in turn, transmits the address over the communication link. This causes the remote extender, which recognizes its own address, to turn on its modem carrier and begin communicating with the control-center extender. Any previously addressed remote extender automatically turns off its modem carrier.

Configuration (e) shows a dial-up system with multiple remote sites capable of being controlled sequentially or on demand from the controller. This type of configuration would be applied where the nature of the monitoring does not warrant dedicated lines or a fast response, where the cost of leased lines is prohibitive, or where on-site data-collection facilities are provided, with the system controller periodically dialing each site and receiving a dump of all the collected information.

In this configuration, the controller sends the telephone number of the appropriate remote site to the local extender. This extender then passes the number to the dialer via the extender RS366 (V.25) interface, and when the call is automatically (or manually) answered, control of the group of instruments at the remote site can begin immediately.

Multiple computers, each controlling their own cluster of measurement stations, can be formed into a distributed system network (see Fig. 8.31). Such a system would be used where more than the maximum number of subsystems controllable from a single computer are needed to provide overall network surveillance or where it is necessary to have two computers controlling the measurement subsystems, with both computers having the same control capability

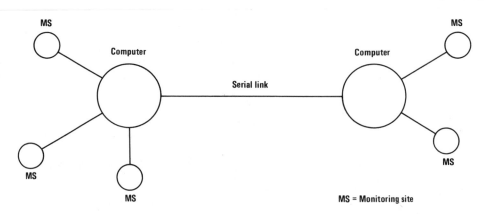

Figure 8.31 Diagrammatic representation of a distributed system network showing the linkages between measuring sites and computer sites.

(for instance, where a cable crosses national boundaries). Distributed computer networks could also be used within an administration or telephone company where computer systems are located on a regional basis with a national computer providing network control.

8.5.3 Control Hardware

Irrespective of the system configuration, the system controller is the heart of the monitoring system. It is this device that holds information about the network under surveillance and the monitoring equipment; it also initiates measurements, either on-demand from an operator or automatically at a scheduled time, and can report or store the results of these measurements. The exact size and type of controller to be employed in a particular system will depend on several factors:

- The number of monitoring sites
- Whether sequential or simultaneous control of the monitoring sites is required
- The memory requirements for the application software, network, and system related databases and quantity of result information to be stored
- The number of system operators and their location(s)
- The data-reduction, -analysis, and -presentation requirements

In general, some form of computer system offers the most realistic control capability, whether it be of the small variety or a minicomputer. (A computer system includes both hardware and software; the hardware is the part that can be physically picked up, whereas the software is the part that provides instructions to the hardware to carry out the appropriate tasks.)

If the monitoring system comprises many stations and all stations are to be operated continuously, a minicomputer having a real-time, multitasking operating system offers the best solution unless relatively slow operating times are acceptable. The operating system (OS) is the software that manages the hardware, including not only the computer processor and memory but also input/output (I/O) devices such as terminals, printers, and instrumentation. A real-time operating system can handle a wide variety of tasks and, as tasks in the real world do not necessarily have equal priorities, the OS permits tasks to be prioritized. This allows tasks whose purpose is computational or display to coexist in the same system with tasks that control and react to measurements from instruments. The multitasking capability effectively allows many programs to be run simultaneously in the computer and is essential for large systems. It also allows multiple access to control programs and result data from user terminals as well as providing the ability to develop programs while the system is actively monitoring remote stations. These features require partitioning of processor memory, a feature which is not available with traditional desktop computers.

If stations are to be accessed sequentially and there is no requirement for multiple user access, a desktop computer may be capable of controlling the system. Figure 8.32 shows a Hewlett-Packard 9816 Desktop Computer, together with an SLMS, access switch, and printer. Some form of mass storage will be needed to hold the system-control software as well as to store the results of measurements and status conditions. Various forms of mass storage are available, data cartridge, floppy disc, hard disc, and mag tape being the most popular. In general, if speed of access is not a factor and the storage requirements are minimal, data cartridges may prove to be satisfactory. Floppy discs have a considerable access speed advantage over cartridges, with hard discs providing the best all-around on-line storage capability. High-speed magnetic tape units are ideal for providing large, off-line, archive storage and are particularly useful for backing up large fixed discs. Table 8.4 provides a comparison of the main specifications between some Hewlett-Packard on-line storage media.

Figure 8.32 HP9816S Desktop Computer, HP9121D Disc Drive, with HP3746A SLMS, HP3754A Access Switch and HP82905A Printer.

Most desktop computers have some form of mass storage built-in, either in the form of cartridge tape drives or floppy disc drives. They are also provided with a hard-copy printer, often of the thermal variety. An external printer may, however, have to be added to provide fast, full-page width (more than 72 characters) output or simultaneous multiple-copy output. A fast, page-width line printer or even a high-speed page printer to log system measurements and list programs that are under development can usually be justified for large monitoring systems using minicomputers.

Many types of user terminals are available, ranging from a simple text keyboard and printer to a combined keyboard, VDU, and printer providing complete graphical capability. Again, the type, or types, of terminals connected to the system controller will depend on the application and, sometimes, on the location. A full-capability terminal would most often be used at the control center and subcontrol centers, with a portable keyboard and printer being ideal for mobile maintenance teams who could manually dial into the system controller using an acoustic coupler to provide the physical interface between telephone and terminal.

TABLE 8.4 Sample Storage-Media Specifications for Some Hewlett-Packard Floppy and Hard Discs

	Floppy disc			Hard disc		
	3½ in.	5¼ in.	8 in.			
	HP9121S	HP82901M	HP9895A	HP7908A	HP7911A	HP7912A
Bytes/drive	270K	270K	1.15M	16M	27M	64M
Transfer rate (bytes/s)	17.8K	6.8K	25.6K	0.5M	1M	1M
Average seek time (ms)	365	187	179	41.6	26.7	26.7

8.6 APPLICATION SOFTWARE

The application software for an FDM monitoring system can be divided into four basic categories:

• Database management
• Measurement routines
• Maintenance and management report routines
• Functional testing and diagnostic routines

This is not to say that every monitoring system will, or must, necessarily have software from all four categories. Certainly, measurement routines and database management are fundamental to any system; functional and diagnostic routines are very desirable, especially in the case of systems having truly remote stations. Whereas maintenance and management reports will be a function of the requirements of the system, in some cases the printout of raw measurement results obtained from the measurement routines may be adequate.

8.6.1 System Database

Information about the FDM network to be monitored, the monitoring system equipment, and the system configuration is stored in the computer system in what is known as a **database.** Typically, the monitoring system information would include details of the following:

• Equipment located at each site
• Controller I/O port assignment for each subsystem
• Type of subsystem communication link (e.g., leased line point-to-point, dial-up, multipoint)

The network information relates to the FDM test points and would typically comprise

- Transmission level (TLP)
- FDM plan type and any modifications
- FDM start and stop values plus skip values relating to unloaded sections of the multiplex
- Test-point flatness characteristic (equalization curve)
- Measurement alarm status

Measurement result and status information can also be held either in the relevant station database or in a separate result database. The storage of such information is a necessity if data analysis and management reporting are to be a function of the system.

The organization of the database may take one of several forms, but for the purposes of this description, suffice to say that, physically, a database is a collection of logically related files containing details of the network and the monitoring system as well as the database structural information.

8.6.2 Measurement Programs

The measurement programs provide control of the equipment by initiating tasks and receiving results. Generally, such programs can be classified as demand or automatic. **Demand measurement programs** are used for investigating faults or for making spot checks, whereas **automatic measurement programs** are used in sequences to provide network surveillance.

Demand measurement programs should be interactive, prompting the operator for information about the measurements to be made. Furthermore, there are advantages in having the programs request information in a manner similar to that in which an operator would manually control the instruments. An example of the sequence of prompts to initiate five measurements at a frequency of 952.08 kHz is shown in Fig. 8.33, together with a sample set of results. Automatic programs, on the other hand, are normally scheduled under computer control, although they may also be initiated manually from a terminal. Complete surveillance sequences can be built up from a set of these programs, as shown in Fig. 8.34, and stored on file for initiating at the appropriate times. Typically, surveillance sequences would be used to accumulate the information necessary to produce fault maintenance reports, network status reports, and trend analysis.

If the monitoring system is to be used for continuous surveillance, it is important that an operator be able to interrupt the surveillance process in some way to make demand measurements. Once a demand-measurement request has been entered from a terminal, the system controller can log it but not process the request until a suitable point has been reached in the surveillance procedure.

```
FREQ, SN1, TP1, P

FREQ(kHz), FILTER (A/P/C/G/W), AVG(S/N/L), # MEASMTS
? 952.08,P,L,5


SN1   TP1   16:03   21 JAN 1983


SINGLE FREQUENCY WITH PILOT FILTER AND LONG AVERAGING

    FREQ(kHz)      LEV(dBm)        LEV(dBmO)
    952.08         -53.04          -20.04
    952.08         -53.05          -20.05
    952.08         -53.04          -20.04
    952.08         -53.04          -20.04
    952.08         -53.03          -20.03
```

where FREQ, SN1, TP1, P is the user-entered run-string to request a level measurement at one frequency, at station SN1, test access point TP1, with the measurement results being output to printer P.

The measurement parameter prompt returned from the system requests the user to enter the frequency (in kHz), SLMS filter type (pilot, channel, group, weighted, or automatic selection), SLMS averaging (short, normal, or long) and the number of measurements to be made.

Figure 8.33 Sample demand measurement prompts and results for level measurements at a single point frequency.

This could take place at the end of one measurement in a sequence or after completing all measurements in a sequence. Alternatively, an operator may be allowed to interrupt a surveillance routine at any time, with the surveillance routine continuing from the point at which it was interrupted. The method chosen is dependent upon the priority assigned by the designer to the demand and surveillance functions. From an operator's point of view, it is best to choose

```
**   GROUP AND SUPERGROUP PILOT STATUS MEASUREMENTS
**   ON TEST POINTS TP1, TP2, TP3 AND TP4
**   AT STATION SN2
**   LAST MODIFIED 23RD FEBRUARY 1983 - RLM
SFDM, SN2, TP1, GRPS, P
SFDM, SN2, TP1, SGPS, P
SFDM, SN2, TP2, GRPS, P
SFDM, SN2, TP2, SGPS, P
SFDM, SN2, TP3, GRPS, P
SFDM, SN2, TP3, SGPS, P
SFDM, SN2, TP4, GRPS, P
SFDM, SN2, TP4, SGPS, P
```

where SFDM is the name of the FDM automatic surveillance program, SN2 is the station name, TP1, TP2, TP3, and TP4 are the test access-point identifiers, which provide the keys to the appropriate FDM related information in the database, GRPS and SGPS identify the group pilot and supergroup pilot status measurements, and P indicates that the results are to be output to the printer.

Figure 8.34 Sample surveillance sequence for measuring group and supergroup pilots.

the second method, because the first method can prove frustrating due to the delay between initiating the request and having it actioned.

Automatic monitoring systems, although providing continuous surveillance, may create a problem in that they can accumulate so many results that it is difficult or even impossible for a user quickly to identify important results. In addition, if all the results are to be retained at the control center, they will occupy a considerable amount of storage. To illustrate this point, consider an FDM monitoring system comprising ten stations, each making one measurement per second and returning every result to the system controller. Each station operating in surveillance mode can take 86,400 readings in a 24-h period, requiring the system controller to process 864,000 readings for the total system. If each result is 10 bytes (characters) in length and every result is to be stored at the control center, 8.6 megabytes of memory would be needed for each day's results. Limiting the results returned from monitoring sites to the system controller to those that are most significant is therefore useful. For many applications, results returned from monitoring sites can be restricted to those that violate predefined limits. Ideally, the intelligence to determine measurement-level violations should reside in the measuring set itself. This will lead to a reduction in the time taken to transfer result information over the data link, reduce the system processing time, and effectively speed up the measurement rate.

With an unintelligent (or "dumb") SLMS, it is necessary for the controller to send the frequency at which a measurement is to be made and then have the set return the level measured. This procedure must be repeated for each measurement in a sequence, as shown diagrammatically in Fig. 8.35(a). The processor and I/O ports of the controller are, therefore, in almost continuous operation, with the further disadvantage that the effective measurement rate is reduced for remote stations because of the time taken to transfer both measurement setup and result information over the data link. Even where the controller is colocated with the measuring equipment, the data-transfer-time overhead can significantly reduce the measurement rate.

With a measuring set having built-in intelligence, the controller need only initiate a measurement sequence—for example, an FDM or frequency spectrum—by sending the setup information (FDM values, test-point equalization curve coefficients, level limits, and measurement type) at the start of the sequence and let the measuring set perform the sequence, returning only those results that violate limit conditions, as shown in Fig. 8.35(b). The system processor is then involved to a considerably lesser degree in handling I/O transfers, thereby allowing more measuring stations to be handled by one controller and yet leaving sufficient time to perform data analysis.

As a benchmark, Fig. 8.36 shows the relative times taken to use the techniques just described to make the same 300 channel measurements and detect two out-of-limits conditions with both "dumb" and "intelligent" SLMSs controlled over various communication links.

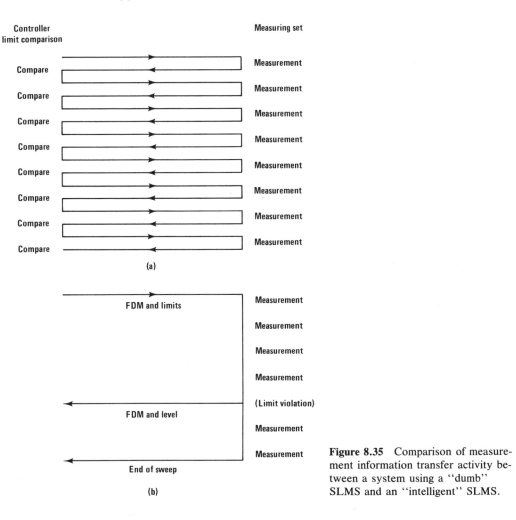

Figure 8.35 Comparison of measurement information transfer activity between a system using a "dumb" SLMS and an "intelligent" SLMS.

8.6.3 Result Reporting

Even with a data-transfer reduction, this measurement may not fully satisfy the needs of a user. Assuming that the parameters in violation of limits remain in violation for an extended period of time, the user will be informed of this situation each time the test equipment measures these parameters. In some cases, this continual identification of violations may be desirable, but generally it is beneficial if the user is presented with information on parameter violations as they are detected and then receives further information only when the parameters return within limits. In other words, the system should provide some form of status

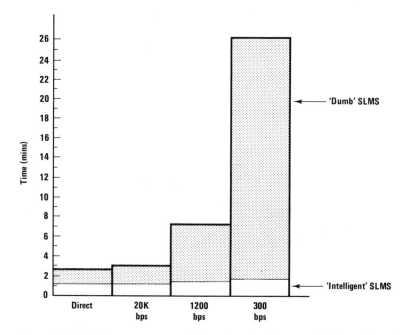

Figure 8.36 Relative measurement times to make the same 300 channel measurements and detect two out of limits conditions using a "dumb" SLMS and an "intelligent" SLMS.

indication. For example, the FDM surveillance software could provide three modes of operation:

- Record
- Limits
- Status

These modes restrict the amount of information returned from the measuring stations and the amount presented to the user to only absolutely necessary information.

The record mode provides a list of every measurement made at a station. In addition, if upper and lower level limits are specified, an out-of-limits message can be printed beside the results of measurements that violate those limits, as shown in Fig. 8.37(a). The limits mode provides a list of only those measurements that violate the level limits and, unlike the record mode, results that are within limits are not returned from the measuring set and hence are not output to the user, as shown in Fig. 8.37(b). The status mode compares the limit conditions of a current set of results with those of the previous set of results for the same measurement type and provides an output only when a change of limit status is detected. In other words, a status change would be indicated only when a measurement that was within limits went out of limits, or vice versa.

```
SN1::12  TP1  16:05  19 JAN 1982  GRPR

NOM LEV = -20.00  UPPER LIM =    2.00  LOWER LIM =    2.00


        FDM      FREQ(kHz)  LEV(dBm)  LEV(dBmO)
      4  1  0     1028.08    -52.60    -19.60
      4  2  0      980.08    -53.50    -20.50
      4  3  0      932.08    -52.60    -19.60
      4  4  0      884.08    -52.60    -19.60
      4  5  0      836.08    -52.90    -19.90
      5  1  0     1276.08    -81.40    -48.40 OUT OF LIMITS
      5  2  0     1228.08    -52.80    -19.80
      5  3  0     1180.08    -52.80    -19.80
      5  4  0     1132.08    -52.90    -19.90
      5  5  0     1084.08    -52.80    -19.80
      6  1  0     1524.08    -52.80    -19.80
      6  2  0     1476.08    -87.90    -54.90 OUT OF LIMITS
      6  3  0     1428.08    -52.80    -19.80
      6  4  0     1380.08    -52.80    -19.80
      6  5  0     1332.08    -52.50    -19.50
      7  1  0     1772.08    -52.70    -19.70
      7  2  0     1724.08    -52.60    -19.60
      7  3  0     1676.08    -52.70    -19.70
      7  4  0     1628.08    -52.70    -19.70
      7  5  0     1580.08    -52.60    -19.60
      8  1  0     2020.08    -52.60    -19.60
      8  2  0     1972.08    -52.60    -19.60
      8  3  0     1924.08    -52.90    -19.90
      8  4  0     1876.08    -52.60    -19.60
      8  5  0     1828.08    -52.60    -19.60




SN1::12  TP1  17:18  19 JAN 1982  GRPL

 NOM LEV = -20.00  UPPER LIM =    2.00  LOWER LIM =    2.00

    FDM      FREQ(kHz)  LEV(dBm)  LEV(dBmO)
  5  1  0     1276.08    -75.50    -42.50 OUT OF LIMITS
  6  2  0     1476.08    -66.80    -33.80 OUT OF LIMITS
```

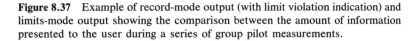

Figure 8.37 Example of record-mode output (with limit violation indication) and limits-mode output showing the comparison between the amount of information presented to the user during a series of group pilot measurements.

The sequence of four sets of measurements on group pilots in Fig. 8.38 serves to explain how the status mode operates. In run 1, the results of all measurements are returned from the measuring set and printed out together with the appropriate status indication. The status conditions are also stored on file in the system controller. In run 2, where pilot 1 goes below the lower limit, pilot 3 comes back within limits, and pilot 5 remains out of limits, only the measurements relating to pilots 1 and 5 are returned from the measuring set. From the previously stored status information (taken on run 1), the system controller identifies a status change on pilot 1 and prints out the measurement details. The next result received is that corresponding to pilot 5. The controller

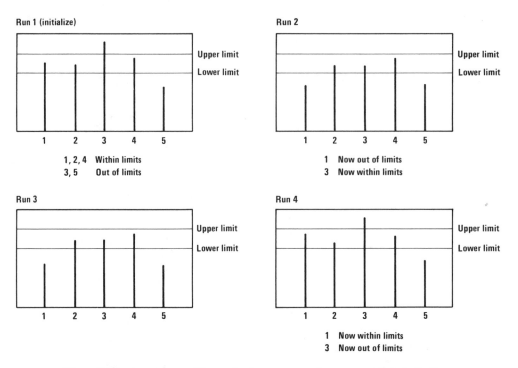

Figure 8.38 A sequence of four sets of measurements on group pilots indicating the information that would be presented for each measurement set.

then scans through the status condition stored for each pilot between 1 and 5 and determines whether any pilots that were out of limits had returned within limits. Any such status changes are printed out. In this particular example, pilot 3 would be identified as having returned within limits.

In run 3, the status of the five pilots does not change, and hence no printout is given. In run 4, pilot 1 comes back within limits and pilot 3 moves out of limits; hence, only these two measurement results are printed out. With a system that provides a status-mode facility and/or the ability to archive results of measurements for future analysis, there are several simple report utilities that can be very useful.

The ability to access the contents of a status file can be used to obtain the current measurement status for any test point for which surveillance measurements (in the status mode) have been made. This can prove beneficial in two ways. By listing on a printer at each station only the out-of-limits measurement conditions held on file, a fault report can be produced for station maintenance staff on a daily basis. Also, a weekly management report can be prepared, giving details of the limit violation percentage for each parameter measured in the network. Such a report might look like this:

Limit Violations	
Group pilots	2.67%
Supergroup pilots	3.33%
Mastergroup pilots	0%
ISG noise	6.67%
Carrier leaks	0.67%

If a network is characterized on a monthly basis by making measurements of all parameters and storing the results, then a utility to list the contents of a set of results is worthwhile. Furthermore, a routine to compare equivalent sets of results will be invaluable, as this can provide an indication of the general trend in the quality of the network. Figure 8.39 shows the output from a typical comparison routine. Pie charts and histograms can also be drawn to represent diagrammatically changes in any particular parameter or overall quality of the network.

8.6.4 System Testing

Diagnosis of faults within a monitoring system itself is of prime importance. A failure in the controller will prohibit the use of the system, leaving only the capability for manual operation of the measuring equipment. A failure in a measurement subsystem, that is, an arm of the system stretching from the controller to a measuring station and including the measuring equipment itself, will prohibit remote operation of that subsystem and, depending on the cause of the failure, may also prohibit manual operation.

An instantaneous failure cannot be predicted, and the only course of action is to initiate a diagnostic procedure to identify the source of the trouble. Equipment degradation can, however, be identified before it affects the system performance by scheduling functional test or verification programs to be run at appropriate times: for example, between measurement sequences, at idle times, or on a periodic basis. If provision is made to store the results of these tests, sequential and absolute deviations can be provided on each test run.

```
                RESULT COMPARISON


1ST RESULT:  SN1     TP2     09:29  27 JAN 1983    SFRQ

2ND RESULT:  SN1     TP2     15:02  27 FEB 1983    SFRQ


                  (1)        (2)      (1)-(2)
FREQ(kHz)      LEV(dBm0)  LEV(dBm0)   DIFF(dB)
27076.00        -11.52     -11.95       0.43
35624.00        -10.81     -10.85       0.04
44424.00        -10.54     -10.65       0.11
```

Figure 8.39 Sample result comparison between two sets of three single point frequency measurements.

The cause of a sudden system or subsystem failure must be identified quickly. To detect the cause of such a failure, a **diagnostic routine,** as opposed to a *functional test routine,* is needed. The diagnostic procedure adopted will depend on the failure symptoms. For example, if a complete surveillance system becomes nonoperational, there is a very high probability that the cause of failure is in the system controller, and this would be the first section of system equipment checked. If only one measurement subsystem is inoperable in a multistation system, the failure could lie in either the interface-communication link or in the measuring equipment. As a rule, the system configuration will play a significant part in defining the diagnostic procedure.

8.7 SAMPLE APPLICATIONS

The operation of most FDM monitoring systems is fundamentally very similar. In the following text, two specific applications are described briefly: a microwave network-monitoring system and an undersea cable-transmission testing system.

8.7.1 Microwave Network Monitoring

Many communications companies using microwave transmission as the backbone of their networks use an FDM-baseband monitoring system as one of the automated systems deemed necessary for network surveillance. The main purpose of such a monitoring system is to provide a means of detecting and controlling high levels in the transmission network.

Before describing how such a system might function, it is important to note one significant difference about the baseband access that may be required when monitoring a microwave network. Unlike a conventional cable transmission system, where access can be made at baseband or lower levels of the multiplex hierarchy, a microwave network may require access to be provided at repeater stations where there is no direct baseband access. At such stations, the information signal is normally brought down to IF (e.g., 70 MHz) before being modulated back up to RF for transmission over the next microwave hop (see Fig. 8.40). In such cases, the FDM baseband is not directly accessible and hence an IF demodulator unit is required. If the microwave system carries several channels, then an equivalent number of demodulators would normally be required to de-modulate the individual signals down to baseband before connecting them to a baseband-access switch. An alternative method using an access switching arrangement having a frequency range up to 90 MHz, for example, is more practical. In this case, the individual radio channel test points are connected to the input ports of the switch; the output of the switch is then fed via one IF demodulator to the input of the measuring set, as shown in Fig. 8.41. (It could be argued that access at IF is not of major importance and that measurements need be made only where channels are brought down to baseband. Although there is some validity to this argument where there are only relatively few hops between

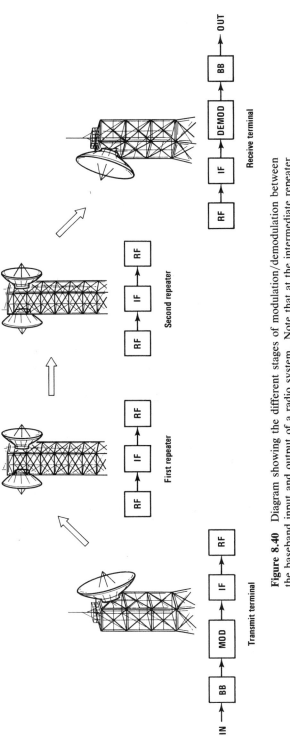

Figure 8.40 Diagram showing the different stages of modulation/demodulation between the baseband input and output of a radio system. Note that at the intermediate repeater stations, the signal is not normally demodulated down to baseband.

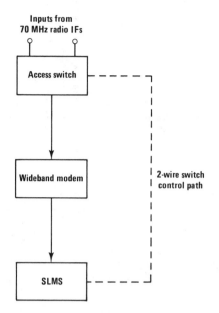

Inputs from
70 MHz radio IFs

Access switch

Wideband modem

2-wire switch
control path

SLMS

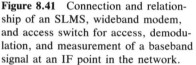

Figure 8.41 Connection and relationship of an SLMS, wideband modem, and access switch for access, demodulation, and measurement of a baseband signal at an IF point in the network.

terminal offices, the need for IF access can be recognized in the case of a radio link traversing many hundreds of miles and where sectionalizing poor transmission performance can be impossible without intermediate monitoring sites.)

The typical operation of a microwave monitoring system is based on several scanning routines, each aimed at providing emphasis on different parameters that are initiated automatically at certain times, depending on their function. For example, a *background scan* might measure all pilot levels, **intersupergroup** (ISG) noise, and baseband spectrum on a continuous basis at each measuring location, with status changes being reported at the control center. A *busy-hour scan,* involving the measurement of group powers (and subsequently channel powers in a group showing a high level), is scheduled by the computer during busy hour, as its name suggests. A *detailed scan* comprising measurement of carrier leakage and channel power can be initiated on a weekly or monthly basis, for instance.

In addition to the three surveillance routines, all measuring sets could be accessed by network-management personnel or technicians at designated plant offices for demand measurements for the purposes of trouble-shooting or maintenance activities. Normally, the measuring sets would be under control of one of the scan routines; only a request for a demand measurement would stop the automated routines and then only for the duration of the demand measurement.

The sequence of the background scan routine for each station is to measure, on the first baseband, all pilot levels and ISG noise, followed by a spectrum search for spurious tones; the scan then advances to the next baseband and performs the same set of measurements. This sequence is repeated until all basebands have been scanned; then the procedure is repeated ad infinitum until interrupted for a demand measurement or one of the other scan routines.

Before leaving this microwave monitoring system example, it is worth noting the importance that has been placed on the ISG noise measurement through practical application. The importance of this measurement stems from the fact that it provides an assessment of the quality of the microwave path and detects deteriorating conditions that will ultimately, if not corrected, seriously affect all traffic, in comparison to other high-level conditions that will normally affect only medium- or high-speed traffic. As a result of ISG noise measurements, several problems have been identified in practice, such as deteriorating traveling wave tubes (TWTs), poor grounding, improper shielding, and even high loading on supergroups. It has also been known to identify misaligned antennas.

8.7.2 Undersea Cable Transmission Testing System

Submarine cable systems are usually laid to carry traffic on international or intercontinental routes. Automatic transmission testing and systems are required to make performance measurements on such cable systems and store the results of the appropriate measurements over the service life of the cables. The following are the main tests to be carried out on undersea cables:

1. Attenuation and frequency response of the transmission system
2. Repeater supervisory response
3. In-station surveillance

Figure 8.42 shows the block diagram of a typical test system, comprised of a computer system and measurement subsystem located at each end of a cable and linked by a high-speed data circuit. Figure 8.43 shows the configuration of the measuring equipment, comprised of a stimulus section (generator, switch controller, and distribution-switching arrangement) and a response section (selective level measuring set, counter, switch controller, and access-switching arrangement).

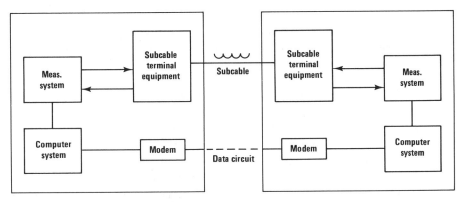

Figure 8.42   Block diagram of a typical undersea cable test system showing the connection of the various elements.

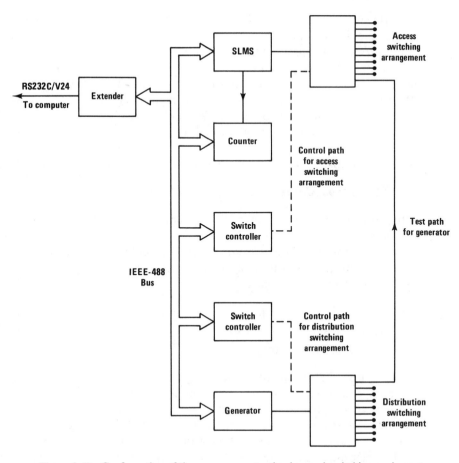

Figure 8.43 Configuration of the measurement, stimulus, and switching equipment used at each end of an undersea cable test system.

Access to a submarine cable system for testing purposes is usually provided at several points. The specific number of test points and their locations will depend on the cable system manufacturer, but, in general, they will correspond to those shown in Fig. 8.44. Testing is normally carried out at the baseband level; however, there may be requirements to test at lower levels in the FDM hierarchy—for example, between hypergroup-in and hypergroup-out. In one direction, the baseband is frequency translated to occupy a different spectrum on the line. In the majority of new systems, high frequencies are transmitted in the B-to-A direction and low frequencies in the A-to-B direction.

The basic attenuation and frequency response measurement is carried out by injecting individual tones at intertraffic frequencies using a generator and measuring the level using an SLMS. Two modes of measurement operation are

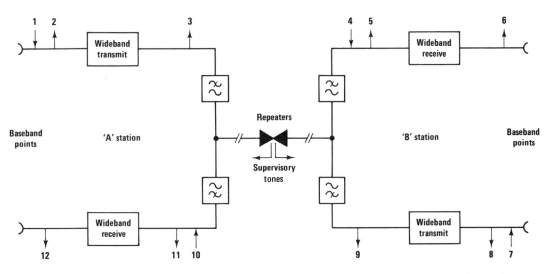

Figure 8.44 Simplified diagram of the structure of an undersea cable system highlighting the location of injection/measurement test points.

required—absolute and relative. In the **absolute mode,** the SLMS simply measures the level of tones output from the generator. For example, by injecting signals at point 1 (Fig. 8.44) and making measurements at point 6, the overall system attenuation and frequency response is obtained. In the **relative mode,** reference measurements are made using the SLMS in the same subsystem as the generator, and final measurements made using the SLMS at the other end of the cable. The results are the difference (in decibels) between the final measurements and the reference measurements. For example, by injecting signals at point 1 and making measurements at points 3 and 5, the loss or gain characteristics of the submerged plant can be established. Similarly, by using test points in the range 7–12, the attenuation and frequency response in the B-to-A direction can be measured.

Each repeater (submerged amplifying unit) in the submarine cable contains two discrete oscillators with frequencies chosen such that the signals are injected in their respective transmission bands and returned to the terminal stations. The automatic test system is required to measure the discrete level and frequency output from each repeater. As the actual frequency of a repeater may drift due to temperature changes in the ocean, the measuring equipment has to tune onto the received signals.

The general method of operation is to tune the SLMS to the last frequency at which a repeater was measured or, in the case of the initial set of measurements, at the ideal repeater frequency adjusted by the commissioning offset frequency. This measurement is made using the pilot (narrow) filter of the SLMS. If the level is within the assigned measurement limits, the frequency of the tone is measured using the frequency counter (see Fig. 8.45).

If the measured frequency is not within ±10 Hz of the frequency, the SLMS is tuned and then another level measurement is made with the SLMS tuned to the offset calculated from the previous frequency plus or minus the off-set frequency. This ensures that the SLMS will capture the tone within the flat top of the pilot filter and not on the skirts of the filter. If the level of the first measurement is not within the assigned limits, a spectrum measurement in 20-Hz steps is initiated around the repeater frequency measured during the last set of repeater-measurement results. When the tone is detected by the measured level being within assigned limits, the exact frequency is measured. If no tone is detected within the measurement limits during the spectrum measurement, a message is output and the repeater measurement program moves on to the next repeater.

```
SN2      TP6      09.30    07 Mar 1983

Reference:  SN1   TP4

FREQ(kHz)          LEV (dBm0)          LEV (dBm0)          GAIN (dB)

  724.00            -23.14              -23.17              -0.03
  628.00            -27.12              -26.08               1.04
  580.00            -23.68              -23.67               0.01
  532.00            -24.02              -24.02               0.00
  972.00            -23.18              -23.18               0.00
  924.00            -24.72              -24.87              -0.15
  876.00            -24.59              -24.38               0.21
  828.00            -24.05              -23.95               0.10
```

Figure 8.45 Sample set of attenuation/frequency response measurements using the relative mode to provide the difference (gain) between the current measurements and the previous measurements.

A typical result output is shown in Fig. 8.46. In addition to the attenuation and frequency response and repeater supervisory response measurements, the test system would be required to measure the following:

1. System pilot levels at test points 6 and 12
2. Traffic pilots at test points 2, 6, 8 and 12—that is, at the inputs and outputs of the submarine cable system
3. The power in the traffic spectrum at the input to the submarine cable system at test points 2 and 8

While the particular system examples mentioned in this chapter relate to FDM network surveillance, similar capabilities could be provided for digital

```
                    37013A REPEATER TONE SUBSYSTEM
                    STATION 1 --- STATION 2 [S/CABLE]
                  RPTR SUPY RESULTS BLOCKSKIP F/R=20 S/R=2

                   ACCESS POINT : TP1     - LOW BAND

                 PRINT MEASUREMENT FOR 10:00   24 MAR 1981
```

		- F R E Q U E N C Y -					SECTION
REPEATER		IDEAL	ACTUAL	DIFF	LEVEL	LOSS	GAINS
CODE	NO.	(kHz)	(kHz)	(Hz)	(dBm)	(dB)	(dB)
A	0	1776.000	1775.999	-1	-64.91	-0.32	1.83
HD	1	1584.800	1584.799	-1	-66.74	-0.53	0.75
HC	2	1584.390	1584.391	1	-67.49	-0.42	-0.03
HB	3	1583.980	1583.987	7	-67.46	-0.39	-0.09
HA	4	1583.620	1583.617	-3	-67.37	-0.39	0.20
GZ	5	1583.200	1583.203	3	-67.57	-0.38	0.33
GY	6	1582.800	1582.799	-1	-67.90	-0.35	0.52
GX	7	1582.380	1582.384	4	-68.42	-0.36	0.24
GW	8	1582.000	1581.999	-1	-68.66	-0.35	0.26
DN	9	1550.000	1549.999	-1	-68.92	-0.35	0.22
DM	10	1549.600	1549.600	0	-69.14	-0.34	-0.19
DL	11	1549.210	1549.208	-2	-68.95	-0.34	1.40
FW	13	1572.400	1572.400	0	-70.35	-0.33	0.99
FU	15	1571.610	1571.616	6	-71.34	-0.35	-0.01
FS	17	1570.800	1570.801	1	-71.33	-0.31	1.71
FQ	19	1570.000	1569.997	-3	-73.04	-0.34	-0.78
FN	21	1569.200	1569.198	-2	-72.26	-0.34	0.39
FL	23	1568.410	1568.406	-4	-72.65	-0.37	0.53
FJ	25	1567.610	1567.608	-2	-73.18	-0.33	-6.72
FG	27	1566.810	1566.813	3	-66.46	-0.32	1.41
FE	29	1565.990	1565.987	-3	-67.87	-0.33	0.27
FC	31	1565.590	1565.594	4	-68.14	-0.32	-0.10
FB	33	1564.800	1564.797	-3	-68.04	-0.34	0.56
FA	35	1564.400	1564.398	-2	-68.60	-0.32	

Figure 8.46 Sample set of repeater measurement results showing the frequency and level values for each repeater together with the gain between each repeater.

surveillance, radio monitoring, and private-line access and test, providing a completely integrated communications surveillance and test system.

APPENDIX 8.1

8.A.1 Gaussian and Chi-Squared Distributions

The Gaussian probability density function is used to describe the statistical behavior of a random variable or noise source with zero mean and mean-squared value of σ^2. It describes the probability of the signal having any specified instantaneous value.

Of more relevance to us is the probability of a noise signal exceeding a threshold value. This function is defined by erfc(x); see Table 8.3. Since the Gaussian distribution is symmetrical and the total probability of a signal across its full amplitude range is 1, then the probability of a noise signal exceeding its mean is erfc(O) = 0.5.

Erfc(x) can be used to find the probability of a noise signal exceeding any value, remembering that σ, the standard deviation of the Gaussian variable, is the rms value of the noise signal. For example, what is the probability of a noise signal exceeding the peak level of a sine wave of the same power? Since the powers are the same, V_{rms} (sinewave) = V_{rms} (noise). Also, V_{peak} (sine) = 1.414 V_{rms}. Therefore, look up erfc(1.414) to find the probability of 7.8%.

Chi-Squared Variables

If we have N independent random variables, each with the same mean and variance, then the composite signal equal to the sum of these variables is the chi-squared variable with N degrees of freedom. The probability density function (pdf) of chi-squared is tabulated in Table 8.5.

To measure the average power of a noise signal, a number of independent estimates of the signal are averaged. If the noise signal had infinite bandwidth, then its autocorrelation function would be an impulse at $t = 0$, so any two consecutive estimates of the noise power would be totally uncorrelated or completely independent. When the noise signal is bandlimited, its autocorrelation function becomes a triangle of finite width. In a time interval T, $2BT$ independent estimates of the power of a noise signal having a bandwidth B can be obtained. So the pdf of chi-squared with $n = 2BT$ degrees of freedom can be used to describe the statistics of the measurement process. For example, we have a weighted filter with noise bandwidth of 1.74 kHz. We measure noise through this filter with an rms detector having an integration time of 35 ms. The uncertainty that can be expected from this measurement is $2BT = 2(1740)(0.035) = 120$ degrees of freedom (approximately). From Table 8.5, we find that 99% of the distribution lies above a chi-squared value of 86.92 and 1% lies above a chi-squared value of 158.9.

Remembering that chi-squared is the sum of the independent variables, each having unit variance or rms value of 1.0, we divide by the number of estimates to obtain a spread of 99% of our readings above 0.724 of the true value and 1% above 1.324 of the true value. So 98% of the results will lie between 0.724 and 1.324 of the true result. Expressed in decibels, we have 98% confidence that our result is correct within +1.22 dB and −1.4 dB, or a spread of 2.6 dB.

The formula quoted earlier was 98% spread within $29/\sqrt{2BT - 1}$, which equals 2.65 dB in this case. For fewer degrees of freedom, the formula is less accurate, so Table 8.5 and the method just given should be used.

TABLE 8.5 Chi-Squared Distribution

Value of $\chi^2_{n;\alpha}$ such that $\text{Prob}[\chi^2_n > \chi^2_{n:\alpha}] = \alpha$

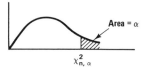

$\chi^2_{n,\,\alpha}$

Area $= \alpha$

					α					
n	0.995	0.990	0.975	0.950	0.900	0.10	0.05	0.025	0.010	0.005
1	0.000039	0.00016	0.00098	0.0039	0.0158	2.71	3.84	5.02	6.63	7.88
2	0.0100	0.0201	0.0506	0.103	0.211	4.61	5.99	7.38	9.21	10.60
3	0.0717	0.115	0.216	0.352	0.584	6.25	7.81	9.35	11.34	12.84
4	0.207	0.297	0.484	0.711	1.06	7.78	9.49	11.14	13.28	14.86
5	0.412	0.554	0.831	1.15	1.61	9.24	11.07	12.83	15.09	16.75
6	0.676	0.872	1.24	1.64	2.20	10.64	12.59	14.45	16.81	18.55
7	0.989	1.24	1.69	2.17	2.83	12.02	14.07	16.01	18.48	20.28
8	1.34	1.65	2.18	2.73	3.49	13.36	15.51	17.53	20.09	21.96
9	1.73	1.09	2.70	3.33	4.17	14.68	16.92	19.02	21.67	23.59
10	2.16	2.56	3.25	3.94	4.87	15.99	18.31	20.48	23.21	25.19
11	2.60	3.05	3.82	4.57	5.58	17.28	19.68	21.92	24.73	26.76
12	3.07	3.57	4.40	5.23	6.30	18.55	21.03	23.34	26.22	28.30
13	3.57	4.11	5.01	5.89	7.04	19.81	22.36	24.74	27.69	29.82
14	4.07	4.66	5.63	6.57	7.79	21.06	23.68	26.12	29.14	31.32
15	4.60	5.23	6.26	7.26	8.55	22.31	25.00	27.49	30.58	32.80
16	5.14	5.81	6.91	7.96	9.31	23.54	26.30	28.85	32.00	34.27
17	5.70	6.41	7.56	8.67	10.08	24.77	27.59	30.19	33.41	35.72
18	6.26	7.01	8.23	9.39	10.86	25.99	28.87	31.53	34.81	37.16
19	6.84	7.63	8.91	10.12	11.65	27.20	30.14	32.85	36.19	38.58
20	7.43	8.26	9.59	10.85	12.44	28.41	31.41	34.17	37.57	40.00
21	8.03	8.90	10.28	11.59	13.24	29.62	32.67	35.48	38.93	41.40
22	8.64	9.54	10.98	12.34	14.04	30.81	33.92	36.78	40.29	42.80
23	9.26	10.20	11.69	13.09	14.85	32.01	35.17	38.08	41.64	44.18
24	9.89	10.86	12.40	13.85	15.66	33.20	36.42	39.36	42.98	45.56
25	10.52	11.52	13.12	14.61	16.47	34.38	37.65	40.65	44.31	46.93
26	11.16	12.20	13.84	15.38	17.29	35.56	38.88	41.92	45.64	48.29
27	11.81	12.88	14.57	16.15	18.11	36.74	40.11	43.19	46.96	49.64
28	12.46	13.56	15.31	16.93	18.94	37.92	41.34	44.46	48.28	50.99
29	13.12	14.26	16.05	17.71	19.77	39.09	42.56	45.72	49.59	52.34
30	13.79	14.95	16.79	18.49	20.60	40.26	43.77	46.98	50.89	53.67
40	20.71	22.16	24.43	26.51	29.05	51.81	55.76	59.34	63.69	66.77
60	35.53	37.48	40.48	43.19	46.46	74.40	79.08	83.30	88.38	91.95
120	83.85	86.92	91.58	95.70	100.62	140.23	146.57	152.21	158.95	163.65

For $n > 120$, $\chi^2_{n;\alpha} \approx n\left[1 - \dfrac{2}{9n} + z_\alpha \sqrt{\dfrac{2}{9n}}\right]^3$, where z_α is the desired percentage point for a standardized normal distribution.

APPENDIX 8.2

8.A.2 Overview of the IEEE 488 Instrumentation Interface Standard

In many areas, the testing of telecommunication networks is moving rapidly toward a system environment. One example of this trend can be seen in the operational maintenance of FDM networks (Chapter 7) where, since the mid-70s, there has been a significant shift toward centralized maintenance systems with many monitoring sites reporting the results of measurements into a network-control center. Such systems cannot function unless the individual elements—instruments, computers, and peripherals—can communicate effectively with one another. There must be a way to tell an instrument (for example, a SLMS) what to do (measure the level of mastergroup pilot 1) and when to do it. There must also be a way for an instrument to tell what it has accomplished (level is -53.4 dBm). This communication, or transfer of information, is accomplished via some form of interface. Although many forms of interface have been developed over the years, one interface in particular has recently gained acceptance as a standard for instrumentation. This is the IEEE Standard 488-1978, *Digital Interface for Programmable Instrumentation*. Because most telecommunications test products aimed at the systems market that have been developed since the mid-70s use this interface, it is worthwhile gaining an understanding of its capability and operation, even if only in broad terms. A condensed description of the interface is provided in this appendix; however, any reader wishing to gain a more detailed understanding should read the references given at the end of the appendix.

Mechanical and Electrical Characteristics

Passive cabling ties the system instruments in parallel to form a common transmission link. Each cable comprises 24 conductors and has both male and female connectors at each end (Fig. 8.47) allowing cables to be stacked in a piggyback arrangement (Fig. 8.48). The total length of cabling in a system is governed by electrical considerations to 2 m times the number of devices in the system or 20 m, whichever is less. Therefore, in a 5-instrument system, the maximum cable length permitted would be 10 m, and in a 12-instrument system, it would be 20 m.

Figure 8.47 IEEE-488 interface connection cable.

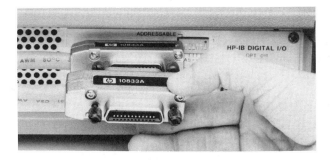

Figure 8.48 Photograph showing how interface cables can be piggybacked on each other.

The interface is a "ground-true" system, which operates at Test Tone Level (TTL) voltages with all signal lines terminated with a resistive divider load that is far greater than the impedance of the interconnecting cable, thereby distributing the load among all the system devices.

Device Roles

To permit communication to take place over the interface, each device connected to the bus has to act in an assigned role. These roles—**talker, listener,** and **controller**—provide the basis for the orderly flow of information on the interface:

- A talker is a device that has the ability to transmit data over the interface.
- A listener is a device that has the ability to receive data over the interface.
- A controller is a device that has the ability to manage the information flow over the interface.

The action of a controller in a system is to direct other devices to talk, listen, or perform other interface functions. Although there can be only one controller in operation at any time, the job of controlling can be passed from one device to another, provided, of course, that these devices contain the controller interface function. It is not, however, essential for a system to contain a controller; a minimum system using the interface may contain only one talker-type device and one listener-type device. For example, the devices could be a SLMS and a printer, with the SLMS set always to talk and the printer set always to listen. In this way, the result of each measurement taken by the SLMS can be sent to the printer without the need for a controller to assign the devices their particular roles.

Addressing

To assign a talker and listeners to become active, the controller uses a technique known as **addressing.** Every talker and every listener has an identifying code called an **address,** and the controller uses these addresses to designate active talkers and listeners; the controller can also address itself to talk or listen, if required.

When assembling a system, each device is assigned an address to distinguish it from other devices on the interface. This is typically done by means of a bank of switches either on the rear panel of an instrument (Fig. 8.49) or inside an instrument. (The address in an interface system is similar to a telephone number in a telephone system. Telephone numbers allow the central switchboard to route calls to the correct telephones; similarly, the controller uses device addresses to select the talker and listeners on the interface). To permit more than one device at any time to function as a listener, a device becomes an active listener when it receives its listen address, and it remains a listener until the unlisten command is transmitted. In the same way, a device becomes an active talker when it receives its talk address but stops functioning as such whenever another talk address or the untalk command is transmitted.

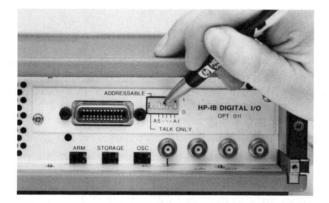

Figure 8.49   Photograph of a typical set of device address switches.

Interface Operation

The interface comprises a set of 16 signal lines, each of which fulfills at least one function. These lines may be classified into three distinct component buses, as shown in Fig. 8.50. The data bus consists of the eight data input/output (DI/O) lines, which are used for sending information between devices. The transfer control, or handshake, bus is made up of the lines data valid (DAV), not ready for data (NRFD), and not data accepted (NDAC). These signal lines are used to coordinate the flow of information on the data lines from a sender to a receiver. The remaining five lines, remote enable (REN), interface clear (IFC), attention (ATN), service request (SRQ), and end or identify (EOI) are used to manage the flow of information over the interface.

The information carried by the interface can be split into two broad categories:

1. Device-dependent messages, which comprise control data, measurement data, and status data

2. Interface-management messages, which are used to manage the operation of the system and control the flow of device-dependent messages

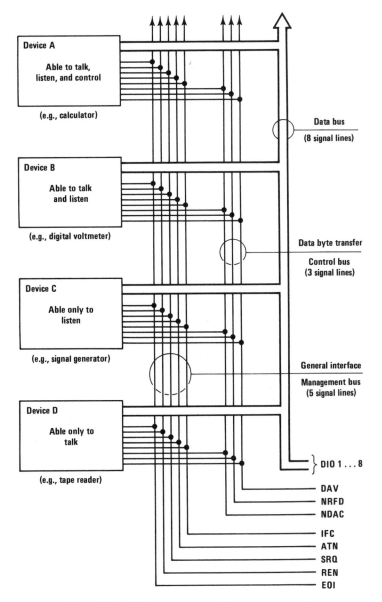

Figure 8.50   IEEE-488 interface signal lines and device functions.

The data bus carries all the coded interface messages and device-dependent messages bidirectionally in a bit-parallel, byte-serial form. Eight lines accommodate the widely used 7-bit American Standard Code for Information Interchange (ASCII) code, leaving one bit that can be used for parity, if required. Also, most computers have word lengths that are multiples of 8 bits, and there are software advantages in having bus bytes and computer words in an integral ratio.

Information on the DI/O lines, whether command or data, is transferred asynchronously under control of a technique known as the **three-wire handshake.** This provides the system with great flexibility, enabling devices of different I/O speeds to be interconnected on the interface, with the speed of transfer adjusting automatically to the rate of the slowest active device. The DAV line is driven by the source of information to indicate that the data on the DI/O lines is valid. The NRFD and NDAC lines are driven by acceptors of information. When the NRFD line goes high, it indicates that all acceptors are ready for data; similarly, when the NDAC line goes high, it indicates that all acceptors have accepted data.

Figure 8.51 illustrates the handshake-timing sequence for one talker and multiple listeners. The NRFD and NDAC signals each represent composite waveform resulting from two or more listeners accepting the same byte of information at different times. This variation in timing is due to the response rates of the individual instruments and variations in the transmission path length.

A listener indicates that it is ready to accept data by letting the NRFD line go high. Listeners are connected to this line in a logic-AND configuration, so the line itself will not go high until all active listeners are ready to receive data. When NRFD does go high, it indicates to the talker that all active listeners are ready for data. The talker then indicates that it has placed a byte on the data lines by setting the DAV line low; this, in turn, enables listeners to accept the message on the data lines. When a listener has accepted data, it lets the output of its NDAC line go high, but, as with the NRFD line, all listeners are connected to the NDAC line in a logic-AND configuration, the line will not go high until

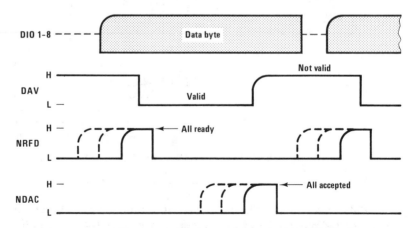

Figure 8.51 Handshake timing sequence showing the relationship between the listener-controlled lines (NRFD and NDAC) and talk-controlled lines (DAV and DIO 1-8).

all the listeners have accepted the data. NDAC remains high until the talker sets the DAV line high. This sequence may then be repeated after NDAC goes low. No action in this series of events can be initiated without the previous step having been completed. In this way, information will be transferred only at the rate of the slowest addressed device.

Both device-dependent (data) and interface-management (command) messages are carried by the eight DI/O lines using this transfer process, and—unless there is to be a restriction placed on the codes used by these types of messages— some means of identification must be provided. The method adopted in this interface is to use the ATN line as a data mode–command mode selector. When ATN is low (true), the codes on the DI/O lines are commands and have the same meaning for all the devices connected on the interface. When ATN is high (false), the codes are pure data and will be understood by devices in their own particular way. The ATN line is driven only by the device in the system which is currently assigned the role of controller. When the controller sets the ATN line true, all other devices must terminate what they are doing and obey the codes that only the controller may then transmit.

The most significant bit of the data bus (DI/O8) is not used when the ATN line is true, and the second- and third-most significant bits are used to gain the meaning of the information contained in the first five bits. Table 8.6 shows the command code sets used on the DI/O lines. Talk and listen address commands select the particular instruments that will transmit and receive data. A specific talk and listen address may be determined by replacing the lower five bits in the address byte with the code for an instrument's device address. For example, an instrument that is capable of functioning as both a talker and a listener and

TABLE 8.6   IEEE-488 Command Sets and Codes Used on the DI/O Lines

8	7	6	5	4	3	2	1	Command type
			DI/O lines					
X	0	0	D_5	D_4	D_3	D_2	D_1	Universal commands
X	0	1	D_5	D_4	D_3	D_2	D_1	Listen addresses
			with the exception of					
X	0	1	1	1	1	1	1	Unlisten command
X	1	0	D_5	D_4	D_3	D_2	D_1	Talk addresses
			with the exception of					
X	1	0	1	1	1	1	1	Untalk command
X	1	1	D_5	D_4	D_3	D_2	D_1	Secondary commands
X	1	1	1	1	1	1	1	Not used

with a device address of 10 would have a talk address of 74 and a listen address of 42 (values in decimal).

Universal commands are interpreted by all devices equipped to identify them in the same manner and result in some specific interface operation being performed. Addressed commands are similar to universal commands, but they are obeyed only by those devices that have previously been addressed. Secondary commands are used to extend, effectively, the number of bits in a message. For example, in the simplest case, 5 bits are available in a device for addressing, allowing a total of 31 different addresses to be given to devices—the 32nd address is associated with the untalk and unlisten commands. However, if a secondary command were to be used to give another 5 bits in the second byte, a total of 961 addresses would be available.

In a multicontroller system, one of the controllers must always be able to take absolute control of the interface without being addressed itself. This controller is designated as the *system controller* and always retains control of two specific bus management lines, namely, REN and IFC. When the REN line is set true, it enables devices to respond to commands from the controller. The IFC line, when set true, terminates all interface activity and causes all devices to unaddress themselves. It is used at the start of a system initialization sequence and also when an operator wishes to abort an automatic sequence.

The EIO line serves a twofold purpose. It can be used by a device functioning as the active talker in a system to indicate the end of a message transmission. When used in conjunction with the ATN line, it is employed to initiate a method of simultaneously checking the status of up to eight devices on the interface. This operation is known as a **parallel poll** and basically allows each instrument in a system to present one status bit to the controller in charge without previously being addressed to talk.

The final interface line is SRQ. This is set low by a device that wishes to gain access to the interface to perform some function that it is not already actively able to accomplish or to indicate some condition, such as a measurement being outside limits. Devices are connected to this line in a logic-OR configuration so that several devices may indicate a request for service at the same time. To determine which instrument or instruments in a system have requested service, the controller responds, at its own discretion, by setting the interface into the serial poll mode.

A **serial poll** is a routine that interrogates each instrument in a system to determine if it requested service. This is achieved by sending the talk address of each instrument, in turn, and receiving from each instrument a single byte (character), known as the **status byte,** which indicates the condition of the instrument to the controller. Once an instrument that has requested service has been polled, it lets the SRQ line go high. However, if more than one device has requested service, the SRQ line will not go high until the last device requesting service has been asked for its status.

The ability for a device to transmit, receive, and act on information from the interface is provided by the implementation of interface functions. Although the electrical, mechanical, and functional elements of the interface are defined in the standard, a designer is given complete freedom to select only those functions required within a particular instrument or device.

Table 8.7 lists the name of the interface functions together with a brief description of each. Various subsets of these functions are available, but for brevity they have been omitted from this description. To summarize the basic features of this interface system, an outline of the specifications is given in Table 8.8.

TABLE 8.7   IEEE-488 Interface Functions and Associated Descriptions

Function	Description
Source handshake	The means of transmitting information on the data bus. Used by a talker when sending device dependent messages and by a controller when sending interface management messages.
Acceptor handshake	The means of accepting information on the data bus. Used by all devices when receiving interface management messages and by listeners when receiving device dependent messages.
Talker	Allows an instrument to send data over the interface.
Listener	Allows an instrument to receive data over the interface.
Controller	Provides an instrument with varying capabilities, which are summarized as follows:
	1. The ability to send the IFC and REN messages.
	2. Provision to respond to service requests by performing the serial poll sequence.
	3. The ability to send interface management messages.
	4. The ability to pass control to another controller and/or receive control from another controller.
	5. The ability to take control, without destroying a transmission of data which may be in progress.
	6. The ability to perform a parallel poll.
Remote-local	Allows control of an instrument to pass from its front panel to the interface. Provision is also made for the locking out of an instrument's manual controls.
Service request	Allows an instrument to indicate some requirement.
Parallel poll	Allows an instrument to return a 1-bit status message to the controller.
Device clear	Provides a means of initializing an instrument to a known condition. All instruments may be cleared together or selectively.
Device trigger	Enables instruments to be triggered into some action, either individually or in a group.

TABLE 8.8 IEEE-488 Interface System Specifications

Function	Description
Number of devices	Any number up to a maximum of 15 on a single contiguous bus.
Signal lines	Sixteen; 8 data lines and 8 management and control lines.
Transmission-path length	20 m
Data rate	Typically 250–500 kb/s over complete path length; 1 Mb/s over limited distances.
Data transfer	Bidirectional in bit-parallel, byte-serial form using three-wire handshake technique for asynchronous operation.
Control	Only one controller at any time. Control may be delegated; it can never be assumed, with a maximum of 1 talker and 14 listeners.
Addressing capability	Primary addresses: 31 talk and 31 listen. Secondary addresses: 961 talk and 961 listen.
Signal levels	TTL compatible. High (false) state >2.4 V. Low (true) state <0.4 V.

REFERENCES

[8.1] Coates, T. "Interfacing to the Interface: Practical Considerations Beyond the Scope of IEEE Standard 488," presented at WESCON Conference, Session 3, September, 1975.

[8.2] Fluke, J. M., Jr. "System Considerations in Using the IEEE Digital Instrument Bus," presented at WESCON Conference, Session 3, September, 1975.

[8.3] *IEEE Standard* 488-1978. "Digital Interface for Programmable Instrumentation." The IEEE, Inc., 345 East 47th St., New York, November, 1978.

[8.4] Knoblock, D. E., D. C. Loughry, and C. A. Vissers. "Insight into Interfacing," *IEEE Spectrum,* May, 1975, pp. 50–57.

[8.5] Loughry, D. C. "IEEE Standard 488 and Microprocessor Synergism," *Proceedings of the IEEE,* Vol. 66, No. 2, February, 1978, pp. 162–172.

[8.6] Myles, R. "IEC Interface: A Microprocessor Based Implementation," presented at IMEKO VII Congress, London, England, Vol. 3, May, 1976.

[8.7] Ricci, D. "IEEE 488: Its Impact on the Design, Building and Automatic Test and Measurement Systems," *WESCON Paper,* May, 1980.

[8.8] Santon, A. "IEEE-488 Compatible Instruments," *EDN,* November 5, 1979, pp. 91–98.

[8.9] Smith, L. "Join 488-Bus Instruments and Efficient Software for Fast, Automatic Tests," *Electronic Design,* Vol. 24, November 22, 1979, pp. 142–148.

[8.10] Young, R. "Implementing an IEEE-488 Bus Controller with Microprocessor Software," *IEEE Transactions on Industrial Electronics and Control Instrumentation,* Vol. IEC-27, No. 1, February, 1980, pp. 10–15.

[8.11] Gauntlett, R. D. "The Improvement of Availability in the U.K. Trunk Transmission Network," *IEEE Conference Publication* 131, September, 1975, pp. 27–31.

[8.12] CCITT Rec. G.232. *Green Book,* Vol. 2, pp. 145–157.

[8.13] CCITT Rec., "Some general observations concerning measuring instruments and measuring techniques," *Green Book,* Volume 4-2, 1973, pp. 463–471.

[8.14] CCITT Rec. G.135. *Green Book,* Volume 3-1, 1973, p. 57.

[8.15] Bradford, K. "Zero-Loss Switching for Communications Test Systems," *Communications International,* Vol. 4, No. 9, September, 1977, pp. 44–51.

[8.16] Zverev, A. I. *Handbook of Filter Synthesis,* John Wiley, New York, 1962.

[8.17] Schiesser, W. E. *Statistical Uncertainty of Power Spectral Estimates,* Weston-Boonshaft and Fuchs—Bulletin 711-C-1.

[8.18] Hamsher, D. H. *Communications System Engineering Handbook,* McGraw-Hill, New York, 1982.

9

ANALOG MICROWAVE DIAGNOSTIC MEASUREMENTS

GUY DOUGLAS

Hewlett-Packard Ltd.

9.1 INTRODUCTION

The white-noise-loading technique is normally used as the final arbiter in determining analog microwave radio performance. White-noise measurements provide a quantitative measure of that performance in order to make a decision to go or not. No alignment or diagnostic information is readily available using this technique. Meanwhile, swept frequency measurement techniques are normally used to align the various subsections of a radio—for example, the modulator or IF equalizers—in terms of their linearity, group delay, and IF amplitude characteristics or to diagnose faults on a previously working system. The objective of such alignment adjustments is to minimize distortion, but the relationship between deviations in these characteristics and the baseband distortions that result under traffic loading conditions is not obvious. For this reason, white-noise loading, which simulates normal traffic, must be used to confirm the validity of the alignment obtained using swept measurements.

The aim of this chapter is to show that (1) some diagnostic information can be obtained from white-noise measurements, particularly with the availability of automatically controllable instrumentation and low cost computing power, and that (2) some qualitative information can be realized from swept-frequency measurements using rule-of-thumb techniques. The benefits of this are, from (1), that trouble-shooting activity can converge more quickly to the source of the problem and, from (2), that the source of the distortion can be identified to the shape of a particular characteristic and that guidelines for realignment can be developed.

9.2 WHITE-NOISE MEASUREMENTS AS A DIAGNOSTIC AID TO MICROWAVE RADIO TROUBLESHOOTING

Noise power ratio (NPR) is defined in Chapter 8, and measurement standards are laid down by CCIR [CCIR, 9.1]. Traditionally, NPR measurements have been used to provide qualitative information about the performance of microwave radio links. If the NPR specification, say 50 dB, is exceeded in (typically) three slots spanning the baseband, then satisfactory performance of the link system under test has been verified. Using this technique, the link can be monitored for NPR degradation on a routine preventive-maintenance basis.

If, however, the NPR falls below specification in any slot, then clearly something is wrong—but what? Little more information is to be gained from single NPR measurements in each slot. Typically, a **microwave link analyzer** (MLA) would be used at this stage to identify the source of distortion, but it is not clear where to start and what to expect on the MLA traces.

More information can be gained by making a series of NPR measurements at various system loading levels and plotting *V-curves*.* These can then be analyzed to indicate the most likely problem area. The V-curve technique has been known for many years but has not been widely used because the manual processes involved in making the measurements, plotting the points, and then analyzing the results are laborious and cumbersome. With the advent of relatively low-cost computing power, the relevant data can be acquired and analyzed with ease, free from operator error, and analyzed more quickly than by the manual method. For this reason, the V-curve technique is worthy of another look, and this section deals with the diagnostic information that it affords. A typical V-curve is shown in Fig. 9.1. A V-curve is obtained by plotting the measured NPR of a system against received load-level variations both above and below the nominal load level, as shown. Well below nominal load, the NPR is dominated by the idle (no-load) noise of the system and improves by 1 dB per decibel increase in loading level. Above nominal load, intermodulation distortion becomes significant and finally dominates. At or slightly above nominal load, the idle and distortion contributions are approximately equal corresponding to the optimum system NPR.

Figure 9.2 shows the V-curve of a well-behaved system that was measured, plotted, and analyzed automatically using the test arrangement shown in Fig. 9.3 together with a V-curve analysis software package [Hewlett-Packard, 9.2].

---

* Also known as noise-loading curves or bucket curves.

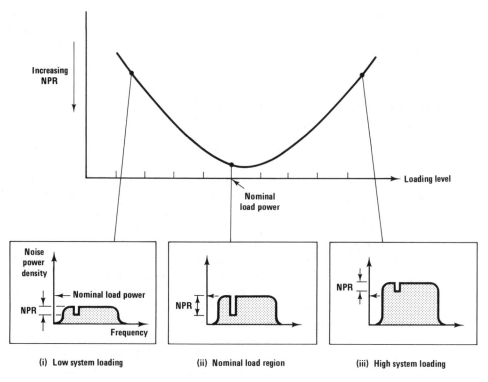

Figure 9.1 Typical V-curve.

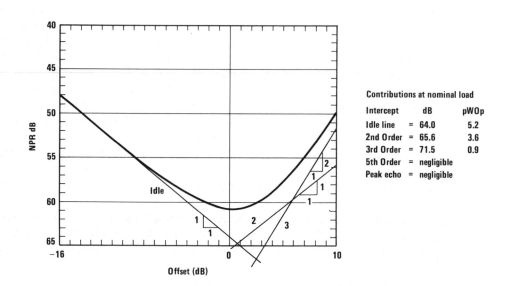

Figure 9.2 Measured V-curve showing idle and intermodulation noise curves.

Figure 9.3 Test set-up for V-curve measurement and analysis.

The software has analyzed the curve to show the various noise contributors in the form of construction lines as follows:

- The idle line, with a slope of -1 dB/dB
- The second-order distortion line, with a slope of 1 dB/dB.
- The third-order distortion line, with a slope of 2 dB/dB.

Each noise contributor line intercepts the nominal load line (0 dB offset), and for a well-behaved radio hop, it has been found by empirical methods that the following statements hold.

- The idle intercept is typically 2 to 3 dB below the system's loaded noise specification.
- The second-order intercept is typically 2 to 3 dB below the idle intercept.
- The third-order intercept is at least 6 dB below the idle intercept.

The basis upon which diagnostic information can be obtained from this analysis is that any noise contributor not meeting the relevant criteria indicates a problem of a very specific type, as shown in the next section.

9.3 EXAMPLE SOURCES OF SECOND- AND THIRD-ORDER DISTORTION

9.3.1 Amplitude (Gain) Imperfections

If second-order noise is dominant or its intercept exceeds the above criterion, then one possibility is that the transfer constant of the modulator or, more generally, of the transmission system is proportional to the input signal voltage.

For a perfect modulator and demodulator pair, the output voltage V_O is directly proportional to the input voltage, V_{in}, that is,

$$V_O = kV_{in} \tag{9.1}$$

where k = the **transfer constant**

If, however, the modulator and demodulator are not perfectly matched such that k is proportional to V_{in}, that is,

$$k = k_1 + e_1 \cdot V_{in} \tag{9.2}$$

where k_1 and e_1 are gain constants, then (9.1) becomes

$$V_O = k_1 \cdot V_{in} + e_1 \cdot (V_{in})^2 \tag{9.3}$$

The undesired distortion term is V_D, where

$$V_D = e_1(V_{in})^2 \tag{9.4}$$

Now if V_{in} consists of two sinusoids of amplitude A and angular frequencies w_1 and w_2, respectively, that is,

$$V_{in} = A \cos(w_1 t) + A \cos(w_2 t) \tag{9.5}$$

then

$$V_D = e_1 A^2 [\cos^2(w_1 t) + 2 \cos(w_1 t) \cdot \cos(w_2 t) + \cos^2(w_2 t)] \tag{9.6}$$

The squared terms yield second harmonics of w_1 and w_2, respectively, and the third term, $2 \cos(w_1 t) \cdot \cos(w_2 t)$, becomes

$$\cos(w_1 + w_2)t + \cos(w_1 - w_2)t \tag{9.7}$$

That is, second-order sum and difference products exist whose amplitudes are proportional to e_1. The distortion power levels are therefore proportional to $(e_1)^2$. A two-tone input has been used, but the result can be generalized to a multitone or white-noise input. This illustrates that a constant slope of e_1 in the gain linearity characteristic of a radio modem or system, as measured on a MLA, generates second-order distortion at a level proportional to $(e_1)^2$. This is revealed in a V-curve analysis by the intercept of the second-order distortion line.

A similar argument can be used to show that a parabolic shape in the linearity characteristic generates third-order distortion products and is revealed in the intercept of the third-order distortion line of a V-curve analysis.

9.3.2 Group Delay Variations in FM and PM Systems

Another possible source of second-order distortion is group delay (GD) variation, which can occur in either the modem section or the carrier section of an FM or PM radio system. Figure 9.4 shows various forms of transmission characteristics for group delay and gain centered on the IF carrier frequency. GD is related to transmission phase by the equation

$$GD = d\phi/d\omega \text{ (seconds)}$$

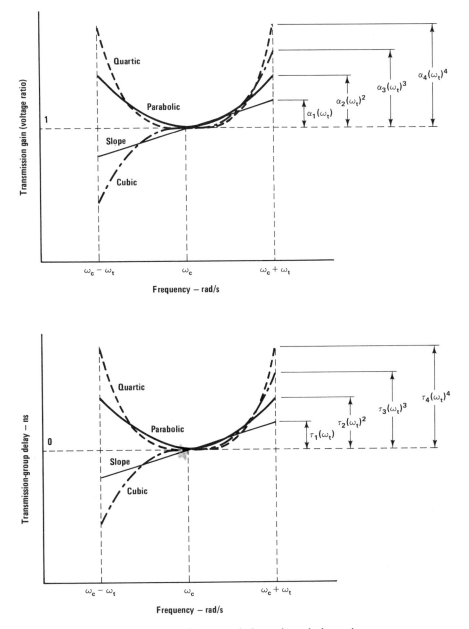

Figure 9.4 Low-order transmission gain and phase shapes

where ϕ is phase and ω is angular frequency. For distortion-free transmission, the time delay through the system should be the same for all frequencies; that is, GD should be constant. This corresponds to a sloping phase characteristic. If, however, parabolic phase, which corresponds to GD slope, is present, then second-order distortion is introduced. This can be reasoned as follows. The normalized transmission characteristic can be derived from Fig. 9.4 as

$$Y_n(w + w_c) = (1 + \alpha_1 w + \alpha_2 w^2 + \alpha_3 w^3 + \alpha_4 w^4)e^{j(\tau_1 w + \tau_2 w^2 + \tau_3 w^3)} \qquad (9.8)$$

where
$$w_c = \text{ carrier frequency in rad/s}$$
$$\alpha_1, \alpha_2, \alpha_3, \alpha_4 = \text{ slope, parabolic, cubic, and quartic gain coefficients, respectively}$$
$$\tau_1, \tau_2, \tau_3 = \text{ slope, parabolic, and cubic GD coefficients, respectively}$$

For FM system applications, it can be shown [Bell Telephone Laboratories, 9.3] that for small deviations,

$$e_2(t) = [1 + P(t)]\cos[w_c(t) + \phi(t) + Q(t)] \qquad (9.9)$$

where $e_2(t)$ = the IF output signal
$\phi(t)$ = the applied phase modulation of the carrier
$P(t)$ = the unwanted amplitude modulation
$Q(t)$ = the unwanted phase modulation

If *GD slope is present*—that is, τ_1 is finite but all other coefficients are zero—then

$$Q(t) = \tau_1 \phi'^2 - \frac{\tau_1^2}{2}\phi'''' \qquad (9.10)$$

where $\phi' = \dfrac{d\phi(t)}{dt}$

and so on. If a limiter can successfully remove the amplitude variations, then $P(t)$ can be ignored. If, however, the amplitude variations reach an AM/PM converter, they become a second source of distortion, and $P(t)$ must be considered.

To find the order of the products generated by GD slope, assume a modulating input signal, $\phi(t) = kV(t)$. Then, from equation (9.10) we have

$$\text{Output distortion} = \phi_0(t) = k^2 \tau_1 [v'(t)]^2 - \frac{k\tau_1^2}{2}V''''(t) \qquad (9.11)$$

The ratio of undesired to desired terms is

$$\frac{\phi_0(t)}{\phi(t)} = k\tau_1 \frac{[V'(t)]^2}{V(t)} - \frac{\tau_1^2}{2}\frac{V''''(t)}{V(t)} \qquad (9.12)$$

The second term has the same frequency components as $V(t)$ and does not vary with deviation sensitivity, k. Therefore, this can be equalized. The first term, however, is proportional to k, and the squared part will generate sum and difference products as in equations (9.5)–(9.7).

This discussion demonstrates that GD slope in FM radio systems *results in second-order distortion* products whose power level is proportional to τ_1^2, where τ_1 is the GD slope.

9.3.3 Distortion Caused by Quartic Gain

If the only transmission deviation now present is the quartic gain, α_4, then equation (9.9) still applies, but this time the unwanted phase modulation becomes

$$Q(t) = \alpha_4\phi'''' - 6\alpha_4\phi''\phi'^2 \tag{9.13}$$

Again substituting $\phi(t) = kV(t)$,

$$\phi_D(t) = \underbrace{k\alpha_4 V''''(t)}_{\text{equalizable}} - 6k^3\alpha_4 V''(t)[V'(t)]^2 \tag{9.14}$$

Ignoring the equalizable term, we have

$$\phi_D(t) = -6k^3\alpha_4 V''(t)[V'(t)]^2$$

That is,

$$\phi_D(t) = -\frac{2k^3\alpha_4 d[V'(t)]^3}{dt} \tag{9.15}$$

and

$$\frac{\phi_D(t)}{\phi(t)} = -2k^2\alpha_4\frac{d[V'(t)]^3/dt}{V(t)} \tag{9.16}$$

The cubed term gives rise to third-order distortion products (e.g., $2w_1 - w_2$ and $2w_2 - w_1$), and this can be directly attributed to the quartic gain present. By examining each case in turn, it can be deduced that: Even- (odd-) order gain and delay transmission deviations cause odd- (even-) order distortion.

9.4 FREQUENCY VARIATION OF DISTORTION CONTRIBUTIONS

The effect of a given transmission deviation, such as GD slope, is not constant with frequency. For example, for a given GD slope, the mid- and high-frequency slots of the baseband of a FDM-FM radio system will exhibit poorer NPR than the lower slots. This is true for a combination of the following reasons:

   1. The effect of pre-emphasis on the baseband.

2. Various powers of frequency terms arising from differentiation, for example, in equations (9.12) and (9.16).

3. Considering second-order products in a radio baseband, more difference products will occur in the low slot than the high slot, whereas more sum products will occur in the high slot than low slot. However, only half of the baseband can produce sum products in the high slot, but all the baseband can produce difference products in the low slot. This gives rise to distortion variations with frequency [Bell Telephone Laboratories, 9.4]. A similar argument can be applied to third-order products.

The typical distribution of intermodulation distortion components in the baseband of a FDM-FM radio system is summarized in Fig. 9.5. When system performance is poor, the following test method suggests itself:

- Measure and plot NPR V-curves for the low, mid, and high slots, respectively.
- Analyze the V-curves in terms of their noise contributors (idle and second- and third-order distortions).

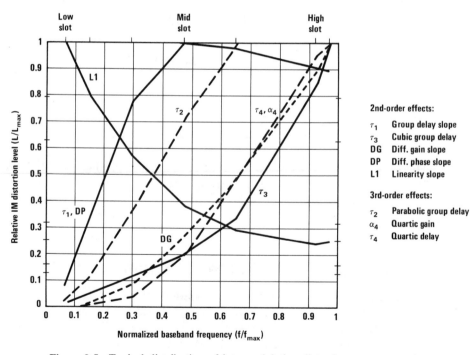

Figure 9.5 Typical distribution of intermodulation distortion components in a FDM-FM radio system.

• Using knowledge of the frequency distribution of and order of distortion caused by transmission parameters, identify the most likely source(s) of the system problem.

Once the curves have been analyzed, the flowcharts in Figs. 9.6 and 9.7 can be used to identify the likely problem source. Information about the mid slot is used in the high-slot flowchart in Fig. 9.7. Note that the cause of idle noise in the high slot may be different from that of idle noise in the low slot. These are usually differentiated as **thermal** and **basic** noise, respectively.

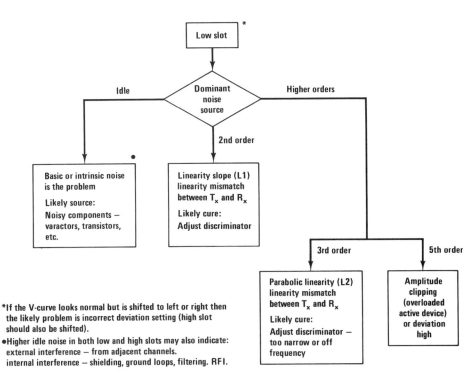

Figure 9.6 Low slot V-curve analysis flow chart in FDM-FM systems.

To measure, plot, and analyze *V*-curves manually is slow, laborious, and prone to operator error; in addition, the analysis requires some level of skill or experience. Automatic test equipment and low-cost controller-calculators are now readily available, however, and the following examples illustrate the use of a commercially available software package that completes the measurement and analysis automatically [Hewlett-Packard, 9.2].

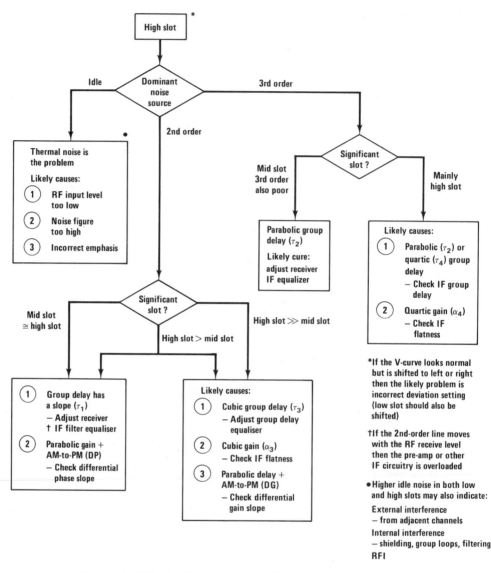

Figure 9.7 High slot V-curve analysis flow chart in FDM-FM systems.

9.5 V-CURVE ANALYSIS EXAMPLES

Example 9.1

Figure 9.8 shows the printout, obtained using a software package [Hewlett-Packard, 9.2], together with the test arrangement shown in Fig. 9.3 for an 1800-channel FDM-FM system. The expected NPR was 60 dB in all slots. Based on the expected NPR and the criteria set out in Section 9.1, the program derives maximum tolerable

(a) **Low slot 534 kHz**

Contributions at nominal load

Intercept	dB	pWOp
Idle line	= 64.9	4.3
2nd order	= 53.9	53.3
3rd order	= 75.7	0.4
5th order	= negligible	
Peak echo	= negligible	

(b) **Mid slot 3886 kHz**

Contributions at nominal load

Intercept	dB	pWOp
Idle line	= 62.3	7.8
2nd order	= 59.7	14.1
3rd order	= negligible	
5th order	= negligible	
Peak echo	= 62.2	

(c) **High slot 7600 kHz**

Contributions at nominal load

Intercept	dB	pWOp
Idle line	= 61.9	8.5
2nd order	= 76.3	0.3
3rd order	= 74.4	0.5
5th order	= negligible	
Peak echo	= 62.6	

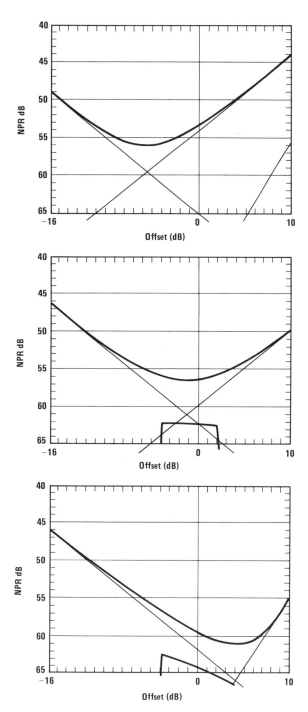

Figure 9.8   V-curve construction for Example 1.

Analysis

Deviation	OK
Thermal noise (8.5pWOp)	HIGH
Low slot distortions	
Linearity slope (53.3pWOp)	HIGH
Para. linearity	OK
Mid/high slot distortions	
2nd-order effects:	
Group delay slope (14.1pWOp)	HIGH
Cubic group delay	OK
Cubic gain	OK
Coupled distortions	OK
3rd-order effects:	
Para group delay	OK
Quartic group delay	OK
Quartic gain	OK
End of analysis	

(d)

Figure 9.8 V-curve construction for Example 1 (continued).

levels for the various noise contributors (idle, second-order, and so on). After plotting the V-curves and identifying the distortion-line intercepts, the program then gives an analysis summary (see Fig. 9.8(d)). Three items are flagged as *high*:

1. Thermal noise, because the high-slot idle intercept occurs at 61.9 dB (should be >62 dB).

2. Linearity slope, because second order in the low slot is well above target, 53.9 against 64 dB.

3. GD slope, because second-order in the mid slot is 59.7 dB against a target of 64 dB.

The noise levels are also given in pWOp. In this example, the most serious distortion source is linearity slope—53.3 pWOp—due to the poor low-slot second-order performance. The analysis indicates, therefore, that the linearity trace, when investigated on a MLA, is expected to exhibit slope. It is possible that the other items flagged are simply manifestations of the linearity misalignment in the other two slots and may disappear once the linearity has been corrected. ∎

9.5.1 Echo

In Fig. 9.8(b) and 9.8(c), the extra curves shown represent the error between the measured V-curve and the power summation of the noise-contributor construction lines. If large amounts of echo are present, this will normally appear as a well-defined inverse parabola. Figure 9.9 gives an example of this. The radio-only V-curve shown here is the power summation of the construction lines. If echo is a significant problem, it should be concluded from the V-curve obtained. It is possible to determine the echo amplitude and approximate delay from the

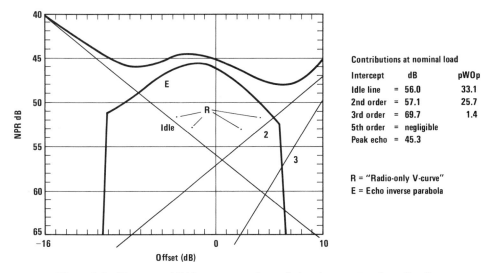

Figure 9.9   V-curve exhibiting severe echo and showing construction of radio-only V-curve and echo inverse parabola.

peak echo NPR, the offset at which it occurs, the slot in which it occurs, and so on [Bennet, 9.5].

Generally, echo in the high slot only indicates an internal feeder echo, possibly caused by a damaged waveguide, whereas far-field long-delay echoes usually also introduce evidence of echo in the low slot and can cause rapidly fluctuating noise readings. This would be indicative of radio-wave reflections from buildings or hills and can be corrected by using larger, more directive antennas.

Example 9.2

Another set of results for an 1800-channel system is given in Fig. 9.10, together with the relevant analysis. The entire program cycle takes 8 to 10 min to complete. In this case, the most significant noise contributor is of second order in both the mid and high slots. From the flowchart of Fig. 9.7, high GD slope is expected. This is confirmed by the calculator analysis (Fig. 9.10(d)), which attributes 225.3 pWOp to GD slope. As in the first example, other contributors flagged in the analysis are consequent to the main problem and will be found to correct themselves when the GD slope is readjusted. ■

Example 9.3

This time the high slot exhibits the poorest performance (Fig. 9.11), which is indicative of a problem in the radio section of the link rather than the modem section. Notice that the high-slot second-order contribution is 5 dB greater than that in the mid slot. This points to one of cubic GD, cubic gain, or parabolic GD + AM/PM as the likely distortion source (see Fig. 9.7). The frequency distribution shown in Fig. 9.5 also makes this clear (it is shown later in the chapter that differential gain (DG) slope is due to cubic gain or the coupled distortion, parabolic GD + AM/PM). In this case, a MLA should now be used to identify which of these three is the source

(a) **Low slot 534 kHz**

Contributions at nominal load

Intercept	dB	pWOp
Idle line	= 64.9	4.3
2nd order	= 58.4	19.1
3rd order	= 76.9	0.3
5th order	= negligible	
Peak echo	= negligible	

(b) **Mid slot 3886 kHz**

Contributions at nominal load

Intercept	dB	pWOp
Idle line	= 62.2	7.9
2nd order	= 47.7	225.3
3rd order	= 62.4	7.5
5th order	= negligible	
Peak echo	= negligible	

(c) **High slot 7600 kHz**

Contributions at nominal load

Intercept	dB	pWOp
Idle line	= 61.5	9.3
2nd order	= 47.9	214.5
3rd order	= 61.4	9.5
5th order	= negligible	
Peak echo	= negligible	

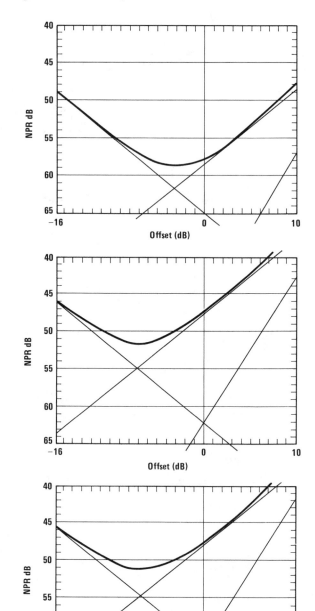

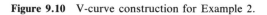

Figure 9.10   V-curve construction for Example 2.

Deviation	OK
Thermal noise (9.3pWOp)	HIGH
Low slot distortions	
Linearity slope (19.1pWOp)	HIGH
Para. linearity	OK
Mid/high slot distortions	
2nd-order effects:	
Group delay slope (225.3pWOp)	HIGH
Cubic group delay	OK
Cubic gain	OK
Coupled distortions	OK
3rd-order effects:	
Para. group delay (9.5pWOp)	HIGH
Quartic group delay	OK
Quartic gain	OK
End of analysis	

(d)

Figure 9.10 V-curve construction for Example 2 (continued).

(a) Low slot 534 kHz

Contributions at nominal load

Intercept		dB	pWOp
Idle line	=	61.7	8.9
2nd order	=	69.6	1.4
3rd order	=	75.0	0.4
5th order	=	negligible	
Peak echo	=	negligible	

(b) Mid slot 3886 kHz

Contributions at nominal load

Intercept		dB	pWOp
Idle line	=	59.3	15.5
2nd order	=	50.0	132.8
3rd order	=	61.4	9.6
5th order	=	negligible	
Peak echo	=	negligible	

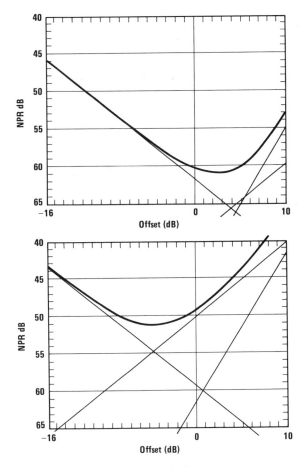

Figure 9.11 V-curve construction for Example 3.

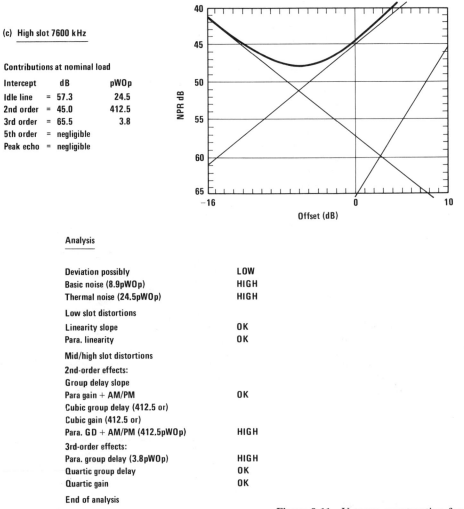

(c) High slot 7600 kHz

Contributions at nominal load

Intercept	dB	pWOp
Idle line	= 57.3	24.5
2nd order	= 45.0	412.5
3rd order	= 65.5	3.8
5th order	= negligible	
Peak echo	= negligible	

Analysis

Deviation possibly	LOW
Basic noise (8.9pWOp)	HIGH
Thermal noise (24.5pWOp)	HIGH
Low slot distortions	
Linearity slope	OK
Para. linearity	OK
Mid/high slot distortions	
2nd-order effects:	
Group delay slope	
Para gain + AM/PM	OK
Cubic group delay (412.5 or)	
Cubic gain (412.5 or)	
Para. GD + AM/PM (412.5pWOp)	HIGH
3rd-order effects:	
Para. group delay (3.8pWOp)	HIGH
Quartic group delay	OK
Quartic gain	OK
End of analysis	

(d)

Figure 9.11 V-curve construction for Example 3 (continued).

of the problem by making measurements in the radio section. The MLA traces corresponding to each of the V-curve examples are given later in the chapter, and their relationship to the V-curves is discussed. ∎

9.6 DIFFERENTIAL PHASE (DP) AND DG

The definition of **differential phase** (DP) is *the difference in phase shift encountered by a low-level, high-frequency sinusoid at two stated instantaneous amplitudes of a superimposed low frequency signal.* Consider an IF signal with center frequency ω_c, modulated by a baseband test tone of frequency ω_m (Fig. 9.12). Then if the FM modulation index is less than unity, only two significant sidebands, $\omega_c + \omega_m$ and $\omega_c - \omega_m$, will be produced. Now consider this IF signal and two sidebands in relation to the phase-frequency characteristic shown in Fig. 9.12. As the carrier, ω_c, is swept over the IF band, the phase shifts encountered by the sidebands vary and are reflected in the changing slope of the chord AB. The variation in slope of AB is displayed on an MLA as a measure of the variation in DP. Figure 9.12 also illustrates the close relationship between DP and GD. Effectively,

$$DP = \frac{\Delta\phi}{\Delta\omega}$$

where $\Delta\omega = 2\omega_m$

If the test-tone frequency, ω_m, is reduced, then A and B will move closer together until we are measuring the tangent at a point on the phase-frequency curve— that is, $d\phi/d\omega$, or GD. The measurement methods for DP and GD are therefore very similar, the only difference being the test-tone frequency used. Normally, to *measure GD, a test-tone between* 80 *and* 500 kHz *is used, whereas DP requires a test-tone frequency greater than* 1 MHz.

Similarly, DG is defined as the difference in gain encountered by a low-level, high-frequency sinusoid at two stated instantaneous amplitudes of a su-perimposed low-frequency signal. It follows that DG can be measured using the same method as for DP but measuring the amplitude difference between the high-frequency sidebands rather than the phase difference. Just as DP is related to GD, so also is DG related to BB linearity in the sense that the measurement

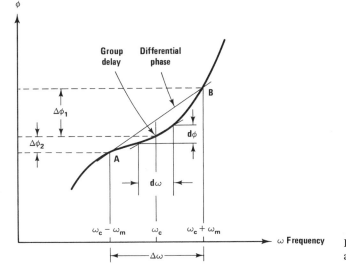

Figure 9.12 Measuring GD distortion and differential phase.

method is identical but for the frequency of the test-tone used. DG can be measured by test tones greater than 1 MHz.

9.6.1 The Need for Measuring DG and DP as Well as GD and Linearity

When microwave radio channel capacities were less than 600, predictions of baseband NPR from linearity and GD were fairly accurate. As system capacities grew larger, however, discrepancies became apparent between swept measurements, using low-frequency test tones, and NPR measurements.

The discrepancies are due to the fact that there are types of distortion arising in the radio section that are important noise contributors whose effects on GD are proportional to test-tone frequency and whose effects on linearity are proportional to the square of test-tone frequency. Therefore, whereas GD and linearity arising in the BB and modem sections of a link remain constant with test-tone frequency, this is not true for the carrier (radio) section (see Fig. 9.13).

The following approximate relations can be derived [Hewlett-Packard, 9.6]:

$$\text{DG at } \omega_c = 1 - [\alpha''(\omega_c) - k\tau'(\omega_c)] \frac{\omega_m^2}{2} \qquad (9.17)$$

$$\text{DP at } \omega_c = [\tau(\omega_c) + k\alpha'(\omega_c)]\omega_m \qquad (9.18)$$

where ω_c is the carrier angular frequency, $\alpha(\omega_c)$ is the IF amplitude response, $\tau(\omega_c)$ is GD response, ω_m is the test-tone angular frequency, and k is the AM-PM conversion coefficient.

Equation (9.17) shows that DG will be exhibited if cubic (or higher-order) terms exist in the IF amplitude characteristic or parabolic (or higher-order) group delay is followed by an AM-PM converter. This second effect is usually associated with the limiters, which remove incidental AM by converting some of the AM to PM at the output. This PM combines with the original FM and usually gives rise to distortion. This combination of a linear distortion GD followed by a nonlinear one, AM-PM, is usually called a **coupled distortion.**

Another coupled distortion, parabolic (or higher-order) amplitude response followed by an AM-PM converter manifests itself as DP (see equation (9.18)), as does GD slope. Figure 9.14(a) and (b) are pictorial representations of equations (9.17) and (9.18), respectively. For example, an MLA DG trace exhibiting a large amount of slope is due either to cubic amplitude response or the coupled distortion. Parabolic content in the MLA trace, however, would suggest quartic amplitude response (quartic gain) or cubic GD plus AM-PM. A similar argument applies to DP. Because DG and DP are functions of test-tone frequency, they are useful measurements for identifying the distortion contributors affecting the high slot of a radio baseband. This is demonstrated in a later example.

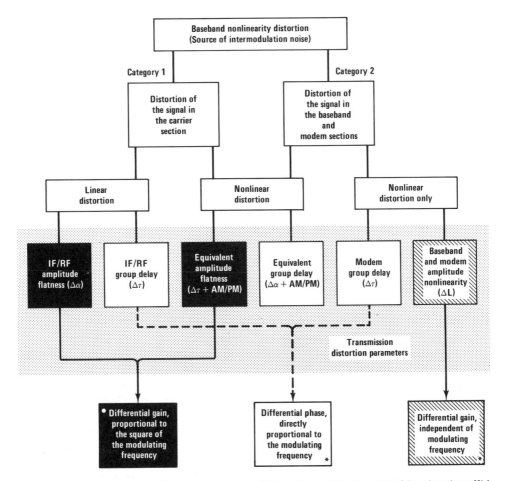

Figure 9.13 Sources of baseband intermodulation noise.

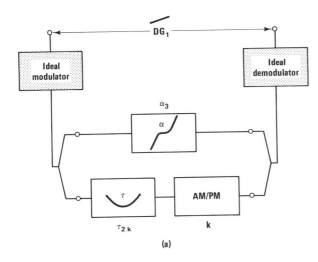

(a)

Figure 9.14 (a) Sources of DG slope (DG₁).

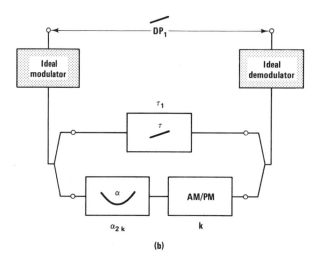

Figure 9.14 (b) Sources of DP slope (DP$_1$).

(b)

9.7 RELATIONSHIPS BETWEEN THE SHAPES OF MLA TRACES AND NPR

We have discussed the fact that BB linearity and group delay measurements reveal the noise contributors affecting the low slot, whereas DG and DP identify the transmission parameters that have most effect on the high slot of a radio. But how can we relate the shapes of these traces on a MLA to the baseband noise distortion measured? If these relationships are understood, then correct compromises can be made when realigning the radio.

Let us assume that we have resolved the relevant MLA traces in terms of their polynomial content, slope, parabolic, and so on, as illustrated in Fig. 9.4. Then Fig. 9.15 shows the relative distorting effects of various trace characteristics in the low slot of an 1800-channel radio. From this we can conclude that slope in the linearity trace is the most important contributor, with slope in the GD trace next. Parabolic and higher-order content is much less significant.

Fig. 9.16 indicates that the same pattern is true for DG and DP slope with respect to the high slot.

9.7.1 Analyzing the Polynomial Content of MLA Traces

We have just seen that the slope of the baseband linearity and of the differential gain have the most significant effect on the NPR performance. Therefore, it is important to be able to identify the amount of slope, in particular, in a given MLA trace. Figure 9.17 outlines one method, which involves measuring the Y-ordinates of 4 frequencies normalized to the center frequency ordinate of the sweep. The required polynomial content can then be calculated using the equations given in the figure. A sweep of ± 10 MHz is assumed in the description, but the same principle applies to any sweep width. In the examples given later, a ± 10 MHz sweep is used throughout, since this adequately covers measurement

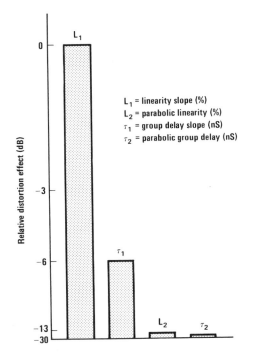

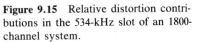

L_1 = linearity slope (%)
L_2 = parabolic linearity (%)
τ_1 = group delay slope (nS)
τ_2 = parabolic group delay (nS)

Figure 9.15 Relative distortion contributions in the 534-kHz slot of an 1800-channel system.

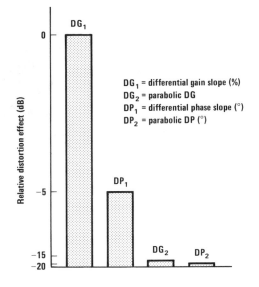

DG_1 = differential gain slope (%)
DG_2 = parabolic DG
DP_1 = differential phase slope (°)
DP_2 = parabolic DP (°)

Figure 9.16 Relative distortion contributions in the 7600-kHz slot of an 1800-channel system.

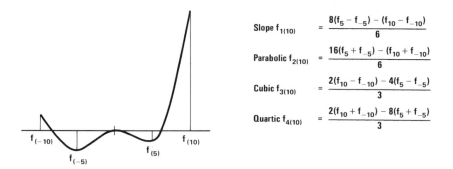

$$\text{Slope } f_{1(10)} = \frac{8(f_5 - f_{-5}) - (f_{10} - f_{-10})}{6}$$

$$\text{Parabolic } f_{2(10)} = \frac{16(f_5 + f_{-5}) - (f_{10} + f_{-10})}{6}$$

$$\text{Cubic } f_{3(10)} = \frac{2(f_{10} - f_{-10}) - 4(f_5 - f_{-5})}{3}$$

$$\text{Quartic } f_{4(10)} = \frac{2(f_{10} + f_{-10}) - 8(f_5 + f_{-5})}{3}$$

Figure 9.17   Formulas for estimating component deviations at 10 MHz for mixed polynomial responses.

of systems up to 2700-channel capacity. These equations are laborious to use, however, so two alternatives to mental arithmetic exist:

1. Read the traces from the MLA into a desktop computer over the IEEE-488 bus using a setup such as that shown in Fig. 9.18. The equations given can then be incorporated in the computer software.

2. Use a rule-of-thumb method to determine the slope content as follows:

$$\text{Slope} \approx f_5 - f_{-5}$$

More generally, the slope content is approximately equal to the difference between the ordinates at plus and minus half-bandwidth points.

The rule-of-thumb method holds provided the cubic content is not high. If the cubic content is less than half the slope content, then NPR predictions based

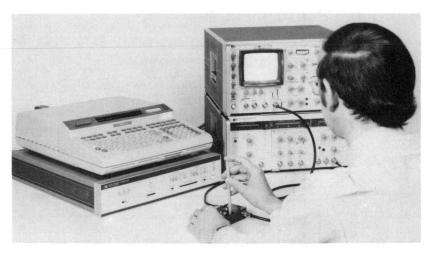

Figure 9.18   High-capacity system measurement setup with a MLA and desktop computer.

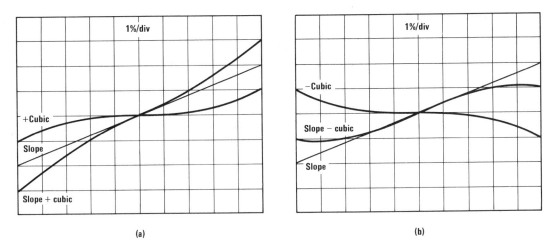

Figure 9.19   (a) 2% slope added to 1% cubic of same sense. (b) 2% slope added to 1% cubic of opposite sense.

on the rule-of-thumb method can be accurate within ± 1 dB. Examples of this condition are shown in Fig. 9.19. If in doubt, use the slope equation in Fig. 9.17. The rule-of-thumb method described in (2) is often sufficient to confirm which MLA trace is the major cause of the poor NPR observed.

Once the polynomial content (particularly slope) has been established, Table 9.1 can be used to calculate the noise contribution (in pWOp).

TABLE 9.1   FDM System Noise Contribution

Measurement trace	Slot for which coefficient is valid	Channel capacity				Units pWO per
		960	1260	1800	2700	
L_1	Low	6	7	9	13	$(\%)^2$
τ_1	Low	<0.1	<0.1	2.2	3.3	$(ns)^2$
τ_1	Mid	7.5	9	28	95	$(ns)^2$
DG_1	High	8	19	115	936	$(\%)^2$
DP_1	Mid	10	12	38	130	$(degree)^2$

where L_1 = BB linearity slope
 τ_1 = GD slope
 DG_1 = DG slope
 DP_1 = DP slope

Notes:
± 10-MHz sweep is assumed.
A 2.4-MHz test tone is assumed for DG and DP measurements.
For different sweep width, x (MHz), divide the coefficient by $(x/10)^2$.
For different test-tone frequency, y (MHz), divide the DG coefficient by $(y/2.4)^4$ and divide DP coefficient by $(y/2.4)^2$.

9.8 EXAMPLES OF MLA TRACE ANALYSIS

Each of the following examples gives the MLA traces relating to the V-curves obtained in the correspondingly numbered V-curve analysis examples.

Example 9.4

Figure 9.20 shows the BB linearity and GD traces for an 1800-channel system. Here GD slope is minimal, and the contribution due to linearity slope can be estimated by evaluating the difference between the ±5-MHz markers, the rule-of-thumb method. By this method, the slope is −2.1%, so the noise contribution, using Table 9.1, is

$$(-2.1)^2 \times 9 = 40 \text{ pWOp}$$

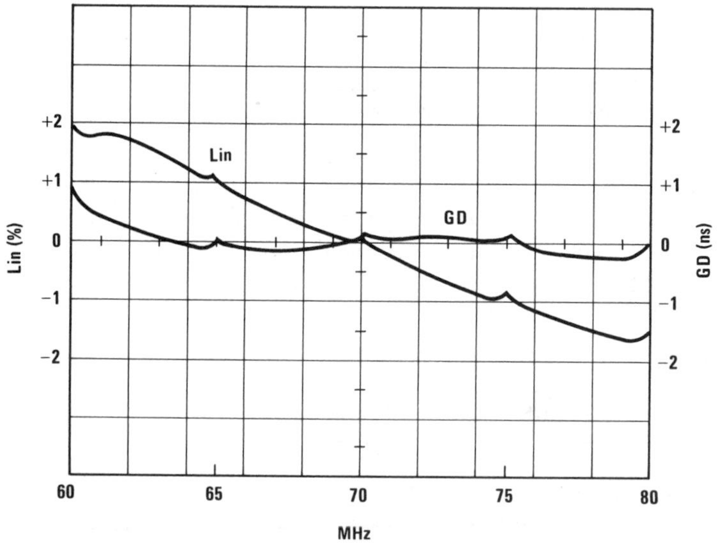

Figure 9.20 MLA traces of linearity and group delay corresponding to the V-curve set of Fig. 9.8.

From Table 9.2, this converts to a NPR of approximately 55 dB. In fact, the linearity trace has some cubic content, which slightly reduces the rule-of-thumb estimate obtained. Using the slope equation given in Fig. 9.17, a slope of 2.3% is derived, which translates to 47 pWOp.

These estimates agree very well with the values of 53 pWOp (53.9 dB NPR) obtained by V-curve analysis. This confirms that linearity slope is the cause of the poor NPR in the low slot. If realignment of the modem section can reduce the difference between ±5-MHz points to less than 0.7%, the second-order contribution will then be approximately equal to the idle noise at nominal load, since

$$(0.7)^2 \times 9 = 4.4 \text{ pWOp}$$ ■

TABLE 9.2   Decibel NPR-to-pWOp Conversions

NPR (dB)	pWOp	S/N (dBp)	dBrnCO	NPR (dB)	pWOp	S/N (dBp)	dBrnCO	NPR (dB)	pWOp	S/N (dBp)	dBrnCO
30.0	13182.6	48.8	42.0	44.0	524.8	62.8	28.0	58.0	20.9	76.8	14.0
31.0	10471.3	49.8	41.0	45.0	416.9	63.8	27.0	59.0	16.6	77.8	13.0
32.0	8317.6	50.8	40.0	46.0	331.1	64.8	26.0	60.0	13.2	78.8	12.0
33.0	6606.9	51.3	39.0	47.0	263.0	65.8	25.0	61.0	10.5	79.8	11.0
34.0	5248.1	52.8	38.0	48.0	208.9	66.8	24.0	62.0	8.3	80.8	10.0
35.0	4168.7	53.8	37.0	49.0	166.0	67.8	23.0	63.0	6.6	81.8	9.0
36.0	3311.3	54.8	36.0	50.0	131.8	68.8	22.0	64.0	5.2	82.8	8.0
37.0	2630.3	55.8	35.0	51.0	104.7	69.8	21.0	65.0	4.2	83.8	7.0
38.0	2089.3	56.8	34.0	52.0	83.2	70.8	20.0	66.0	3.3	84.8	6.0
39.0	1659.6	57.8	33.0	53.0	66.1	71.8	19.0	67.0	2.6	85.8	5.0
40.0	1318.3	58.8	32.0	54.0	52.5	72.8	18.0	68.0	2.1	86.8	4.0
41.0	1047.1	59.8	31.0	55.0	41.7	73.8	17.0	69.0	1.7	87.8	3.0
42.0	831.8	60.8	30.0	56.0	33.1	74.8	16.0	70.0	1.3	88.8	2.0
43.0	660.7	61.8	29.0	57.0	26.3	75.8	15.0	71.0	1.0	89.8	1.0

Example 9.5

In this case, V-curve analysis indicates GD slope as the likely source of the problem. Figure 9.21, therefore, displays group delay paired with linearity. By rule-of-thumb and using Table 9.1,

$$\tau_1 = -1 - (+2) = -3 \text{ nS}$$

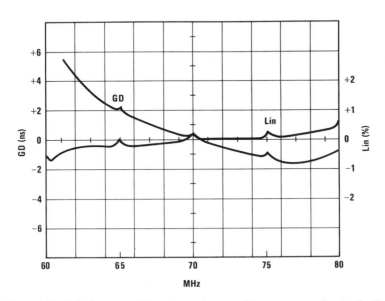

Figure 9.21   MLA traces of linearity and group delay corresponding to the V-curve set of Fig. 9.10.

so noise contribution is $(-3)^2 \times 28 = 252$ pWOp in the mid slot of an 1800-channel radio.

From Table 9.2, 252 pWOp converts to about 47 dB NPR. The linearity contribution in the low slot is negligible:

$$\text{Noise from } L_1 = (0.3)^2 \times 9 = 0.8 \text{ pWOp}$$

The parabolic content of the GD trace can be shown to be approximately 1.7 pWOp (see [Hewlett-Packard, 9.6] for coefficient), which is negligible. The realignment goal here is, therefore, to minimize the difference between the ± 5-MHz markers of the GD trace. The parabolic content, which will cause third-order distortion, can be tolerated. ■

Example 9.6

V-curve analysis of this 1800-channel radio revealed a poor high slot (7600 kHz). DG and DP measurements should be used to investigate such a problem (see Fig. 9.22).

For the DG trace, rule-of-thumb gives -2.3% slope, which gives 608 pWOp or 43.5 dB NPR (Table 9.1). A significant amount of cubic DG is present in this case, so a better estimate of slope is obtained from the equation in Fig. 9.17:

$$DG_1 = \frac{8 \times (-1 - 1.3) - (-3 - 3.6)}{6} = 1.96\%$$

and noise contribution is $(1.96)^2 \times 115 = 442$ pWOp (44.5 dB NPR). This agrees well with the V-curve prediction. The DP slope contribution is about 70 pWOp in the mid slot.

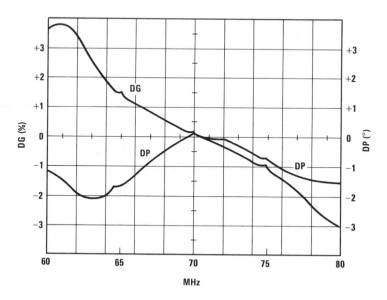

Figure 9.22   MLA traces of differential gain and phase corresponding to the V-curve set of Fig. 9.11.

V-curve analysis points to one of cubic group delay, cubic gain, and parabolic gain plus AM-PM conversion as the source of the problem. MLA trace analysis reveals DG slope as the major contributor, and from Fig. 9.14(a), this is caused either by cubic gain or the coupled distortion. Thus cubic group delay is eliminated. (This would have appeared as cubic DP.)

Having obtained confirmation of the V-curve analysis, the next step is to check for cubic content in the gain trace of the carrier section. In this case, the problem is, in fact, due to the coupled distortion, τ_2 + AM-PM. The IF circuit is as shown in Fig. 9.23. A limiter converts AM, from the unequalized parabolic group delay of the 20-MHz bandpass filter, to PM. The GD response of the overall network can be equalized by the GD equalizer, and the coupled distortion that is present shows up only by examining the DG trace. This situation could be cured either by equalizing the GD before the limiter or by improving the AM-PM coefficient of the limiter. ■

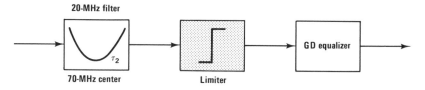

Figure 9.23   The IF circuit producing DG slope and high-slot distortion.

PROBLEMS

9.1. If the NPR specification of a radio hop is 50 dB at nominal load, what ought the NPR contributions be from idle, second- and third-order distortion types, respectively? What are the equivalent objective values in pWOp and dBrnCO?

9.2. If the second- and third-order distortion contributions are 55 and 65 dB (NPR) at nominal load, what will be the offset loading at which the second- and third-order distortion levels are equal? What will the system NPR be at this offset level (assume no other contributions are significant)?

9.3. At high-noise loading, clipping may occur in amplifiers and introduce a fifth-order distortion component. If it is acceptable for the fifth-order contribution to be equal to that of third-order at ±10 dB offset, what should the fifth-order nominal load intercept level be relative to the idle intercept?

9.4. Figure 9.24 is a V-curve measured on an 1800-channel radio in the low slot whose NPR specification is 60 dB. Use the flowchart in Fig. 9.6 to identify the likely cause. Which section of the radio is indicated? Which MLA trace and polynomial will show deviation from the ideal? What percentage deviation is expected on the MLA trace? (Use Table 9.1.)

9.5. Figure 9.25 shows a high-slot V-curve (1800 channels). Using the appropriate flowchart, what transmission problems are indicated if (a) the mid-slot V-curve is very similar to this one, and (b) the second-order intercept in the mid slot is 53 dB? Describe the main features expected of the relevant MLA trace for each case.

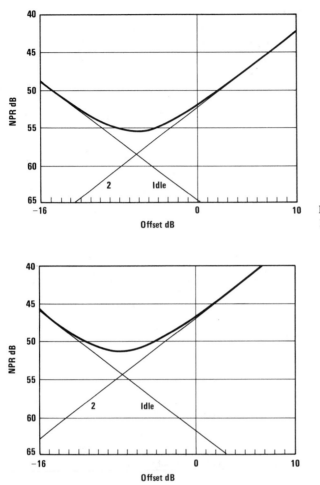

Figure 9.24 Low slot V-curve of an 1800-channel radio. See Problem 9.4.

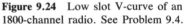

Figure 9.25 High slot V-curve. See Problem 9.5.

9.6. Which slot will be most affected by the deviations shown in Fig. 9.26? Evaluate the distortion contributions in pWOp and dB NPR from the slope and parabolic content of each trace in the affected slot. For parabolic content use coefficients:

$$L_2 = \frac{0.5 \text{ pWOp}}{(\%)^2} \quad \text{and} \quad \tau_2 = \frac{0.02 \text{ pWOp}}{(\text{nS})^2}$$

Compare the rule-of-thumb results for slope with the equation method of Fig. 9.17.

9.7. Which slot will be most affected by the deviations shown in Fig. 9.27? Evaluate the slope content of the DP and DG traces, comparing rule-of-thumb and equation results. Based on the dominant contributor, estimate the NPR at nominal load in the affected slot. What order of distortion would a V-curve exhibit? What transmission distortions are indicated by this dominant contributor?

9.8. If realignment of the radio alters the traces in Fig. 9.27 to eliminate DG slope at the

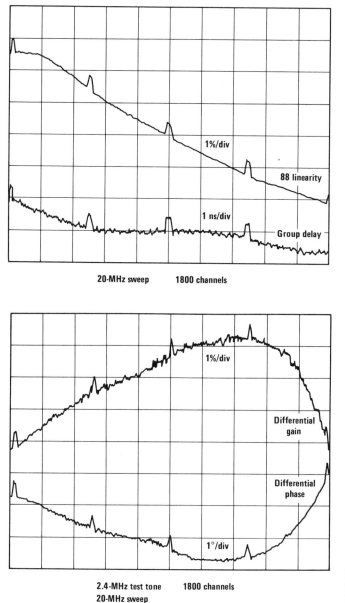

1%/div

88 linearity

1 ns/div

Group delay

20-MHz sweep   1800 channels

Figure 9.26 Linearity and group delay deviations. See Problem 9.6.

1%/div

Differential gain

Differential phase

1°/div

2.4-MHz test tone   1800 channels
20-MHz sweep

Figure 9.27 Differential gain and differential phase measurements. See Problem 9.7.

expense of increasing parabolic DG to 2%, what will be the new NPR at nominal load? (Remember the DP trace remains unchanged; the coefficient for DG_2 is 3.1 pWOp/(%)².) Estimate the positions of the second- and third-order construction lines for the new V-curve and so plot the V-curve between 0 and +10 dB offset (neglect parabolic DP).

REFERENCES

[9.1] CCIR Rec. 399-3. "Measurement of Noise Using a Continuous Uniform Spectrum Signal on FDM Telephony Radio Relay Systems," CCIR Recommendation 399-3, Geneva, 1978.

[9.2] Hewlett-Packard Application Note AN320. "Automatic V-curve Analysis of Microwave Radio Systems," part No. 5953-5401, describing software package part No. 03724-10101/2, December, 1982.

[9.3] Bell Telephone Laboratories, Inc. "Transmission System for Communications," Western Electric Company, Inc. *Technical Publications,* Chapter 21, 1971.

[9.4] Bell Telephone Laboratories, Inc. "Transmission Systems for Communications," Western Electric Company, Inc. *Technical Publications,* Chapter 10, 1971, p. 258.

[9.5] Bennet, W. R., H. E. Curtis, and S. O. Rice. "Interchannel Interference in FM and PM Systems Under Noise Loading Conditions," *Bell System Technical Journal,* Vol. 34, May, 1955, pp. 601–636.

[9.6] Hewlett-Packard Application Note 175-1, "Differential Phase and Gain at Work," part No. 5952-3164, November, 1975.

[9.7] CCITT Rec. G.223. "Assumptions for the Calculation of Noise of Hypothetical Reference Circuits for Telephony," CCITT *Yellow Book,* Vol. 3, Fascicle III.2, Geneva, 1981.

[9.8] Laine, R. E. "The Bucket Curve: Indispensable Noise Analysis Tool for Microwave Links," *GTE Lenkurt,* San Carlos, California, 1976.

[9.9] Tant, M. J. "The White Noise Book," Marconi Instruments Ltd., St. Albans, England, 1974.

[9.10] Urquhart, R. "A New Microwave Link Analyzer with High Frequency Test Tones," *Hewlett-Packard Journal,* September, 1972, pp. 8–16.

[9.11] Hewlett-Packard, "MLA Measurement Concepts," Hewlett-Packard Publication 5952-3126, October, 1972.

[9.12] Hewlett-Packard, "Introducing the BBA," Hewlett-Packard Publication 5953-6659, September, 1981.

[9.13] Hewlett-Packard, "Introductory Operating Guide for the 3724A/3725A/3726A Baseband Analyzer with the 85A Personal Computer," Hewlett-Packard Publication 5953-6671, October, 1981.

[9.14] Hewlett-Packard, "Transmission Distortions," Hewlett-Packard Publications 5952-3136, July, 1973.

[9.15] "Annual of the Research Institute for Telecommunication 1973," TKI, Budapest, 1973, pp. 57–89.

10

ENHANCED MICROWAVE RADIO MEASUREMENTS USING A TRACKING DOWNCONVERTER

ROBERT EASSON and ROBIN SHARP

Hewlett-Packard Ltd.

10.1 INTRODUCTION

Conventional microwave link analyzer (MLA) measurements using upconverters and downconverters as required to measure distortions in a microwave transmitter or receiver or in a microwave path or subsystem are limited by the test equipment in two major aspects: first, **residual distortions** (which limit measurement accuracy) and, second, **bandwidth** (which limits the usefulness of such measurements in waveguide testing and in some wideband satellite systems). In this chapter, the use of a tracking downconverter to overcome both limitations is described, and a number of applications are highlighted.

10.2 WHERE RESIDUALS OCCUR

For the measurement systems shown in Fig. 10.1, the corresponding distribution of residual error contributions from individual test instruments is shown in Table 10.1. Two observations can be made concerning MLA contributions. First, the MLA generator can be omitted, since it is normally used only to generate single baseband tones and a low-frequency (typically 70 Hz) sweep. In the one case where its IF output is used with an IF upconverting transmitter, the MLA generator and receiver IF residual errors can be calibrated out by a separate

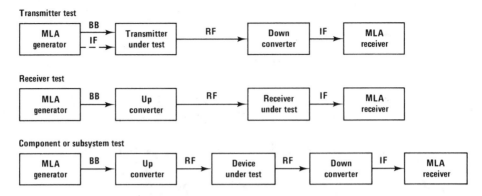

Figure 10.1   Testing different portions of microwave radios.

back-to-back measurement. Second, in microwave path or subsystem testing, the upconverter and downconverter residuals can, in principle, be calibrated out in a similar way. However, practical difficulties such as cable effects frequently introduce further uncertainties in measurement accuracy.

What is required is to identify separately the residuals due to the upconverter, microwave cable assemblies, and the downconverter. Three methods can be used, each of which enables the residual performance of a different part of the measurement chain to be isolated. By combining these methods, individual instrument contributions can be identified and, furthermore, provide a check on the accuracy achieved.

First, if it is possible to simulate a "perfect" upconverter, we can measure the combined residual errors of the downconverter and MLA receiver directly. These errors can then be calibrated out by storage-normalizer techniques. The perfect-upconverter concept is illustrated in Fig. 10.2, where the upconverter is modulated only with test tone, and it is the downconverter local oscillator (LO) that is swept. This eliminates the upconverter residual errors and also the errors

TABLE 10.1   Contributions to Residual Errors

Unit under test	Test system component				
	MLA generator	Up-converter	RF cables	Down-converter	MLA receiver
Transmitter (BB input)	—	—	X	X	X
Transmitter (IF input)	(X)	—	X	X	(X)
Receiver	—	X	X	—	X
RF component or subsystem	—	X	X	X	X

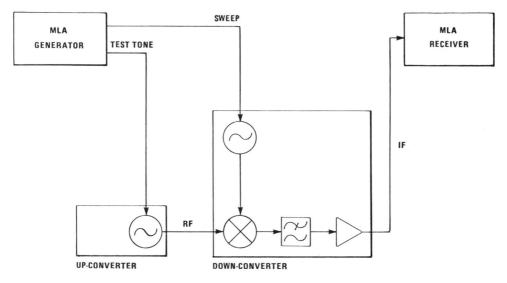

Figure 10.2   "Perfect upconverter" simulation.

introduced by the interconnecting microwave cable, since effectively it is transmitting only a CW signal (the test-tone deviation is negligible). *The MLA receiver
now measures directly the downconverter IF section residual error plus the MLA
receiver error, with no incoming-signal contribution.*

In the second method, the downconverter operates in a tracking mode (Fig.
10.3). To achieve this, an automatic frequency control (AFC) circuit holds the

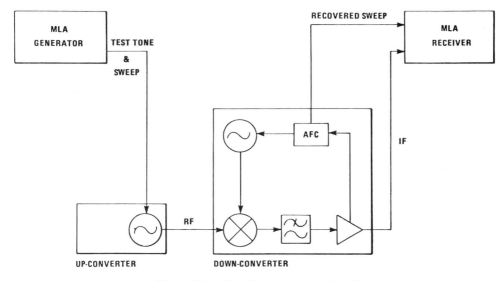

Figure 10.3   Tracking mode operation of downconverter.

LO frequency at a constant difference (equal to the IF center frequency) from the incoming swept RF signal. Provided the loop gain is sufficiently high, the bandwidth of the resulting IF signal is so small that the IF residual error contributed by the downconverter and MLA receiver is negligible. For example, if the RF input is swept over a 50-MHz bandwidth and the loop gain is 50, then the IF bandwidth is reduced from 50 MHz to 1 MHz. In consequence, *the incoming RF distortions are measured directly, with no contribution from the downconverter and MLA receiver*.

These two methods are complementary. In the first method, the downconverter and MLA receiver residual errors were directly measured. In the second method, they were eliminated. Therefore, if the residual errors due to the microwave cable assemblies can be either characterized or eliminated, the upconverter residuals can be uniquely identified. This is the basis of the third method of isolating residual errors.

This third method eliminates the RF interconnection contribution to RF residual errors. While this contribution is in part due to cable imperfections (loss, impedance variation, and so on) in most measurement systems where only short lengths of cable are used, the most serious source of error is poor return loss. For example, a 2-m length of coaxial cable with 0.5 dB/m attenuation at its operating frequency and a return loss of 20 dB at each end, connected to a source and load with 15 dB output and input return losses, respectively, gives

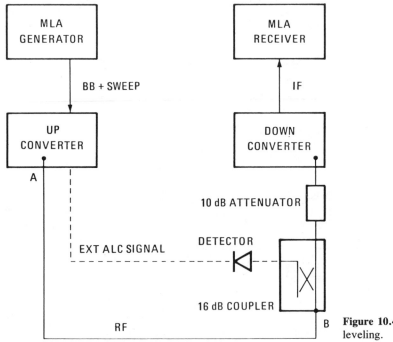

Figure 10.4 Test system—external leveling.

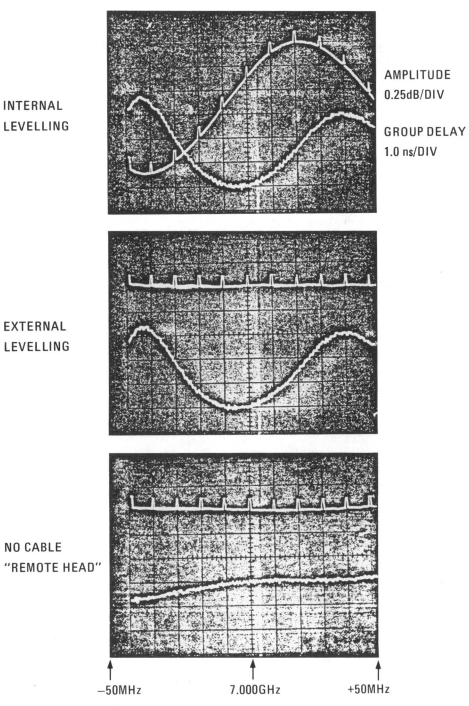

INTERNAL
LEVELLING

AMPLITUDE
0.25dB/DIV

GROUP DELAY
1.0 ns/DIV

EXTERNAL
LEVELLING

NO CABLE
"REMOTE HEAD"

−50MHz   7.000GHz   +50MHz

Figure 10.5 RF cable effects.

rise to a group delay ripple of 1.4 ns and a corresponding amplitude ripple of 0.6 dB (both peak-to-peak). Amplitude flatness can be improved through external leveling of the upconverter, but this will not affect group delay (GD) variations. Only under ideal conditions (all return losses at least 25 dB) can GD ripple be reduced to around 0.25 ns. These conditions are unlikely to be met within normal operating environments.

The preferred approach, therefore, is to eliminate interconnecting microwave cables altogether. This is feasible if the downconverter is capable of *remote head* operation, that is, if the front end can be removed from the mainframe on an umbilical cable and interfaced directly with the system under test. The IF cabling, which now effectively replaces the microwave cable, does not cause any significant distortion. A basic test system shown in Fig. 10.4 allows internal or external leveling of the upconverter, or removal of the 1.2 m microwave cable between *A* and *B*. Figure 10.5 dramatically shows the effects on flatness and GD of internal and external leveling and of direct connection between *A* and *B*. Thus, by eliminating interconnecting microwave cabling and combining this technique with the foregoing methods of separating RF and IF residual errors, the contributions of all components can be individually identified.

10.3 ANALYSIS OF RESIDUAL DISTORTION COMPONENTS

In this section, the relationships between theory and residual distortions as measured by the foregoing three methods are derived, and from these it is shown how they can be checked against each other for consistency.

$$R = (R_{\text{upconverter}} + R_{\text{cable}})$$
$$+ (R_{\text{downconverter}} + R_{\text{MLA receiver}}) \qquad (10.1)$$
$$= R_{\text{RF}} + R_{\text{IF}}$$

where R_{RF} = RF residual
R_{IF} = IF residual

R may be any distortion parameter: amplitude flatness, GD, and so on. This relationship assumes that the downconverter has no significant RF residual error, which will be justified later.

With the downconverter in its normal (nontracking) mode, the residual distortion as measured by the MLA receiver is R_{N}, where

$$R_{\text{N}} = R_{\text{RF}} + R_{\text{IF}} \qquad (10.2)$$

When the tracking mode is used, the measured residual distortion is R_T, where

$$R_{\text{T}} = R_{\text{RF}} \qquad (10.3)$$

When the downconverter LO is swept instead of the upconverter (i.e., perfect-upconverter simulation), the measured residual distortion is R_{S}, where

$$R_{\text{S}} = R_{\text{IF}} \qquad (10.4)$$

Hence

$$R_N - R_S = (R_{RF} + R_{IF}) - R_{IF}$$
$$= R_{RF} \qquad (10.5)$$
$$= R_T$$

Since R_N, R_S, and R_T are measured quantities, equation (10.1) enables the assumptions of this analysis to be quantitatively checked. The chief assumption is that the downconverter contributes only to the IF component of residual error—that is, all input circuits and mixer assembly can be neglected. Since the mixer operates under different conditions depending on whether R_N, R_S, or R_T is measured (i.e., whether the RF input, the LO input, or both are swept), equation (10.1) provides a comprehensive check on the validity of this assumption.

10.4 IDENTIFYING RESIDUAL ERRORS WITH THE 3730B DOWNCONVERTER

With the test setup using a particular downconverter—Model 3730B, shown in Fig. 10.6—measurements of R_N, R_T, and R_S for amplitude flatness and GD were made as shown in Fig. 10.7. By using the storage normalizer, $R_N - R_S$ was

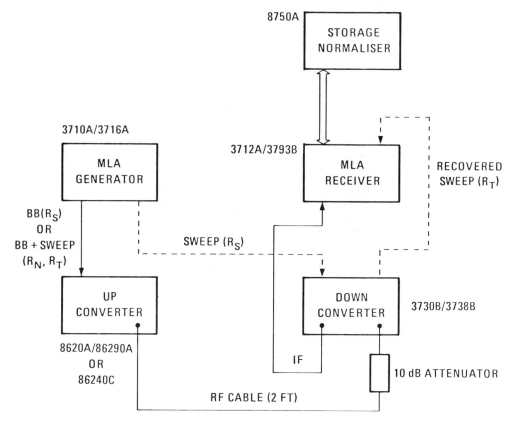

Figure 10.6 Test setup measures R_N, R_T, and R_S.

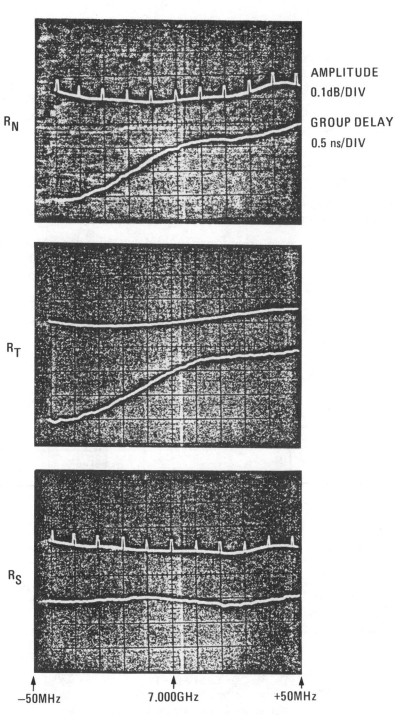

Figure 10.7 Measured residuals (1).

measured as shown in Fig. 10.8. Comparison with R_T shows that equation (10.1) is satisfied within 0.02 dB for amplitude flatness and 0.1 ns for group delay over a sweep width of 100 MHz. Measurements were repeated at different microwave center frequencies with LO RF instead of a reference source, and in all cases consistent results were obtained.

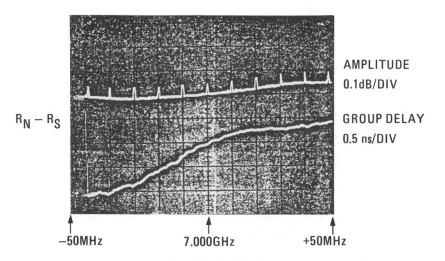

Figure 10.8   Measured residuals (2) (using storage normalizer).

Measurements were also made of baseband linearity and differential gain and phase, with similar results.

The analysis of residual distortions has shown how these may be separated and identified. Measurements have shown that uncertainties in measured values can be reduced to 0.02 dB in amplitude flatness and 0.1 ns in GD, with correspondingly minimal uncertainties in other distortion parameters. These results confirm that the omission of downconverter RF residual errors in the analysis is justified in practice.

Three ways of identifying residual errors have been established:

1. By sweeping the downconverter LO to simulate a perfect upconverter, downconverter plus MLA receiver residuals are directly measured.

2. By using the tracking mode of the downconverter, RF distortions alone are directly measured.

3. By using the remote head facility of the downconverter, interconnecting microwave cable residuals are eliminated.

10.5 EXTENDING MEASUREMENT BANDWIDTH

It has been shown how the tracking mode, by compressing the incoming RF sweep bandwidth into the downconverter by a factor of 50, effectively eliminates any residual error contribution from the test receiver (i.e., downconverter plus MLA receiver). The usual MLA maximum bandwidths of 50 MHz (at 70 MHz IF) and 100 MHz (at 140 MHz IF) are thus reduced to 1 and 2 MHz, respectively. The corollary of this is that the tracking-mode downconverter enables the MLA receiver to handle a much larger effective RF bandwidth. The standard 3730B will handle an incoming RF bandwidth of 125 MHz (with 70 MHz IF) and 250 MHz (with 140 MHz IF). Since the IF bandwidths are only 2.5 and 5 MHz, respectively, residual errors are still negligible (see Appendix 10.1). Special versions have been produced that will accept RF bandwidths of 500 MHz and 1 GHz (with 70 and 140 MHz IFs, respectively), thereby enabling MLA measurements over standard communication bands (e.g., 3.7–4.2 GHz or 10.7–11.7 GHz) to be made.

It is, of course, not necessary to use an MLA receiver as the measurement tool. A swept-amplitude analyzer can be substituted if only amplitude measurements are required. The advantage of using it in conjunction with a tracking-mode downconverter is the recovered sweep output; this avoids the usual necessity of a cable from the sweep output of the generator to the analyzer X-input as all the information is transmitted via the microwave path. Remote measurements (over a microwave link, for example) are thus practical.

10.6 APPLICATIONS

10.6.1 Production Test

Normally network analyzers provide an ideal measurement tool in the manufacture of microwave hardware, as long as input and output frequencies are in a common range. However, the submodules of a microwave radio system are usually configured such that some frequency translation takes place between input and output—for example, modulators, demodulators, upconverters, and downconverters. Measurement thus requires a proven system designed to interface at all sections of the microwave link—that is, BB, IF, and RF—as shown schematically in Fig. 10.9. This is achieved by an MLA plus upconverter and downconverter, whose accuracy can be determined by measurement of residuals as described earlier.

Also, the same system can be used as an economical RF-RF test system, capable of performing swept-response amplitude and GD measurements across RF components such as RF branching filters. Furthermore, the extended RF measurement bandwidth, using the tracking mode of the downconverter, enables wideband components to be evaluated just as easily.

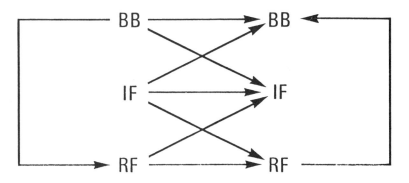

Figure 10.9   Measurement configurations.

10.6.2 Terrestrial Systems

On conventional *n*-for-1 protected radio systems, it is common practice to perform transmission distortion measurements using back-to-back transmitter-receiver combinations, thus avoiding any need for RF interface. However, for systems using hot standby protection, it is impossible to perform end-to-end measurements without taking the complete route out of service. The only practical solution is to evaluate each transmitter and receiver on an individual basis. A test downconverter is used to measure transmitter distortion, whereas an upconverter is used to measure receiver distortions. With the techniques described earlier, errors can be minimized and accuracy limits can be determined.

10.6.3 Satellite Systems

The use of test upconverters and downconverters for transmission distortion analysis is common practice. This allows the transmit and receive sections of the ground station to be fully evaluated separately before interfacing with the satellite transponder, thus minimizing end-to-end alignment time through the transponder.

 Most satellite transponders have a relatively large bandwidth, and hitherto systems have had to be evaluated in subbands due to the restricted sweep width of the MLA. However, by using the tracking mode of the downconverter, characterization can be done with a single swept measurement. A proposed test system for GD measurement over 250 MHz for a WESTAR satellite transponder is outlined in Appendix 10.1.

10.6.4 Waveguide and Antenna Checks

The testing of waveguide and antenna systems has traditionally involved single port tests, such as time domain reflectometry. This technique allows waveguide return loss problems to be easily identified and located, but it does not highlight problems associated with multimode or multipath propagation effects. To highlight these effects, end-to-end tests of GD and amplitude response are required. By using an MLA equipped with an upconverter and a tracking downconverter, it is now possible to evaluate the waveguide and antenna system on an end-to-end basis over the full communications bandwidth (e.g., 500 MHz or 1 GHz).

As noted earlier, the upconverter and tracking downconverter have also found applications with other test equipment, such as a scalar network measurement system. One interesting application has been tested for assessing the crosspolar discrimination of a dual-band antenna system, as described in Appendix 10.2.

10.7 CONCLUSIONS

The applications highlighted earlier indicate the wide range of measurements that become practical through the use of a tracking downconverter such as the 3730B. Coupled with the methods to identify residual errors that are also described, it is clear that such an instrument makes possible rapid, accurate measurements on microwave systems on both a local and an end-to-end basis.

APPENDIX 10.1
GD TEST SET MEASURING OVER 250 MHz CENTERED AT 1.7 GHz

Using the tracking mode of the 3730B downconverter, the swept RF carrier signal is reduced to 2.5 MHz. The FM modulation (250 kHz or 500 kHz) on the carrier required for GD measurement is unaffected. However, the MLA receiver markers are affected, and an independent marker on the RF carrier is required. This is obtained using a cavity wavemeter and crystal detector via a directional coupler on the RF input. The detected signal is fed to the EXT Y-input of the MLA receiver, providing a swept-amplitude response trace with frequency marker. The storage normalizer is used to remove any system residual error. See Fig. 10.A.1.

APPENDIX 10.2
ANTENNA CROSS-POLAR DISCRIMINATION MEASUREMENTS

An antenna cross-polar discrimination measurement is required for the alignment of the antennas for a digital radio system operating over the 10.7–11.7-GHz band. To increase system capacity, transmission takes place on orthogonal polarizations at the same frequency, and thus it is essential that discrimination between the polarizations is maintained across the complete 1-GHz wideband. Since antenna adjustments affect both channels, a swept real-time measurement of the ratio between copolar and cross-polar channels is, therefore, desirable to enable rapid antenna alignment. This is achieved as shown in Fig. 10.A.2.

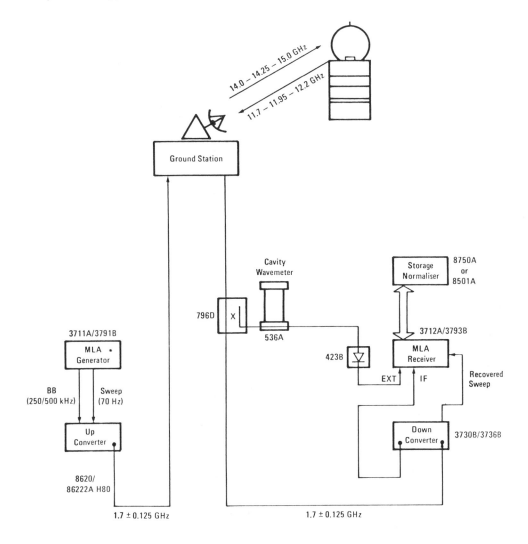

Figure 10.A.1   Group delay test set measuring over 250-MHz bandwidth.

*Alternative: 3707A BB + Sweep Generator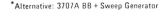

The RF signal is generated by the 8620C/86245A combination. The standard 8620C sweeper is capable of generating only a ramp sweep voltage; this is not suitable for end-to-end measurements where a synchronous X-deflection signal must be recovered at the remote receiver to display the results on a swept basis. To achieve this, a sinusoidal sweep voltage is required—this is supplied by the 3707A BB plus sweep generator and the 86245A. FM drive circuits were modified to allow a 1-GHz sweep width. The marker system of the standard 8620C uses

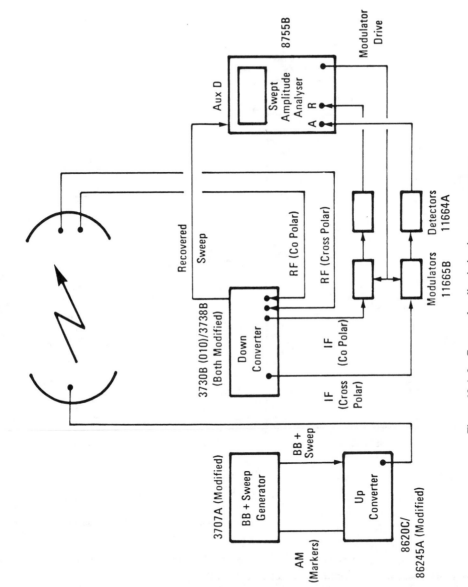

Figure 10.A.2 Cross-polar discrimination measurements.

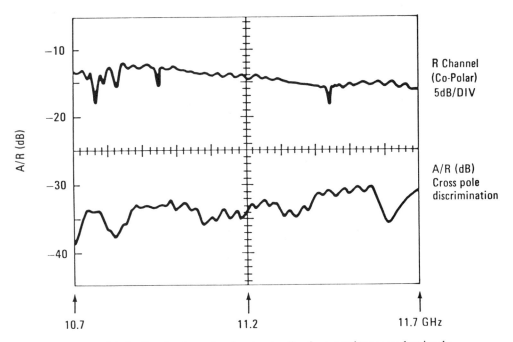

Figure 10.A.3   Received copolar signal and ratio of cross-polar to copolar signals over a 40-km microwave link. Copolar trace shows amplitude markers at ±250 MHz.

a *Z*-modulation approach, which is acceptable on local measurements. However, for end-to-end applications it is not usable; therefore, the 3707A BB plus sweep generator was modified to include a marker generator based on amplitude modulation of the RF carrier.

The RF receiver comprises the 3730B/3738B RF downconverter. The tracking AFC facility of the 3730B is essential for remote end-to-end measurement capability, allowing a synchronous sweep voltage to be recovered from the incoming RF test signal. The standard product is intended for downconversion of a single RF channel; however, for cross-pole discrimination measurements, dual-channel capability is required. To facilitate this, a second RF channel was added, the local oscillator drive being obtained from the main RF module via a splitter. Option 010 on the 3730B provides an additional 25 dB of IF gain; this effectively increases the input sensitivity of the signal path to better than -70 dBm. (Noise performance of the 3730B is discussed in detail in [Easson, 10.1].)

The detection and display instrumentation employed standard Hewlett-Packard *Frequency Response Test System*—the 8755B, 182T display and associated modulator and detectors 11665B and 11664A, respectively. These products allow detection of the two IF signals and subsequently display the ratio of the two detected IF signals on a swept basis. Figure 10.A.3 shows the received copolar signal and the ratio of the cross-polar to copolar signals as displayed on the swept-amplitude analyzer.

REFERENCES

[10.1] Easson, R. M. "Making Difficult Measurements of Noise Power Ratios on Transmitters," *Microwave Systems News*, February, 1981, pp. 88–89.

11

PHASE-NOISE
MEASUREMENTS

CATHARINE M. MERIGOLD

Hewlett-Packard Co.

11.1 INTRODUCTION

As the performance of radio communications systems improves, the phase-noise requirements of the signal sources often become the limiting factor for the overall system. In this chapter, the first part, on phase-noise basics (Sections 11.1 to 11.5), defines phase noise and its importance in modern communications systems. Then in the second part, on phase-noise measurements (Sections 11.6 to 11.8), the most common methods for phase-noise measurements are described, with a comparison of the methods, their performances, and their applicabilities.

11.2 WHAT IS PHASE NOISE?

11.2.1 Fundamental Concept of Phase Noise

The commonly used term *phase noise* is really a subset of the broader category of **frequency stability.** Frequency stability is the degree to which an oscillating source produces the same frequency value throughout a specified period of time [Lance, 11.14]. The stability of a signal source decreases if the signal is anything other than a perfect sine function.

To understand better frequency stability and phase noise, we first gain an intuitive feel. A perfect sinusoidal signal can be represented by

$$V(t) = V_0[\sin 2\pi f_0 t] \qquad (11.1)$$

where $V(t)$ = instantaneous output voltage of the signal
V_0 = nominal peak amplitude of the signal
f_0 = average (nominal) frequency of the signal

This signal appears in the time domain as a perfect sinusoid (Fig. 11.1) and in the frequency domain as a single spectral line (Fig. 11.2(a)).

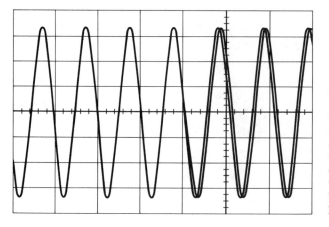

Figure 11.1 Phase noise is very difficult to see in the time domain; it would appear as changes in the 0 crossings relative to a "perfect" signal with no phase noise. Since the noise is at rates much, much lower than the carrier frequency, it takes many cycles of the carrier to observe the change in the zero crossings. (One way to think of how phase noise would appear in the time domain is to picture a signal being frequency modulated with a very low modulation index.)

Unfortunately, in the practical world all signals exhibit unwanted amplitude and frequency or phase fluctuations, represented by

$$V(t) = [V_0 + \varepsilon(t)] \sin [2\pi f_0 t + \phi(t)] \qquad (11.2)$$

where $\varepsilon(t)$ = instantaneous amplitude fluctuations
$\phi(t)$ = instantaneous phase fluctuations

The $\phi(t)$ fluctuations are added to the phase-angle term of the equation. But since frequency and phase are related (frequency is the time derivative of phase), these instantaneous instabilities can be treated equivalently as unwanted frequency or phase fluctuations.

This chapter is about the characterization of these phase fluctuations, commonly called *phase noise*. The phase fluctuations appear in the time domain as changes in the 0 crossings (Fig. 11.1). Since phase noise arises from a composite of low-frequency signals (much lower than the carrier frequency), in the time domain many cycles of the carrier will pass before the change in the 0 crossings occurs.

In the frequency domain (as seen on a spectrum analyzer), the phase fluctuations appear as noise sidebands, extending above and below the nominal

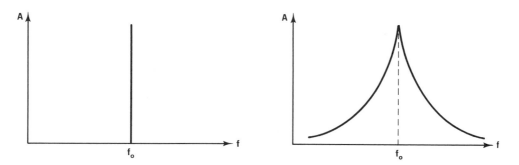

Figure 11.2   An "ideal" signal and a "real-life" signal with phase noise in the frequency domain as seen on a spectrum analyzer. The spectrum analyzer shows the envelope of power vs. frequency. The phase fluctuations appear as noise sidebands, extending above and below the nominal frequency.

frequency, as shown in the RF sideband spectrum of Fig. 11.2(b). Phase noise in the frequency domain can be thought of as simply an infinite number of phase modulation sidebands, each arising from a low-frequency modulation signal.

11.2.2 Conceptual Measurement of Phase Noise

To help gain an intuitive understanding of phase noise measurements, Fig. 11.3 shows a simple direct cancellation technique. The device under test (DUT) with phase noise is summed with an ideal reference source without phase noise. If the two input signals have the same frequency, the same amplitude, and are 180° out of phase, the carrier frequency is *nulled*, or cancelled out, leaving only the noise power. The resulting noise signal can then be measured with a wideband power meter, giving total noise, or—with a selective level meter—yielding noise energy as a function of frequency.

Unfortunately, this conceptually simple method is not practical for a number of systems applications. Phase-noise measurement systems are commonly needed

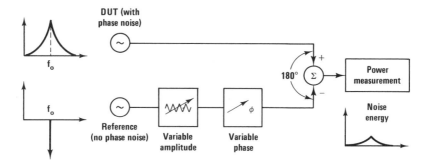

Figure 11.3   Direct cancellation method of measuring phase noise. If the two input signals are the same frequency and amplitude and 180° out of phase, the carrier is nulled, leaving the phase noise to be measured.

to test sources from 5 MHz to 26 GHz, with increasing need for measurements to 40, 60, and 100 GHz. Though the summation device in the direct cancellation method could be as simple as a resistive power splitter, maintaining a 180° phase relationship between two sources that draft independently is a difficult task. A critical feedback path to one of the sources is needed; for good cancellation of the carrier, the 180° relationship would need to be maintained with less than 5° of deviation! This "tight" feedback loop is very difficult to design, especially over a broad range of frequencies. This chapter (Sections 11.6, 11.7, and 11.8) describes measurement methods with better sensitivity which either do not require a feedback loop or require a phase feedback loop only to within 15°.

11.3 WHY MEASURE PHASE NOISE: THE EFFECT OF PHASE NOISE ON COMMUNICATIONS SYSTEMS

The phase noise of signal sources is a major concern in frequency-conversion applications, where the input signal levels span a wide dynamic range. The necessary phase-noise performance may vary greatly for different RF and microwave systems. But, in general, sideband phase noise can be the cause of interference into the information bandwidth and can limit the overall system sensitivity.

11.3.1 Phase Noise on a Receiver Local Oscillator

For example, suppose two signals are connected to a frequency converter, where they are mixed with a local oscillator (LO) signal down to an intermediate frequency (IF) for processing (Fig. 11.4). The local oscillator phase noise will be directly translated onto the mixer products. Although the receiver's IF filtering may be sufficient to remove the (larger) interfering signal's main mixing product, the (smaller) desired signal's mixing product is masked by the downconverted phase noise of the LO. The noise on the LO thus degrades the receiver's sensitivity and selectivity.

11.3.2 Phase Noise in a Digital Communications System

Figure 11.5 shows the state diagram (or IQ diagram, for in-phase versus quadrature phase) of a digitally modulated 16-QAM (quadrature amplitude modulated) system. The integrated phase noise in rms degrees (see Section 11.4.2) over a typical voice channel moves the radial position of a given state. (A good microwave source at 10 GHz has integrated phase noise of 0.25° rms over the 300-Hz to 3-kHz voice channel, and a typical system will have many voice channels.) Given sufficient phase noise on the local oscillator, the "dot" corresponding to a given state will actually move into the quadrant of another state, giving a false

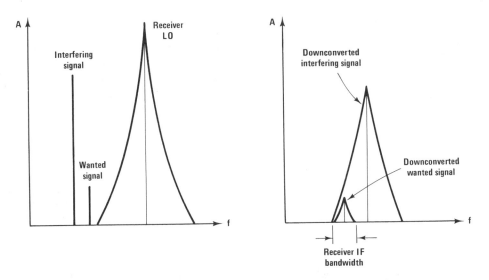

Figure 11.4 Phase noise can degrade receiver selectivity in a multisignal environment. The phase noise translated onto the IF (intermediate frequency) signals can mask a low-level signal.

bit and increasing the bit error rate (BER). The more states in the system, the more susceptible the system is to phase noise.

In digital communications systems, phase noise on the carrier signal also affects carrier-tracking accuracy. For example, LOs in the satellite-earth terminal upconverter, the satellite frequency translator, and the receive earth terminal downconverter can all contribute to the system phase noise. As stated, this directly affects the bit error probability of a digitally modulated system.

In general, the degradation caused by phase noise increases with the number of carrier signal sources between regenerative states in the system, unless phase-cancellation techniques are used. Similar phase noise accumulation problems may also occur in terrestrial microwave and other modulated systems.

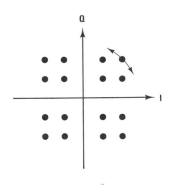

Figure 11.5 Integrated phase noise moves the radial position of the "dot," indicating a state in a 16-QAM modulated system. Excessive phase noise on the receiver LO will increase the bit error rate of the system.

11.4 QUANTIFYING PHASE NOISE

The noise on a frequency source is a superposition of causally generated signals (i.e., related to a definable cause) and random, nondeterministic noise (see Fig. 11.6). Some causal effects—for example, the aging process of components in the resonator—give rise to **long-term instability** (commonly called *frequency drift*). Long-term instability describes frequency variations that occur over time periods greater than a few seconds and is usually expressed in parts per million per time.

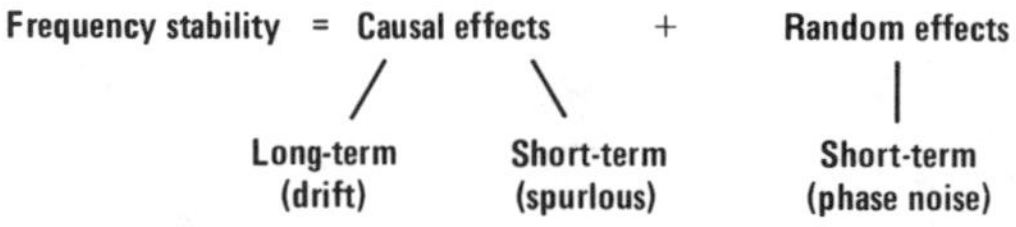

Frequency stability = Causal effects   +   Random effects

        /   \ |

Long-term   Short-term   Short-term
  (drift)   (spurious)   (phase noise)

Figure 11.6 Frequency instability is caused by the superposition of causally generated signals and random, nondeterministic noise.

Other causal effects—for example, power supply fluctuations or vibration—give rise to **short-term instability,** or frequency variations occurring over time periods less than 1 s. These causal effects are deterministic (systematic, discrete) signals, which appear as distinct components on an RF sideband spectrum (Fig. 11.7). These signals are commonly called **spurious.**

The random or power-law noise causes only random, short-term instability, or *phase noise* (Fig. 11.7). The random noise sources include thermal noise, shot noise, and flicker noise. This chapter focuses only on the measurement of this random noise on signal sources.

11.4.1 Phase Noise Described by a Power Distribution

Phase noise can be quantified in both the time and frequency domains, depending on the application. For example, a time domain description helps to determine the accuracy of clocks from a clock timing-error standpoint, whereas a frequency domain description can be used to determine the required carrier-tracking loop bandwidth. Most measurement techniques measure phase noise in the frequency domain, but the data can be converted to the time domain if desired [Fischer, 11.8].

In the frequency domain, the characteristic randomness of phase noise is best described by a power spectral density function, a distribution of relative power versus frequency. The most common units of phase noise describe the *power in the phase fluctuations*, the *power in the frequency fluctuations*, or the *RF sideband power*. These units are all mathematically equivalent and exist primarily because different measurement methods produce different units directly.

Figure 11.8 contrasts an RF signal spectrum and a demodulated spectral density distribution. The power spectral density function, $S(f_m)$, is a postdetection distribution of power as a function of offset frequency from the carrier. Essentially, the carrier is coherently demodulated to 0 Hz, and the power in the frequency or phase fluctuations is described as a function of the rate at which this phase

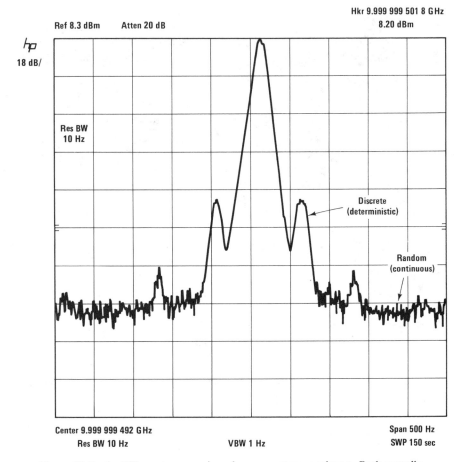

Figure 11.7   An RF spectrum as viewed on a spectrum analyzer. Both causally generated short-term instability (discrete spurious at the line frequency) and random short-term instability (phase noise) is seen.

modulation occurs. The rates of the modulating signals, or equivalently the offset frequency from the carrier, are designated by f_m (for modulation frequency, offset frequency, baseband frequency, or Fourier frequency).

Power spectral densities are typically described in a normalized 1-Hz bandwidth, though sometimes the spectral density is integrated over a bandwidth related to the application. Also, because of the large magnitude variations of the phase noise on an oscillator, phase noise is often conveniently described in logarithmic units (10 log of the spectral density of function used).

Spectral Density of Phase Fluctuations, $S_\phi(f_m)$

The most basic representation of phase noise is the spectral density of phase fluctuations, $S_\phi(f_m)$, where $S_\phi(f_m)$ is defined as the mean squared power in the

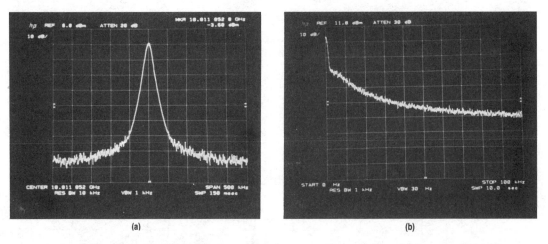

(a)   (b)

Figure 11.8   An RF spectrum as viewed on a spectrum analyzer (a) is contrasted with a demodulated phase noise power spectral density (b). Phase noise is usually described as power vs. offset frequency from the carrier, with the carrier translated to dc.

phase fluctuations per unit bandwidth:

$$S_\phi(f_m) = \frac{\Delta\phi^2_{\text{rms}}(f_m)}{\text{BW}} \quad \text{rad}^2/\text{Hz} \tag{11.3}$$

where BW (measurement bandwidth) is negligible with respect to any changes in S_ϕ versus the offset frequency f_m.

Expressed logarithmically, $S_\phi(f_m)$ is $10 \log S_\phi(f_m) = 10 \log \frac{\Delta\phi^2_{\text{rms}}}{\text{BW}}$, or

$$S_\phi(f_m) \text{ [dBr/Hz]} = 20 \log \frac{\Delta\phi \text{ (rad)}}{1 \text{ (rad)}} \text{ per Hz} \tag{11.4}$$

where dBr/Hz is decibels relative to 1 rad per hertz bandwidth.

$S_\phi(f_m)$ is measured directly when using a phase demodulator to measure phase noise (Section 11.8). $S_\phi(f_m)$ data are useful for analysis of phase-noise effects of systems with phase-sensitive circuits such as digital FM, PSK (phase-shift keying), and QAM communication links [Feher, 11.6].

Spectral Density of Frequency Fluctuations, $S_{\Delta f}(f_m)$

$S_{\Delta f}(f_m)$—the spectral density of frequency fluctuations on a per-hertz basis—is another common term for quantifying phase noise. As shown in Section 11.7, $S_{\Delta f}(f_m)$ data are obtained when measuring phase noise with a frequency discriminator. Analogous to $S_\phi(f_m)$, $S_{\Delta f}(f_m)$ is defined as

$$S_{\Delta f}(f_m) = \frac{\Delta f^2_{\text{rms}}(f_m)}{\text{BW}} \quad [\text{Hz}^2/\text{Hz}] \tag{11.5}$$

or the mean squared power in the frequency fluctuations per unit bandwidth. As in $S_\phi(f_m)$, the measurement bandwidth must be negligible with respect to any changes in $S_{\Delta f}(f_m)$. Expressed logarithmically, $S_{\Delta f}(f_m)$ is $10 \log \Delta f_{rms}^2/\text{BW}$, or

$$S_{\Delta f}(f_m) \quad (\text{dBHz/Hz}) = 20 \log \frac{\Delta f \,(\text{Hz})}{1 \,(\text{Hz})} \text{ per Hz} \qquad (11.6)$$

where dBHz/Hz is decibels relative to 1 Hz per hertz bandwidth.

Since frequency and phase are related, phase noise can be treated equivalently in terms of phase fluctuations ($S_\phi(f_m)$) or frequency fluctuations ($S_{\Delta f}(f_m)$). $S_{\Delta f}(f_m)$ can be directly derived from $S_\phi(f_m)$ using Laplace transforms [Rutman, 11.19], resulting in the relationship

$$S_{\Delta f}(f_m) = f_m^2 \, S_\phi(f_m) \qquad (11.7)$$

Logarithmically, this relationship is expressed as

$$S_{\Delta f}(f_m) \quad (\text{dBHz/Hz}) = S_\phi(f_m) \quad (\text{dBr/Hz}) + 20 \log \frac{f_m}{1 \,(\text{Hz})} \qquad (11.8)$$

Figure 11.9 illustrates this relationship graphically. The phase noise of a 10-GHz source is shown in terms of both frequency and phase fluctuations versus offset from the carrier. Note that the two plots differ by the expected f_m^2.

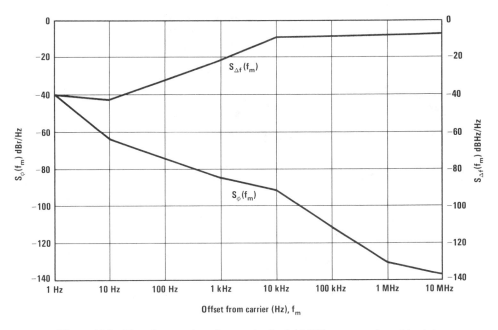

Figure 11.9 The phase noise of a synthesized 10-GHz source plotted both in terms of phase fluctuations and frequency fluctuations. On equal scales of $S_\phi(f_m)$ and $S_{\Delta f}(f_m)$, the two plots differ by a slope of f_m^2.

Single Sideband-Noise-to-Carrier-Power Ratio, $\mathscr{L}(f_m)$

The most common expression of phase noise, $\mathscr{L}(f_m)$ (script L) is not directly phase noise at all. $\mathscr{L}(f_m)$ is defined as the ratio of the power in one phase modulation sideband per hertz of bandwidth, P_{ssb}, to the total signal power, P_s (Fig. 11.10).

$$\mathscr{L}(f_m) = \frac{P_{ssb}}{P_s} = \frac{\text{power density (one phase-modulation sideband)}}{\text{total carrier power}} \quad \frac{\text{W}}{\text{W}} \quad (11.9)$$

$\mathscr{L}(f_m)$ has remained popular primarily because it is easily related to the RF power spectrum observed on a spectrum analyzer, a common measurement tool. Like the spectral densities, $\mathscr{L}(f_m)$ is usually presented logarithmically as a plot of sideband level expressed *in decibels relative to the carrier per hertz of bandwidth* (dBc/Hz):

$$10 \log \mathscr{L}(f_m) \quad (\text{dBc/Hz}) = 10 \log P_{ssb} - 10 \log P_s \quad (11.10)$$

$\mathscr{L}(f_m)$ is an indirect measure of noise energy that can sometimes be directly related to the fundamental unit of phase noise—$S_\phi(f_m)$—by a simple approximation. If the modulation sidebands of the sources are such that the peak phase deviation at a given rate (offset frequency) is much less than 1 rad ($\Delta\phi_{pk} \ll 1$ rad), then

$$\mathscr{L}(f_m) \cong \frac{1}{2} S_\phi(f_m) = \frac{S_{\Delta f}(f_m)}{2f_m^2} \quad (11.11)$$

In other words, because $\mathscr{L}(f_m)$ is defined as the power in a single phase-modulation sideband, it is only a valid expression of phase noise if there is insignificant energy in the higher-order sidebands. The relationship between $\mathscr{L}(f_m)$ and $S_\phi(f_m)$ can be derived from basic modulation theory, as shown in [Hewlett-Packard,

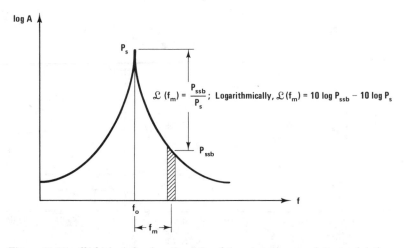

Figure 11.10 $\mathscr{L}(f_m)$ is defined as the ratio of the power in one-phase modulation sideband on a per-hertz basis to the total power in the carrier. On a spectrum analyzer, power is measured in dBm, a logarithmic scale, so $\mathscr{L}(f_m)$ in dBc (dB below the carrier) is P_{ssb} (in dBm) $-$ P_s (in dBm).

11.10]. The approximation of equation 11.11 holds for a modulation index $\beta = \Delta\phi_{pk}$ of approximately less than 0.2 rad.

In logarithmic units, the relations between $\mathscr{L}(f_m)$, $S_\phi(f_m)$, and $S_{\Delta f}(f_m)$ are

$$\mathscr{L}(f_m) \quad (\text{dBc/Hz}) = S_\phi(f_m) \quad (\text{dBr/Hz}) - 3 \quad \text{dB} \tag{11.12}$$

and

$$\mathscr{L}(f_m) \quad (\text{dBc/Hz}) = S_{\Delta f}(f_m) \quad (\text{dBHz/Hz}) - 20\log f_m - 3 \quad \text{dB} \tag{11.13}$$

For most phase-locked sources, the peak phase deviation at any modulation rate will be less than 0.2 rad. However, for free-running sources, the large-amplitude, low-rate phase fluctuations correspond to a high modulation index at low rates, and $\mathscr{L}(f_m)$ cannot be used to describe the phase noise close to the carrier.

11.4.2 Integrated Phase Noise

Some applications need a description of short-term instability integrated over a bandwidth, or **integrated noise power.** For example, in an FM mobile radio system, the integrated **frequency noise** (measured in rms hertz) over the receiver bandwidth (typically 20 Hz to 15 kHz) is important.

In a digital communications system, the integrated phase fluctuations in rms degrees can be useful for analyzing system performance. The receiver designer may be interested in the phase fluctuations integrated over the carrier-tracking loop bandwidth. This integrated noise will show if the chosen loop bandwidth is sufficient to track the phase fluctuations of the carrier with enough accuracy to produce high correlation in coherent demodulation [Spilker, 11.23]. Or, the designer may be interested in the effect of the carrier phase noise on the BER of the system. In this case, the integrated phase noise added over the voice channels—300 Hz to 3 kHz—is often the desired data. In other applications, the area of integration is the double-sided Nyquist bandwidth around the carrier frequency.

Integrated noise data over any bandwidth of interest is easily obtained from the spectral density functions. **Integrated frequency noise,** commonly called *residual FM*, can be found by

$$\text{Residual FM} = \left[\int_a^b S_{\Delta f}(f_m)\, df_m \right]^{1/2} \quad \text{[rms Hz]} \tag{11.14}$$

Integrated phase noise is determined similarly:

$$\text{Residual } \phi m = \left[\int_a^b S_\phi(f_m)\, df_m \right]^{1/2} \quad \text{[rms deg]} \tag{11.15}$$

11.4.3 Units in the Time Domain

There are several ways to express phase noise in the time domain as well. Specifically, phase noise can be expressed in terms of Allan variance or in terms of a probability density function (pdf).

Except for a short definition, a discussion of these units is beyond the scope of this book. Briefly, an Allan variance is a statistical calculation of fractional frequency differences. A phase pdf is a statistical distribution of a given phase, in rms radians, versus the number of occurrences of that phase. Further information on time domain units can be found in [Fischer, 11.8].

11.5 PHASE NOISE AND SOURCES

Choice of a phase-noise measurement technique is largely dependent on the level of phase noise to be measured. Correspondingly, the level of phase noise on a source is a function of the oscillating element used and the design of the source. This section examines two typical sources and their characteristic phase-noise plots. Then, these two sources will be used throughout the chapter to compare and contrast the various phase noise measurement techniques.

11.5.1 Slopes of Phase Noise

The spectral density plot of an oscillator signal is a combination of different power-law processes. On a log frequency scale, these processes produce five different characteristic slopes, from f_m^{-4} to f_m^0, labeled as shown in Fig. 11.11. To an oscillator designer, these slopes tell much about how to change the source design to reduce noise over a certain range of offset frequencies.

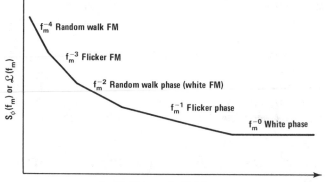

Figure 11.11 Noise processes of an oscillator, viewed on a phase noise plot in the frequency domain, yield different slopes as a function of log offset frequency.

11.5.2 Phase-Locked Versus Free-Running Sources

In terms of typical phase-noise power spectral densities, sources can be divided into two main categories, phase-locked and free-running. Figure 11.12 shows the typical phase noise of these sources. A free-running oscillator is typified by a power spectral density that follows an f_m^{-3} slope at offset frequencies close to the carrier and then an f_m^{-2} slope until the broadband noise floor, f_m^0, is reached.

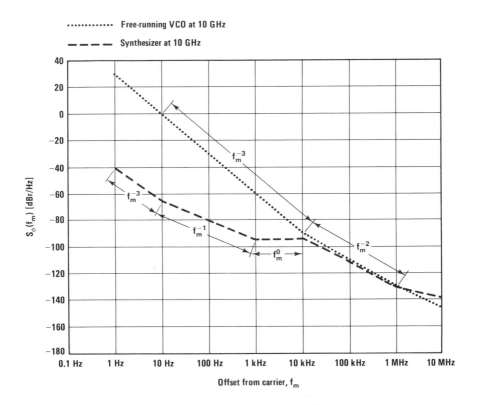

Figure 11.12 Comparing the phase noise of two 10-GHz sources, one phase-locked and one free-running. Compared on a scale of phase fluctuations, both have similar noise floors. However, the free-running source has higher noise close to the carrier, the noise increasing as f_m^{-2} and f_m^{-3}. The noise of the phase-locked source eventually slopes up as f_m^{-2} and f_m^{-3} but much closer to the carrier.

A source phase-locked to a stable reference (such as a crystal) has lower noise close to the carrier than a free-running oscillator; it will also typically have an f_m^{-1} or f_m^0 or flat region (synthesizer knee) before hitting the broadband noise floor. It is this difference in close-in noise performance that usually determines if a free-running source or a locked source will be used in a communications system. The close-in noise is also often the principal determinant of which phase-noise measurement method to use.

11.5.3 Multiplied Sources and Phase Noise

In many modern communications systems, the carrier frequency is derived from (or multiplied from) a stable reference source. This multiplication process increases the phase noise of the signal. Thus *when comparing the phase noise of sources, it is important also to state the carrier frequency.*

The same is true when comparing the sensitivities of phase-noise measurement techniques. Because some measurement methods use a source in addition to the test source, the sensitivity of the method will vary with carrier frequency. Sometimes the sensitivity or noise floor of a measurement system is given at a carrier frequency other than the desired test frequency. Knowing the effect of multiplication on phase noise, this noise-floor information can be translated to the equivalent noise floor at the desired frequency.

The effect of multiplication on the phase noise of a signal is a corollary of the effect of multiplication on a signal with FM modulation. If a signal f_0 with a single tone of FM modulation at rate x and with y hertz of deviation is doubled, the *new signal $2f_0$ has FM at the same rate x but with twice the deviation*. Thinking of phase noise as an infinite number of phase or frequency modulation sidebands, doubling the frequency of a signal doubles the amount of deviation at any offset (modulation) frequency. Since phase noise is related to the frequency or phase deviation squared (equations (11.3) and (11.5)), doubling the carrier frequency increases the phase noise by 6 dB. In the general case, when the carrier is multiplied, the phase noise at any offset frequency is increased by 20 log of the multiplication factor [Hewlett-Packard, 11.10].

Example 11.1

A good 10-MHz reference source has phase noise of -140 dBc/Hz at 10 Hz and -158 dBc/Hz at 100-Hz offsets. If this reference source is multiplied to 10 GHz, then the noise on the resulting signal will be 20 log(10 GHz/10 MHz) = 60 dB higher. The 10-GHz signal will have $\mathscr{L}(10\ \text{Hz})$ of -80 dBc/Hz and $\mathscr{L}(100\ \text{Hz})$ of -98 dBc/Hz. ∎

Sections 11.6 to 11.8 describe three methods of phase noise measurements. For each method, the basic concepts, the sensitivity and limitations, and the measurement procedure are examined.

The applicability of the method is also shown, by comparing the system sensitivity to the noise of the typical oscillators of Section 11.5. For any phase-noise measurement method, the *key specification is noise floor, or sensitivity*. The phase noise of the test source can be measured only if it is higher than the noise floor of the measurement system.

Measuring the phase noise on an oscillator, particularly a low-noise oscillator, requires measuring very low-level signals (down to less than 150 dBm), so careful system setup is important for all three methods. For low-noise measurements, the source (device) under test should be warmed up and properly heat sunk to reduce frequency drift. Also, oscillators are very sensitive to their environments; a careless measurement setup can induce phase noise. Signals can be induced into the measurement from the power supply, from microphonic vibrations, or from other signal sources in the environment.

Several things can be done to reduce instability induced by power supply. The power supply to the source might need to be regulated to avoid large line-frequency sidebands from being induced in the oscillator. For very unstable sources with large power supply, spurious (such as free-running yttrium-iron-garnet (YIG) oscillators) running the test source from a battery might be necessary. Also, short lead lengths from the power supply often help reduce instability induced by power supply.

Induced microphonic signals can also cause problems. These can be minimized by tight connections; an isolator might be needed at the output of the DUT to reduce frequency pulling from movement of the connecting cable. Isolation from induced microphonics can be improved by placing the DUT on a vibration absorbing pad.

Other signals can be induced into the measurement from signals in the environment. For very low noise measurements, an RFI shielded room might be necessary. High-power radio stations have been known to appear as spurious signals in the noise spectrum!

11.6 DIRECT SPECTRUM MEASUREMENT OF PHASE NOISE

11.6.1 Basic Concepts

The simplest and most straightforward method of phase noise analysis of sources is the direct spectrum method. The source under test is connected directly to a spectrum analyzer tuned to the carrier frequency, where the phase noise is measured in terms of $\mathscr{L}(f_m)$, ($\mathscr{L}(f_m) = P_{ssb}/P_s$, equation 11.9). On a spectrum analyzer, $\mathscr{L}(f_m)$ data can be calculated directly from the display. Since the analyzer's amplitude scale is logarithmic, $\mathscr{L}(f_m)$ is simply $10 \log P_{ssb}/P_s$, or $10 \log P_{ssb} - 10 \log P_s$.

11.6.2 Sensitivity and Limitations

The fundamental limitation of the direct spectrum method is noise floor, or sensitivity. Figure 11.13 shows the noise floors of typical spectrum analyzers, compared to the phase noise of the two sources from Section 11.5. Spectrum analyzers typically do not have sufficiently low noise floor to measure at offsets greater than about 10 kHz from the carrier.

In addition to poor sensitivity, the direct spectrum method has other limitations. It cannot be used on sources with high AM noise or for close-to-the-carrier measurements on drifting sources.

A spectrum analyzer with no phase information does not distinguish amplitude modulation from angular modulation—it measures combined noise. For $\mathscr{L}(f_m)$ to be a valid representation of phase noise, the noise power measured must contain only phase-modulation sidebands. If the source has high AM noise, the spectrum analyzer will also measure this AM noise, giving an incorrectly high phase-noise result.

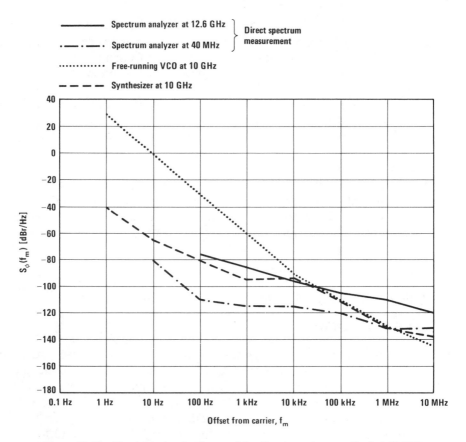

Figure 11.13 The typical noise floors of the direct spectrum method at 40-MHz and 12-GHz carrier frequencies are compared to the noise of the phase-locked and free-running sources. The direct spectrum measurement does not have sufficient sensitivity to measure the broadband noise floors of these sources, nor can it measure the close-in noise of the phase-locked source.

Also, for offsets close to the carrier, a spectrum analyzer uses a narrow resolution bandwidth. Narrowing the resolution bandwidth increases the sweep or span time. If the source drifts more than about $\frac{1}{20}$ of the total analyzer frequency span during the span time, the resultant measured data can be significantly in error.

11.6.3 Making a Measurement

First the carrier level or signal level, P_s, is measured, and then the power at an offset frequency, P_m, is measured (P_m for power measured in bandwidth B), as shown in Fig. 11.14. Since $\mathscr{L}(f_m)$ is defined in a 1-Hz bandwidth, the measured sideband power must be converted to the equivalent power in a 1-Hz noise

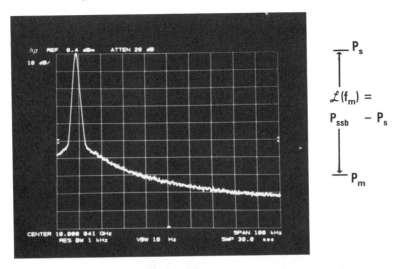

Figure 11.14   Computing $\mathscr{L}(f_m)$ from a spectrum analyzer display. $\mathscr{L}(f_m) = P_{ssb} - P_s$. On the display, the measured sideband level P_m must be normalized to a 1-Hz bandwidth to obtain P_{ssb}.

bandwidth. This is a simple power normalization process, where the noise bandwidth normalization scale factor, SF_{bw}, is

$$SF_{bw} = 10 \log \frac{B_m}{B_M} \qquad (11.16)$$

where B_m = the measurement bandwidth
$\quad\ B_M$ = the equivalent 1-Hz noise bandwidth

With P_{ssb} = equivalent single sideband power in 1-Hz bandwidth,
$\quad\ \ P_m$ = measured noise power in bandwidth B_m

then

$$P_{ssb} = P_m - SF_{bw} \qquad (11.17)$$

The measured noise must also be corrected for any inaccuracies that the measurement system might have when measuring random noise. For example, *noise* data read from an analog spectrum analyzer (which is optimized for measurements on sine waves) is too low due to the effect of the log amplification/peak detection circuitry of the analyzer. A general formula, then, for obtaining $\mathscr{L}(f_m)$ from a spectrum analyzer display is given by

$$\mathscr{L}(f_m) \text{ (dBc/Hz)} = 10 \log P_{ssb} - 10 \log P_s = P_m - SF_{bw} + C_m - P_s \qquad (11.18)$$

where P_m = measured noise level in bandwidth B_m (dBm)
$\quad SF_{bw}$ = noise bandwidth normalization (dB)
$\quad\ \ C_m$ = correction for the measurement system (dB)
$\quad\ \ P_s$ = signal or carrier level (dBm)

Example 11.2

The spectrum analyzer photo of Figure 11.15 shows a measurement on a 10-GHz synthesized source. The signal or carrier level is +8.4 dBm. The measured noise power, P_m, at 10.00001 GHz (a 10-kHz offset from the carrier) is −52 dBm, in a 1-kHz measurement bandwidth. For this source, the measured $\mathscr{L}(f_m)$ is

$$
\begin{array}{lrl}
P_m & -52 & \text{dBm} \\
-SF_{bw} & -30 & \text{dB (10 log 1 kHz/1 Hz)} \\
+C_m & +2.5 & \text{dB (for the HP 8566 spectrum analyzer used)} \\
\underline{-P_s} & \underline{-8.4} & \text{dBm} \\
\mathscr{L}(10 \text{ kHz}) = & -87.9 & \text{dBc/Hz}
\end{array}
$$

C_m was obtained from [Hewlett-Packard, 11.12]. This value is typical of analog spectrum analyzers within a range of about ±0.2 dB. ∎

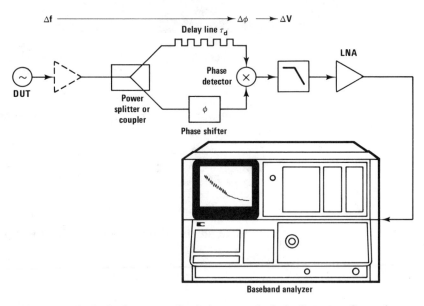

Figure 11.15 In the frequency discriminator method, the frequency fluctuations of the device under test (DUT) are translated to low frequency voltage fluctuations that can be measured by a low frequency analyzer. The delay line/mixer implementation of a discriminator is shown here; the delay line translates frequency fluctuations into phase fluctuations, and the phase detector translates phase fluctuations into voltage.

11.6.4 Summary of the Direct Spectrum Method

In summary, the direct spectrum method of phase-noise measurement is simple, and $\mathscr{L}(f_m)$ can be calculated directly from the display. Unfortunately, it cannot be used to measure low noise sources, nor can it measure "noisier" sources that have either high AM noise or drift. This technique could not be used to measure the entire offset frequency range of the two sources from Fig. 11.12. The direct spectrum method is a good technique for measuring stable (phase-locked) sources with relatively high phase-noise sidebands.

11.7 FREQUENCY DISCRIMINATOR MEASUREMENT OF PHASE NOISE

11.7.1 Basic Concepts

The frequency discriminator method of phase-noise measurements obtains increased sensitivity by nulling or demodulating the carrier and then measuring the noise of the resulting baseband signal. In this method (also called the one-source method), the short-term *frequency fluctuations* of the source under test are translated to low-frequency voltage fluctuations, which can be measured by a low frequency analyzer (see Fig. 11.15). The analyzer must be able to measure power as a function of frequency; selective voltmeters, wave analyzers, or spectrum analyzers can be used. The fundamental output of a frequency discriminator is $S_{\Delta f}(f_m)$, the spectral density of frequency fluctuations.

11.7.2 Overall Performance

First, we put the frequency discriminator in perspective. Figure 11.16 shows typical sensitivities of this method and the noise of the two sources in units of $S_\phi(f_m)$. The frequency discriminator method offers excellent sensitivity at high offset frequencies, but the sensitivity degrades as $f_m^{-2} f_m^{-3}$ as the offset frequency approaches 0 Hz. Notice, however, that this slope follows the same noise characteristic as the free-running source. Because of this correlation and because of the relative simplicity of this method, the frequency discriminator method is optimized for measurement of free-running sources—sources with large-amplitude, low-rate phase instabilities. With careful design, this technique can be used to measure free-running sources at offsets as low as 100 Hz or less. But because of the poor close-in sensitivity, it is *usually not very useful for measuring phase-locked sources close to the carrier* (offsets less than 1 kHz).

There are several common implementations of frequency discriminators. Ashley [Ashley, 11.3] and Ondria [Ondria, 11.16] discuss the use of a cavity resonator. A cavity resonator provides high sensitivity, good AM noise rejection, and good suppression of the carrier. However, cavity discriminators typically have very narrow input bandwidth. At very high frequencies (VHF), good sensitivity can be obtained using slope detectors and ratio detectors as discriminators. For VHF through millimeter frequencies, a delay line/mixer can be used, as described by Halford [Halford, 11.9] and Ashley [Ashley, 11.3]. For best sensitivity at higher frequencies, the test signal can first be downconverted with a low phase-noise reference source, and the resultant IF signal can then be analyzed with a discriminator in the VHF range (Section 11.7.5).

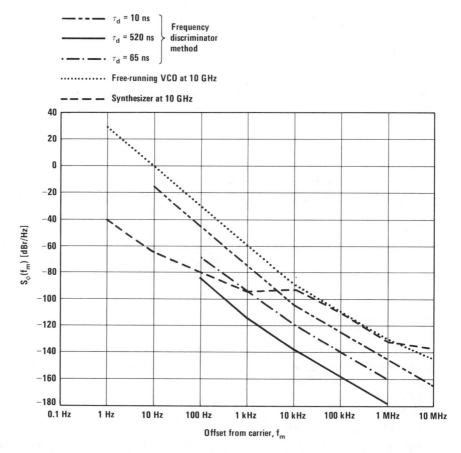

Figure 11.16 Typical sensitivities of the frequency discriminator method compared to two typical sources. This method has a very low broadband noise floor, but as the offset frequency approaches 0 Hz, the noise floor rises as f_m^{-2} and f_m^{-3}.

Each implementation of a discriminator has its own set of advantages and disadvantages, depending on such factors as the input frequency range, sensitivity, and cost. This section examines a delay line/mixer implementation to illustrate the basic principles of the frequency discriminator method. The delay line/mixer implementation is widely used and has the advantage of being able to accept a fairly broad range of input frequencies.

11.7.3 Delay Line/Mixer as Discriminator

Figure 11.15 shows the delay line/mixer implementation of a frequency discriminator. The device under test is split into two channels. One channel (sometimes called the nondelay, reference, or LO channel) is applied to the LO port of a double-balanced mixer. The other channel is delayed through a time element and then applied to the R port of the mixer.

The delay line works as a frequency-to-phase transducer $(f - \phi)$. That is, the delay line yields a nominal phase shift relative to the reference channel,

$$\phi_0 = 2\pi\tau_d f_0 \qquad (11.19)$$

where ϕ_0 = the nominal phase shift
  τ_d = the propagation delay of the line
  f_0 = the nominal carrier frequency

Around this nominal value, the instantaneous frequency fluctuations give rise to linearly proportional phase shifts:

$$\Delta\phi = 2\pi\tau_d\Delta f \qquad (11.20)$$

where $\Delta\phi$ = the instantaneous phase change, and
  Δf = the instantaneous frequency fluctuation

The delay element uncorrelates the noise of the source arriving at the mixer inputs via the two paths.

At the output of the delay line and the reference channel, a double-balanced mixer is used as a phase detector, converting the phase fluctuations into voltage equivalents for measurement by the low-frequency analyzer. To operate the mixer as a phase detector, the two signals (from the same source and, therefore, at identical frequencies) to the mixer are maintained nominally in phase quadrature (90° out of phase; $\phi_0 = 2\pi\tau_d f_0 = 90°$). The mixer sum frequency is filtered off, and the difference frequency is 0 Hz. Riding on this dc signal are small ac voltage fluctuations. For small phase deviations, ΔV is linearly proportional to the fluctuating phase difference between the two signals, $\Delta\phi$ (see Section 11.8.3). The conversion between phase and voltage is defined as K_ϕ, the phase detector constant, in V/radian:

$$\Delta V(f_m) = K_\phi\Delta\phi(f_m) \qquad (11.21)$$

After the phase fluctuations are converted to baseband voltages, a high-gain, low-noise amplifier (LNA) is used to provide additional system sensitivity. The LNA improves the noise figure of the front end of the analyzer. Typically, the LNA has about 40 dB of gain and a noise figure of about 2 to 3 dB.

The system analyzer is typically a spectrum analyzer covering the *offset frequency range* of interest. If measurements close to the carrier are desired, an analog analyzer should have a synthesized LO, or an FFT analyzer can be used. These stable analyzers will also provide narrow resolution bandwidths, which allow the resolution of spurious signals from the random noise.

The noise floors of these stable analyzers are usually not critical. Because the frequency discriminator method improves sensitivity by (1) removing the carrier and (2) using the LNA to improve the noise figure of the analyzer, the overall sensitivity of the system is set by the discriminator transfer function.

11.7.4 Sensitivity and Limitations

Transfer Function of a Discriminator

A rigorous derivation of the delay line/mixer used as a frequency discriminator yields a transfer function of

$$\Delta V(f_m) = K_\phi 2\pi\tau_d \Delta f(f_m) \frac{\sin(\pi f_m \tau_d)}{(\pi f_m \tau_d)} \tag{11.22}$$

where $\Delta V(f_m)$ = the voltage out of the discriminator
$\quad\quad\quad \Delta f(f_m)$ = the short-term frequency fluctuations
$\quad\quad\quad\quad\quad\quad$ of the DUT [Hewlett-Packard, 11.11]

The output follows a $\sin x/x$ response, is periodic in $\omega = 2\pi f_m$, and has peaks and nulls, with the first null at $1/\tau_d$. This is shown in Fig. 11.17, a swept calibration display showing the magnitude of the transfer response of a system with 500 ns of delay. The first null occurs at $1/\tau_d = \frac{1}{500}$ ns, corresponding to an offset frequency of 2 MHz.

It is possible to measure at offset frequencies out to and beyond the nulls by scaling the measurement results with the $\sin x/x$ response. However, since the system has very poor sensitivity in the region of the nulls, typically the discriminator method is used only to offset less than $1/2\pi\tau_d$, where the system response is at a maximum and is linear. In this region,

$$\Delta V(f_m) \simeq K_\phi 2\pi\tau_d \Delta f(f_m) = K_d \Delta f(f_m) \tag{11.23}$$

K_d, the frequency discriminator constant in volts per hertz, is the overall translation factor between the instantaneous frequency fluctuations at the input to the discriminator and the corresponding voltage output. For offset frequencies less than $1/2\pi\tau_d$,

$$K_d = K_\phi 2\pi\tau_d \quad\quad [V/Hz] \tag{11.24}$$

The reduced transfer function of equation (11.23) is *linearly proportional to the instantaneous frequency fluctuations*, and is also a function of K_ϕ and τ_d.

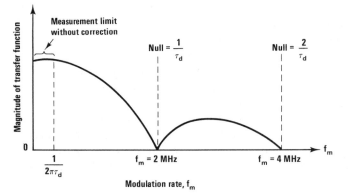

Figure 11.17 The transfer function of a frequency discriminator system with a 500-ns delay time. The transfer function follows a $\sin x/x$ response, with the first null at $1/\tau_d$.

This implies that the sensitivity of the discriminator can be increased simply by increasing the delay τ_d or by increasing the phase detector constant K_ϕ. This assumption is not completely correct, and there are some trade-offs associated with increasing these constants.

First, the transfer function shows that increasing τ_d increases the sensitivity of the system. However, increasing τ_d also decreases the offset frequencies that can be measured without compensating for the sin x/x response. For example, a 200-ns delay line will have better close-in sensitivity than a 50-ns line but will not be useable beyond 2.5 MHz without compensating for the sin x/x response; the 50-ns line is useable to offsets of 10 MHz.

Second, K_ϕ is dependent on the level of the input signals to the mixer, but K_ϕ has a mixer-dependent maximum. Given sufficient LO drive to the mixer, increasing the level to the R port of the mixer (the output of the delay line) will increase K_ϕ until the mixer reaches **compression.** At compression, K_ϕ is at a maximum. Above the compression point of the mixer, increasing the R port input level will not increase system sensitivity.

Finally, in actual use the delay line/mixer is a power-limited system, and K_ϕ and τ_d are not independent. Increasing the time delay (length of line), τ_d, will decrease the input level of the delayed signal to the mixer and thus reduce K_ϕ. Maximizing system sensitivity thus involves a trade-off between the length of the line and the attenuation induced in that length.

The optimum length of line occurs where the decrease in the output voltage is equally compensated by the increase in $2\pi\tau_d$. This condition occurs when the attenuation of the line is 8.686 dB [Hewlett-Packard, 11.11]. Of course, the length of line yielding 8.68 dB of attenuation is a function of the carrier frequency and the delay element used. Also note that the transfer function of a discriminator is *not a function of carrier frequency*, f_0. Given a constant input level into the mixer such that K_ϕ is not reduced, a 50-ns line yields the *same system sensitivity whether operating at* 10 *MHz or* 10 *GHz.*

Overall Noise Floor and Conversion of Units

The voltage output of the discriminator is linearly proportional to the input frequency fluctuations by the discriminator constant, K_d. Substituting equation (11.23) into the definition of $S_{\Delta f}(f_m)$ yields

$$S_{\Delta f}(f_m) = \frac{\Delta f_{\mathrm{rms}}^2}{\mathrm{BW}} = \frac{\Delta V_{\mathrm{rms}}^2(f_m)}{K_d^2} = \frac{\Delta V_{\mathrm{rms}}^2(f_m)}{(K_\phi 2\pi\tau_d)^2} \tag{11.25}$$

where $\Delta V_{\mathrm{rms}}^2(f_m)$ is normalized to a unit bandwidth.

Equation 11.25 shows the sensitivity of the discriminator in units of frequency fluctuations. However, to compare the discriminator method directly with other measurement methods, the output of the discriminator must be described in terms of phase fluctuations. This is easily done using the transformation developed in Section 11.4, $S_{\Delta f}(f_m) = f_m^2 S_\phi(f_m)$ [equation (11.7)].

In units of phase fluctuations, the output of the discriminator is *a function of offset frequency*. Figure 11.16 shows the sensitivity of the discriminator method described in units of $S_\phi(f_m)$. At high offset frequencies, the delay/line mixer discriminator has a noise floor equivalent to the combined noise of the mixer and LNA at an offset frequency of $1/2\pi\tau_d$. The offset frequency squared (f_m^2) term of equation (11.7) "tips up" by 20 dB per decade the noise floor set by the mixer and LNA as the offset frequency decreases. Close to the carrier, the flicker characteristic of $S_\phi(f_m)$ translates into an f_m^{-3} slope in units of $S_\phi(f_m)$.

11.7.5 Making a Measurement

Phase-noise measurements using the frequency discriminator method can be broken into four simple steps:

1. Setup
2. Calibrate
3. Measure
4. Convert to phase-noise units

Setup

System setup is shown in Fig. 11.15. First, the power levels into the mixer should be verified. This (1) ensures that the discriminator will operate at the desired sensitivity (sufficient K_ϕ) and (2) allows use of a substitution source for calibration.

Attenuation through the delay line is a function of delay element and carrier frequency. Figure 11.18 shows that a 50-ns delay line (type N cable, RG 214) has more than 17 dB of attenuation at 10 GHz. Testing a 10-GHz source would require approximately +23 dBm of power at the input to the power splitter in order to provide 0 dBm into the R port of the phase detector (assuming a 6 dB loss through the splitter).

Constructing a 520-ns delay (the lowest noise floor shown in Figure 11.16) would require over 346 ft (more than 105 m) of cable! (Cable with a polyethelene dielectric, with a delay of approximately 1.5 ns/ft.) Clearly, delay line discriminator measurements at microwave frequencies require low-loss cable, high power, or alternative block diagrams.

If there is insufficient power at the output of the desired delay line length, one alternative is to add a low-noise amplifier *before* the splitter, as shown as a dashed line in Fig. 11.15. The noise added by the amplifier must be below the noise of the test source in order not to degrade system sensitivity.

Another alternative at microwave frequencies is shown in Fig. 11.19. Here, the high-frequency device under test is first downconverted with a low-noise reference source. The downconverted intermediate frequency now contains the noise of the (higher-frequency) test source but can be connected to a delay line/discriminator operating at the IF (typically less than 1 GHz). At the IF

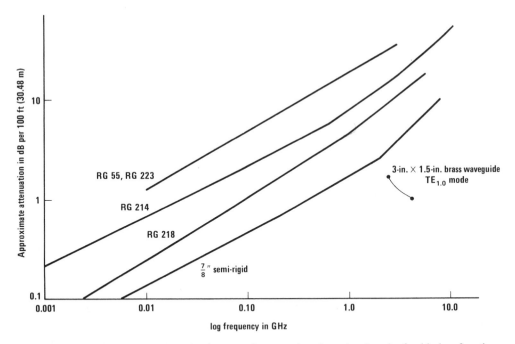

Figure 11.18 The amount of attenuation through a length of cable is a function of the type of cable (i.e., the dielectric and the inner conductor) and the frequency passing through the cable.

frequency, there is much less loss through a length of delay line than there is at the carrier frequency. With an easily achievable IF amplitude of $+7$ dBm (at a frequency of 500 MHz), over 100 ft (30.5 m) of simple RG 223 cable can be used as a delay line, for a delay time greater than 150 ns.

Calibrate

The calibration procedure determines the discriminator constant K_d to use in the transfer response:

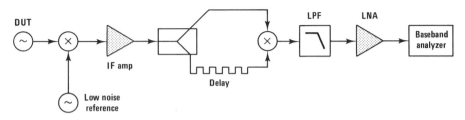

Figure 11.19 Downconverting and then discriminating at the IF (intermediate frequency). Since the IF is a lower frequency than the carrier, the delay line can be made longer for more sensitivity with less loss than would be achievable directly at the carrier frequency.

$$\Delta V(f_m) = K_d \Delta f(f_m) \tag{11.23}$$

The discriminator constant K_d can be obtained by

1. Measuring K_ϕ and τ_d individually [Hewlett-Packard, 11.10].
2. Measuring the overall K_d by analyzing the response of the system to a known input.

Usually the easiest way to determine K_d is by measuring the system response to a known signal. A signal with a single FM tone is ideal. The modulation index β should be kept below 0.2 dB radians so that the power in the higher-order sidebands is negligible. Often the source under test itself can be frequency modulated to produce this calibration signal. If not, *an alternate source at the same frequency and power* can be substituted for the test source.

Since the calibration procedure characterizes the system in its operation as a discriminator, the mixer must be acting as a phase detector during calibration. The input signals to the mixer must be set in phase quadrature (90° out of phase). There are several ways to establish quadrature at the phase detector (usually determined by monitoring for 0 V dc on an oscilloscope). First, since the nominal phase arriving at the phase detector is $\phi_0 = 2\pi\tau_d f_0$ (equation 11.19), the phase can be changed by varying the source frequency f_0 slightly. This is often the easiest way to establish quadrature, since a 50-ns line requires only a few megahertz of frequency change to produce 90° of phase shift.

If the source frequency cannot be changed, a line stretcher can be added to the fixed delay, changing τ_d, until quadrature is established. Typically, less than a few inches of variable length are needed. If both of these techniques are awkward, either analog or digital phase shifters could also be used.

Figure 11.20 shows a sample calibration signal, a signal with a single FM tone at a 1-kHz rate with 100 Hz of peak FM deviation, for a modulation index β of 100/1 kHz = 0.1. With the phase detector in quadrature, the system demodulates the carrier, and the power output at 1 kHz represents the response of the discriminator to 100-Hz peak frequency fluctuations at a 1-kHz rate. Rearranging equation (11.23) and substituting for the calibration values gives:

$$K_d = \frac{\Delta V_{\text{cal pk}}}{\Delta f_{\text{cal pk}}} = \frac{\sqrt{2}\,\Delta V_{\text{cal rms}}}{\Delta f_{\text{cal pk}}} \tag{11.26}$$

Measure

If another source has been substituted during calibration, the test source is reconnected and quadrature is set. With the mixer in quadrature, the phase noise of the DUT can be measured. Typically, the delay line has sufficient bandwidth such that the two inputs to the phase detector will remain in quadrature for the duration of the measurement even if the test source drifts. However, quadrature should be monitored throughout the measurement in case the source frequency changes significantly.

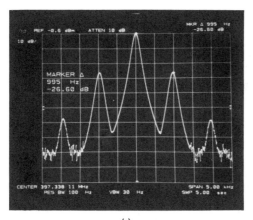

(a)

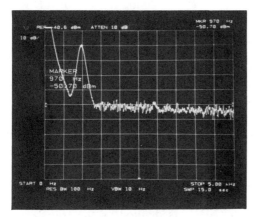

(b)

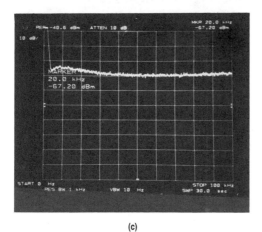

(c)

Figure 11.20   A frequency discriminator can be calibrated by measuring the response of the system to a known input signal. (a) The input calibration signal with ΔSB_{cal} of -26 dBc, and $f_{m\ cal}$ of 1 kHz. (b) The response of the discriminator to the calibration signal. (c) The detected phase noise of the test signal, in voltage proportional to the instantaneous frequency fluctuations of the test source.

Convert to Phase-Noise Units

Equation 11.25 describes the spectral density of the frequency fluctuations of the source under test in terms of the detected power output ($\Delta V_{rms}^2(f_m)$) and the discriminator constant K_d.

Substituting the calibration constant K_d from equation (11.26) into equation (11.25), the detected noise is expressed in units of $S_{\Delta f}(f_m)$:

$$S_{\Delta f}(f_m) = \frac{\Delta V_{rms}^2(f_m)}{\dfrac{2\Delta V_{rms\ cal}^2}{\Delta f_{pk\ cal}^2}} \tag{11.27}$$

where $\Delta V_{rms}^2(f_m)$ is the normalized power in the frequency fluctuations measured at any offset frequency and $\Delta V_{rms\ cal}^2$ is the power measured in the system response calibration signal.

Expressing $S_{\Delta f}(f_m)$ logarithmically yields

$$S_{\Delta f}(f_m) = P_m - P_{cal} + 20 \log \Delta f_{cal} - 3 \quad \text{dB} \tag{11.28}$$

If units of $\mathscr{L}(f_m)$ are desired and the small angle criterion is valid ($\Delta\phi_{pk} \ll 1$ rad), then the $S_{\Delta f}(f_m)$ data could be converted using $\mathscr{L}(f_m) = S_{\Delta f}(f_m)/2f_m^2$ [equation (11.11)], or, logarithmically, $\mathscr{L}(f_m) = S_{\Delta f}(f_m) - 20 \log f_m - 3$ dB [equation (11.13)].

However, if the desired unit is $\mathscr{L}(f_m)$, the system response can be calibrated directly with a known sideband-to-carrier ratio ΔSB_{cal} at a known rate f_{mcal}. For example, in Fig. 11.20(a), ΔSB_{cal} is read as -26 dBc, at f_{mcal} of 1 kHz. The power measured at the discriminator output in response to ΔSB_{cal} represents the response of the overall system to noise at that level and rate (Fig. 11.20(b)). Rate information must be included because the relationship between $\mathscr{L}(f_m)$ and $S_{\Delta f}(f_m)$ is a function of f_m. The measured noise power at any offset must be scaled by the ratio of the measured offset to the calibration offset.

The system scale factor, SF_{sys}, is defined as the transfer function of the discriminator to a known noise level at a known rate:

$$SF_{sys} = \Delta SB_{cal} - P_{cal} \tag{11.29}$$

Then,

$$\mathscr{L}(f_m) = P_m + SF_{sys} - 20 \log \frac{f_m}{f_{mcal}}$$

More completely,

$$\mathscr{L}(f_m) = P_m + \Delta SB_{cal} - P_{cal} - 20 \log \frac{f_m}{f_{mcal}} - SF_{bw} + C_m \tag{11.30}$$

where P_m = measured noise power (dBm)

ΔSB_{cal} = calibration sideband-to-carrier ratio (dBc)

P_{cal} = measured calibration power (dBm)

$20 \log f_m/f_{mcal}$ = conversion to $\mathscr{L}(f_m)$ at measured offset (dB)

SF_{bw} = conversion to 1-Hz noise bandwidth (dB)

C_m = correction for measurement system (dB)

and $\mathscr{L}(f_m)$ is in dBc/Hz.

The measured noise power is normalized to the power in a 1-Hz noise bandwidth. This requires normalizing the resolution bandwidths in the spectrum analyzer, as discussed in Section 11.5. Also, any corrections for the inaccuracies of the measurement system while measuring random noise must be taken into account.

Example 11.3

Figure 11.21 shows measurement results on a 10-GHz cavity oscillator. The test signal was first downconverted with a low phase-noise reference source, and the delay line discriminator operated at 400 MHz. A 68-ft (20.73-m) section of RG cable yielded 100 ns of delay.

Figure 11.21 Typical hardware used to make phase noise measurements. A free-running 10-GHz Gunn oscillator is placed in a vice for heat sinking.

Figure 11.20(a) shows the input calibration signal, with ΔSB_{cal} of -26 dBc. The discriminator response to the calibration signal is shown in Fig. 11.20(b). Figure 11.20(c) shows the detected phase noise power. The measured noise power at 100 kHz is -69 dBm in a 1-kHz bandwidth. The phase noise of this 10-GHz source at a 100-kHz offset from the carrier is:

$$
\begin{array}{rll}
\mathscr{L}(100 \text{ kHz}) = & -69 \quad \text{dBm} & P_m \\
& -26 \quad \text{dBc} & + \Delta SB_{cal} \\
& -(-51) \quad \text{dBm} & - P_{cal} \\
& -40 \quad \text{dB} & - 20 \log f_m/f_{mcal} = (20 \log 100k/1k) \\
& -30 \quad \text{dB} & - SF_{bw} \\
& \underline{+2.5 \quad \text{dB}} & + C_m \text{ (for the analyzer used)} \\
& -111.5 \quad \text{dBc/Hz} &
\end{array}
$$

A typical frequency discriminator test setup is shown in Fig. 11.21. Measuring the 10-GHz cavity source used in the example above, the system of Fig. 11.21 produced the hard-copy output shown in Fig. 11.22.

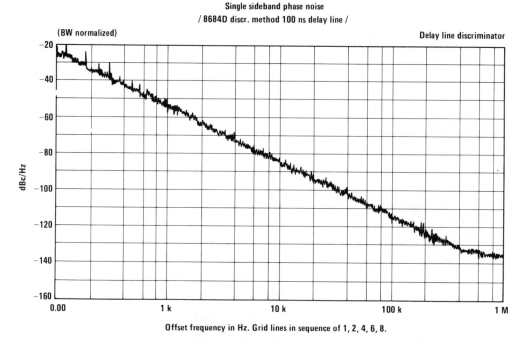

Figure 11.22 Typical output of a frequency discriminator system, showing the phase noise of a free-running 10-GHz cavity oscillator.

11.7.6 Summary of the Frequency Discriminator Method

The frequency discriminator method has a low broadband noise floor, and its sensitivity matches the free-running source characteristic. However, the discriminator method is not very useful for measuring close-in on phase-locked sources, unless very high Q cavities or very long delay lines are used.

The discriminator measurement is ideal for measuring unstable sources. The delay line/mixer implementation is not very sensitive to drift in the source under test and can measure sources with very high close-in phase noise or spurious.

11.8 PHASE-DETECTOR MEASUREMENT OF PHASE NOISE

11.8.1 Basic Concepts

As does the frequency discriminator, the phase-detector method (or two-source method) increases system sensitivity by demodulating the carrier and then measuring the noise on the resulting baseband signal.

The basic phase-detector method is shown in Fig. 11.23. In this method, the **short-term phase fluctuations** of the source are translated directly to baseband voltage fluctuations, which can be measured by a low frequency analyzer. K_ϕ, the phase-detector constant in volts per radian, is the proportionality constant between the phase noise of the test source and the corresponding voltage output:

$$\Delta V(f_m) = K_\phi \, \Delta\phi(f_m) \qquad (11.31)$$

The fundamental output of the phase detector is voltage proportional to phase, or $S_\phi(f_m)$, the spectral density of phase fluctuations.

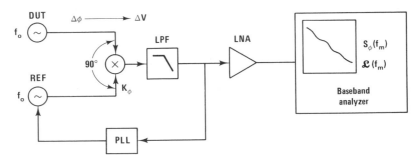

Figure 11.23 In the phase-detector method, two sources at the same frequency and 90° out of phase are connected to a double-balanced mixer. The mixer acts as a phase detector, translating the phase fluctuations of the source under test to low frequency voltage fluctuations. A PLL is needed to keep the two sources 90° out of phase.

11.8.2 Overall Performance

Figure 11.24 shows the typical sensitivity of the phase detector with the phase noise of the two sources from Section 11.5. The phase-detector method has a low noise floor over the entire offset frequency range. The **low broadband noise floor** (offsets greater than 100 kHz) enables the phase-detector method to be used to measure free-running sources. And the *low noise close to the carrier* means that stable or locked sources can be measured as well. Since most new communications systems use locked sources, the phase-detector method is becoming increasingly widespread.

Though the phase-detector method offers the best sensitivity, it is not always the optimal solution. Sources with high drift or very high close-in phase noise may be difficult to measure with the phase-detector method. Also, the phase-detector method is more complex than either the direct spectrum or frequency discriminator methods. If the extra sensitivity is not needed, the other techniques are simpler.

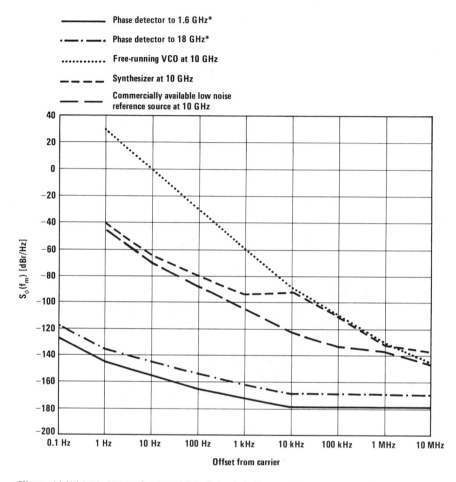

─────────── Phase detector to 1.6 GHz*

·──·──·── Phase detector to 18 GHz*

············ Free-running VCO at 10 GHz

── ── ── Synthesizer at 10 GHz

── ── Commercially available low noise
　　　　reference source at 10 GHz

*Note: with high level mixers, performance shown above requires
　　　　a second source of comparable phase noise.

Figure 11.24 Typical sensitivity of the phase detector method compared to two typical test sources. In a practical system, the sensitivity of the phase detector system is limited by the reference source used. Also shown is the noise of a commercially available low noise reference source at 10 GHz.

11.8.3 Components of the Phase-Detector Method

Phase Detector, Noise-Signal Processing, and Analyzer

At the heart of the phase detector method is the double-balanced mixer used as a phase detector (Fig. 11.23). Two *separate* sources at the same frequency are connected to the mixer. The sum frequency is removed with a low-pass filter (LPF). When operated in quadrature (90° out of phase), the difference frequency is 0 Hz, with an average voltage output of 0 V. Riding on this dc signal are small ac voltage fluctuations directly proportional to the *combined phase noise* of the two input signals. For small phase deviations, the mixer operates in the linear region around 0 V (Fig. 11.25), where the output voltage is related to the input phase by the constant K_ϕ.

Note. When mixers are operated around quadrature, there is a small dc offset on the output (instead of exactly 0 V). This small dc offset usually contributes only a very small error to the measurement.

As in the frequency discriminator method, following the detector are an LNA and a baseband analyzer. Because of the low system-noise floor of the phase-detector method, the LNA is critical. The LNA must have sufficient gain to amplify the detected noise above the noise floor of the analyzer.

To achieve the low system-noise floor of the phase detector method, in particular close to the carrier, a stable analyzer (a synthesized analog analyzer or an FFT analyzer) must be used. With a stable analyzer and the gain added by the LNA, the overall sensitivity of the phase-detector method will be set by the phase detector and LNA and not by the noise of the analyzer.

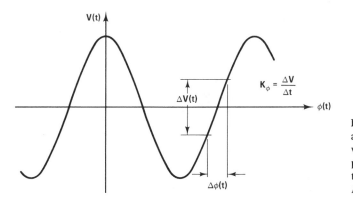

Figure 11.25 The mixer is operated around quadrature (90° out of phase), where the output voltage is linearly proportional to the input phase fluctuations by the phase detector constant K_ϕ.

Reference Source

The most critical component of the phase-detector method (and usually the most costly) is the reference source. Since the output of the detector represents the *rms sum of the noise of both oscillators*, the most important criterion for choosing a reference source is that its phase noise must be less than the noise of the test source (device under test). A margin of 10 dB is usually sufficient to ensure the measurement results are not significantly affected. The measured combined noise will be in error (always high) according to the relationship

$$\text{Error (dB)} = 10 \log \left(1 + \text{antilog} \frac{\mathscr{L}_{\text{ref}} - \mathscr{L}_{\text{dut}}}{10} \right) \qquad (11.32)$$

where $\mathscr{L}_{\text{ref}}$ = noise of the reference source
$\mathscr{L}_{\text{dut}}$ = noise of the DUT

If a reference source with noise below that of the test source is not available, a common alternative is to use two "identical" sources under test (that is, with noise properties assumed identical). With assumed equal noise contributions, 3 dB is subtracted from the measured value to yield the noise of any one source.

For accurate measurements of sources with comparable noise (within about 3 to 6 dB of each other), three sources can be used in three pairwise measurements. This yields three equations in three unknowns, from which the noise of any one source can be calculated.

In addition to phase noise, there are other requirements on the reference source. It is desirable to have the reference source be electronically tunable (the phase-detector method requires one of the sources to be tunable; see Section 11.8.3). With a tunable reference source, the system can make measurements on (1) nontunable test sources and (2) test sources whose noise changes if voltage is applied to the frequency control port. The reference source should also have sufficient tuning range to track the drift of the source under test.

If measuring a very low noise source, the reference source should have low AM noise. Though operating the mixer in quadrature suppresses the AM noise by about 20 dB, low AM noise on the reference source lends increased confidence to the measurement.

Phase-Locked Loop (PLL)

Although both the phase detector and frequency discriminator methods use a phase detector, the input signals to the phase detector are quite different. In the frequency discriminator method, the source (device) under test is split, and one path is delayed relative to the reference path. Since both paths are from a common source, the two channels will track in nominal phase. Thus, for a given time delay and once set in quadrature, the signals will *stay in quadrature* unless the source drifts significantly.

In the phase-detector method, two *different* sources are connected to the phase detector. And since no two sources will stay unaided in a 90° phase relationship for very long, phase tracking must be forced with a feedback path to one of the oscillators (see Fig. 11.26). A second-order PLL provides an error voltage to the frequency-control port of one of the oscillators. The PLL forces one oscillator to track the other in phase.

Note. In rare cases, the PLL may not be necessary. For example, two low-frequency (less than 100 MHz) sources with very good long-term stability (such as cesium beams) will remain in a set phase relationship long enough to make a measurement. Also, since the output of the phase detector varies as the cosine of the phase difference input, even a 15° offset from quadrature will contribute only 0.3 dB of error (error = 20 log(deviation from quadrature), [Hewlett-Packard, 11.10]). If this error can be tolerated, the PLL can sometimes be eliminated.

In most cases, a PLL must be used. The action of the PLL forces one source to track the other in *phase*. However, the desired phase noise information

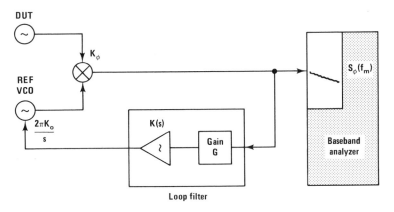

Figure 11.26 A second-order PLL is used to maintain the two sources in quadrature at the phase detector inputs. The bandwidth of the PLL is a function of the gain G and the RC filtering in the loop.

out of the phase detector is related to the *phase difference* between the two input sources. Obviously, if the sources are *tracking* in phase, the output of the mixer will no longer accurately represent the phase *difference*, or the combined phase noise, of the two sources.

The PLL directly affects the data phase fluctuations measured. There is a time constant, or loop bandwidth, associated with the PLL (adjusted by varying the gain or RC filtering within the loop). This loop bandwidth determines the *rates* of change of phase that the loop can track. Outside of this loop bandwidth, the phase of the reference source and the test source are not correlated (phase changes are not tracked). Inside the loop bandwidth, the sources track in phase, and the PLL suppresses (cancels) some of the detected phase noise. Thus the measured noise data accurately represents the noise of the source under test (without correction) *only* for offset frequencies *greater* than the phase lock loop bandwidth.

Figure 11.27 shows two phase-detector measurements made on the same 10-GHz synthesized source, with 1-Hz and 1-kHz measurement system loop bandwidths. Clearly, the PLL directly affects the measured data. The phase noise within the 1-kHz measurement loop bandwidth is suppressed by the action of the PLL, and the measured data is invalid for offsets less than 1 kHz. When the test source is locked in a 1-Hz measurement loop bandwidth, the detected phase noise is valid for all offset frequencies greater than a few hertz.

If measurements at offsets inside the measurement loop bandwidth are desired, it is possible to characterize the noise suppression effect of the loop. Flat random noise (flat with frequency) can be injected into the loop, and the response of the loop, or the transfer function, can be measured. The loop transfer function can then be used to correct the measured detected phase noise for the effect of the loop [Hewlett-Packard, 11.10].

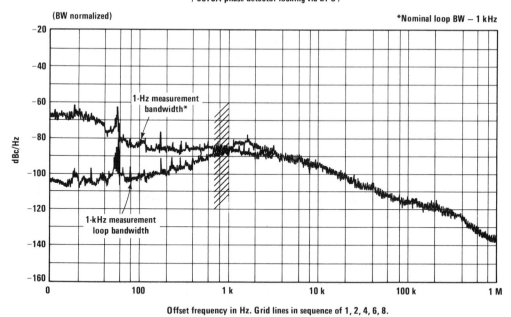

Single sideband phase noise
/ 8673A phase detector locking via EFC /

Offset frequency in Hz. Grid lines in sequence of 1, 2, 4, 6, 8.

Figure 11.27 Two-phase noise measurements on the same 10-GHz synthesized source. The upper trace shows the results when the test source was locked in a 1-Hz measurement loop bandwidth, and is an accurate measure of the noise. In the lower trace, the test source was locked in a 1-kHz measurement loop bandwidth; for offsets less than 1 kHz, the phase noise is suppressed.

What measurement system loop bandwidth must be chosen for a given source? The purpose of the PLL is to keep the input signals at the same frequency and nominally in quadrature, in the linear range of the phase detector. The loop bandwidth must be chosen large enough to suppress any phase fluctuations that would drive the phase detector out of its linear range (phase deviations greater than 0.2 rad). Thus a low-noise, stabilized 10-GHz source can be locked up with less than a 1-Hz measurement system loop bandwidth.

But if the close-in noise on the test source is high (such as on a free-running source or if there are many close-in spurious signals, such as line-related spurs), a wider loop bandwidth must be chosen to suppress this noise. In general, the noisier the source (the higher the residual FM), the wider the loop bandwidth needed.

11.8.4 Sensitivity and Limitations

With perfect input sources, the fundamental noise floor of the phase-detector method is limited to the noise added by the mixer and LNA, as shown in Fig. 11.24. However, since the phase-detector output is the rms sum of noise of the two inputs, the *practical* noise floor, or sensitivity, is set by the noise of the reference source. Also shown in Fig. 11.24 is typically achievable system sensitivity based upon the noise floor of a commercially available low-noise reference signal.

11.8.5 Making a Measurement

There are many similarities between making a measurement with the phase detector method and making a measurement with the frequency discriminator method. As in the frequency discriminator method, the phase-detector method can be broken down into four simple steps:

1. Setup.
2. Calibrate.
3. Establish quadrature and measure.
4. Convert to phase-noise units.

Setup

System setup is shown in Fig. 11.23. Since the phase detector method (1) is very sensitive and (2) depends on keeping the low-rate phase fluctuations less than 0.2 rad, careful system setup (as discussed in the introduction to the second part) is very important.

Calibrate

A rigorous derivation of system calibration determines the phase detector constant K_ϕ used in the transfer function $\Delta V(f_m) = K_\phi \Delta\phi(f_m)$ [equation (11.31)], and there are several ways to determine K_ϕ. For clarity and consistency, a technique similar to the calibration method discussed for the frequency discriminator will be described.

Since the output of the phase detector is proportional to $S_\phi(f_m)$, and since $\mathscr{L}(f_m) = \frac{1}{2}S_\phi(f_m)$ (the small angle criterion will be forced since the PLL will suppress all phase deviations greater than 0.2 rad), the output of the phase detector can be calibrated with a known sideband-to-carrier ratio.

As in the frequency discriminator method, the effective K_ϕ is determined by finding the system response to a known signal with a single FM or ϕM tone. (For single tone modulation, FM and ϕM are essentially equivalent in the frequency domain.) The system response can be calibrated by applying the FM or ϕM tone to the *test source*, the *reference source*, or a *calibration source* at the same frequency and power.

For example, a signal with a 1-kHz FM tone with sidebands -26 dBc (ΔSB_{cal}) (Fig. 11.28) is sent through the phase detector. With the phase detector in quadrature, the system demodulates the carrier, and the power output at 1 kHz (P_{cal}) represents the response of the system to phase noise -26 dB below the carrier.

The system scale factor, SF_{sys}, is defined as the transfer function of the

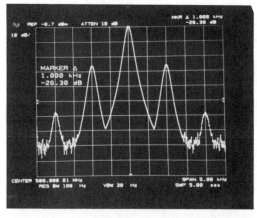

(a)

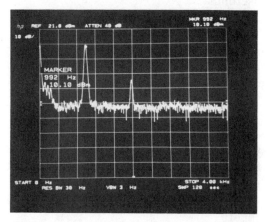

(b)

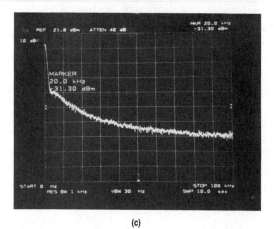

(c)

Figure 11.28 A phase detector can be calibrated by measuring the response of the system to a known input signal. (a) The input calibration signal with ΔSB_{cal} of -26 dBc. (b) The response of the phase detector to the calibration signal. (c) The phase-detected rms sum of the noise of the DUT and the reference source.

system to a known amount of phase noise:

$$\mathrm{SF}_{\mathrm{sys}} = \Delta\mathrm{SB}_{\mathrm{cal}} - P_{\mathrm{cal}} \tag{11.33}$$

Note. As in the frequency discriminator method, the mixer must be acting as a phase detector during calibration. To establish the necessary quadrature condition, the PLL is enabled, paying attention to the rate and level of the calibration sideband. First, in order to ensure that the PLL is not suppressing the system response to the calibration tone, the modulation rate on the calibration signal must be greater than the PLL bandwidth. Second, the calibration tone must be set small enough not to drive the phase detector or the LNA out of their linear ranges.

During calibration, the phase detector is usually operated at high input levels for maximum system sensitivity. The same input levels must also be maintained during the measurement for the calibration constant to be valid.

Establish Quadrature and Measure

If another source has been substituted during calibration, the test source is reconnected and locked in quadrature. The sources are usually locked up in the minimum bandwidth necessary to maintain quadrature. This allows the broadest range of offset frequencies to be measured without correction for the suppression of the PLL.

Convert to Phase-Noise Units

The measured noise is linearly related to the spectral density of phase fluctuations, $S_\phi(f_m)$. Since the response of the phase detector is not a function of offset frequency, the phase noise at any offset frequency is determined from the measured noise and the system scale factor, $\mathrm{SF}_{\mathrm{sys}} = \Delta\mathrm{SB}_{\mathrm{cal}} - P_{\mathrm{cal}}$ (equation 11.32). The phase noise described in units of $\mathscr{L}(f_m)$ is simply

$$\mathscr{L}(f_m) = P_m + \mathrm{SF}_{\mathrm{sys}} \tag{11.34}$$

More completely,

$$\mathscr{L}(f_m) = P_m + \mathrm{SF}_{\mathrm{sys}} - \mathrm{SF}_{\mathrm{bw}} + C_m \tag{11.35}$$

where P_m = (dBm) measured noise level
$\mathrm{SF}_{\mathrm{sys}}$ = (dB) transfer function of system
$\mathrm{SF}_{\mathrm{bw}}$ = (dB) noise bandwidth normalization
C_m = (dB) correction for measurement system

and $\mathscr{L}(f_m)$ is measured in dBc/Hz.

The measured noise power is (1) normalized to the power in a 1-Hz bandwidth and (2) corrected (if necessary) for the measurement circuitry, as discussed in Section 11.5.

Example 11.4

Figure 11.28 shows actual measurement results on a 10-GHz synthesized source. Figure 11.28(a) shows the input calibration signal, with 1-kHz sidebands down 26 dB from the carrier; Fig. 11.28(b) shows the system response to the calibration signal. The measured P_{cal} is $+10$ dBm. Thus the $SF_{sys} = \Delta SB_{cal} - P_{cal} = -26 - (-10) = -36$ dB. Figure 11.28(c) shows the detected phase noise of the source. At a 20-kHz offset, the measured noise power in a 1-kHz resolution bandwidth is $P_m = -31$ dBm. This yields $\mathscr{L}(20 \text{ kHz})$ of

$$
\begin{array}{lll}
\mathscr{L}(20 \text{ kHz}) = & -31 & \text{dBm} \quad P_m \\
& -36 & \text{dB} \quad +SF_{sys} \\
& -30 & \text{dB} \quad -SF_{bw} \\
& \underline{+ \ 2.5} & \text{dB} \quad +C_m \text{ (for the analyzer used)} \\
& -94.5 & \text{dBc/Hz}
\end{array}
$$

Figure 11.27 shows an entire phase-noise plot of a 10-GHz source (measurements made on the same system as Figure 11.22, except now used in the phase-detector mode). The sharp spikes on the phase noise plot are spurious, many related to the line frequency, and are not part of the random noise of the test oscillator. This measurement was made to a 10-Hz offset frequency; the phase detector method is commercially offered to offset frequencies as low as 0.02 Hz. ■

11.8.6 Summary of the Phase-Detector Method

The phase-detector method of phase-noise measurements features a low noise floor over the entire offset frequency range. It can be used to measure both free-running and phase-locked sources.

The low system-noise floor is paid for with an increase in complexity. A second-order PLL is used to keep the source under test and the reference source in quadrature at the phase detector. Free-running sources with very high phase noise or spurious close-to-the-carrier or sources with high drift rates may be difficult to measure with the phase-detector method without automated correction techniques.

Though the fundamental sensitivity of the phase detector method is set by the noise of the mixer and LNA, the useable system noise floor is limited by the noise of the reference source. Since the reference source sets system sensitivity, choice of a reference source will determine which test sources can be measured.

11.9 COMPARISON OF METHODS AND SUMMARY

Figure 11.29 presents a summary of the phase-noise performance of the two typical sources and the achievable sensitivities of the three measurement methods. The sensitivity of the direct cancellation method is not given, for though it is conceptually simple, it is more difficult to implement and is not as sensitive as other measurement techniques.

The direct spectrum method is optimal for measuring stable sources with relatively high noise. It can also be used to give a quick estimation of the measure of the source noise; if the noise measured is equal to the known noise of the spectrum analyzer, then it is known that the noise of the test source is

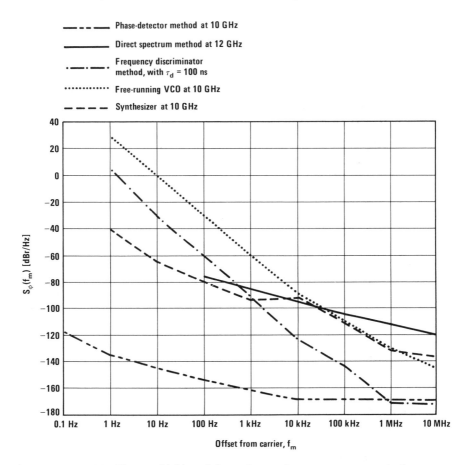

— - - — Phase-detector method at 10 GHz

———————— Direct spectrum method at 12 GHz

· — · — · Frequency discriminator method, with τ_d = 100 ns

············ Free-running VCO at 10 GHz

— — — Synthesizer at 10 GHz

Figure 11.29 The sensitivities of three phase noise measurement methods are compared against the noise of typical sources to be measured. Choice of a method is dependent on the noise of the source to be measured and the offset frequency range of interest. For example, the frequency discriminator method could be used to measure the free-running source over the entire range of offset frequencies or could be used to measure the synthesized source at offsets greater than about 50 kHz.

at least that good (but maybe significantly better). The direct spectrum method is not useful for measuring low-noise sources, sources with high AM, or sources with high drift.

The frequency discriminator method is optimized for free-running sources with high close-in noise sidebands. A delay line/mixer discriminator has less sensitivity to drift in the test source than either the direct spectrum or the phase-detector methods. The discriminator method is not useful for measuring close-in on stabilized sources, nor is it typically used to measure low carrier frequency signals.

Of the many implementations of discriminators, the delay line/mixer implementation has the largest input-frequency range. However, due to higher attenuation in the delay line, it may be difficult to achieve a low noise floor for a delay line/mixer discriminator operating directly at the carrier frequency. For best sensitivity with a delay line/mixer discriminator at high carrier frequencies, the source under test can first be downconverted with a low phase-noise reference signal.

The phase-detector method is the most versatile of all three methods. Featuring low close-in noise and low broadband noise, the phase detector method can be used to measure both free-running and phase-locked sources. In actual practice, the sensitivity of this method is limited by the noise of the reference source used. It can be used for measurements at millihertz offsets from the carrier or offsets of tens of megahertz.

In practice, the input frequency range of a phase-detector system is also determined by the reference source. Since two like-frequency sources are compared in a phase detector, any frequency—even through millimeters—can be measured with this method, given a clean reference source and a dc-coupled mixer at the carrier frequency of interest.

Because the sources in telecommunications systems are usually stabilized—locked to a stable reference—the most widely used method of phase noise measurement in this application is the phase-detector method. In a research and development, production, or service environment, the distribution of the noise (the actual spectral density on a per-hertz basis) would be the most useful data, to determine how to change or improve the noise spectrum. In an on-site or operational readiness-testing application, the desired data would probably be integrated phase noise (rms degrees) over some bandwidth or perhaps a probability density function. The detected phase noise could either be input to a receiver with the bandwidth of interest or the spectral density could be measured and then either integrated to yield residual phase noise or transformed into an equivalent probability density.

REFERENCES

[11.1] Allen, D. W. "The Measurement of Frequency and Frequency Stability of Precision Oscillators," NBS Technical Note 669, U.S. National Bureau of Standards (NBS), Boulder, Colo., 1975.

[11.2] Ashley, J. R., T. A. Barley, and G. J. Rast. "Measurement of Noise in Microwave Transmitters," *IEEE Transactions on Microwave Theory and Technique*, Vol. MTT-25, No. 4, 1977, pp. 294–318.

[11.3] Ashley, J. R., C. B. Searles, and F. M. Palka. "The Measurement of Oscillator Noise at Microwave Frequencies," *IEEE Transactions* on MTT-16, No. 9, 1968, pp. 753–760.

[11.4] Barnes, J. A., A. R. Chie, and L. S. Cutler. "Characterization of Frequency Stability," U.S. NBS Technical Note 394, National Bureau of Standards, Boulder, Colo., 1970.

[11.5] Baghdady, E. J. et al. "Short-Term Frequency Stability: Characterization, Theory, and Measurement," *Proceedings of the IEEE*, Vol. 53, 1965, pp. 704–722.

[11.6] Feher, K. *Digital Communications: Satellite/Earth Station Engineering*, Prentice-Hall, Englewood Cliffs, N.J., 1983.

[11.7] Fischer, M. "An Overview of Modern Techniques for Measuring Spectral Purity," *Microwaves*, Vol. 18, No. 7, 1979, pp. 66–75.

[11.8] Fischer, M. "Frequency Stability Measurement Procedures," Proceedings, Eighth Annual PTTI Meeting, Goddard Space Flight Center, Code 250, Greenbelt, Md., 1976, pp. 575–617.

[11.9] Halford, D. "The Delay Line Discriminator," notes from NBS Technical Note 10, 1975, pp. 19–38.

[11.10] Hewlett-Packard. "RF and Microwave Phase Noise Measurement Seminar," 1984.

[11.11] Hewlett-Packard. "Phase Noise Product Note 11729C-2, Characterization of Microwave Oscillators, Frequency Discriminator Method," 1985.

[11.12] Hewlett-Packard. "Spectrum Analysis . . . . Application Note 150-4, Noise Measurements," 1974.

[11.13] Howe, D. "Frequency Domain Stability Measurements: A Tutorial Introduction," U.S. NBS Technical Note 679, National Bureau of Standards, Boulder, Colo., 1976.

[11.14] Lance, A. L., W. D. Seal, and F. Labarr. "Phase Noise and AM Noise Measurements in the Frequency Domain," Chapter 7 of *Infrared and Millimeter Waves*, Vol. 2, Academic Press, New York, 1984.

[11.15] Lance, A. L., W. D. Seal, and N. W. Mendoza. "Automating Phase Noise Measurements in the Frequency Domain," *31st Frequency Control Symposium Proceedings*, 1977, pp. 347–358.

[11.16] Ondria, F. G. "A Microwave System for Measurements of AM and FM Noise Spectra," *IEEE Transactions on Microwave Theory and Techniques*, Vol. MTT-16, 1968, pp. 767–781.

[11.17] Robbins, W. P. "Phase Noise in Signal Sources," *IEE Telecommunications Series* No. 9, 1982.

[11.18] Rutman, J. "Relations between Spectral Purity and Frequency Stability," *Proceedings of the 28th Annual Frequency Control Symposium*, 1974, pp. 160–165.

[11.19] Rutman, J. "Characterization of Phase and Frequency Instabilities in Precision Frequency Sources: Fifteen Years of Progress," *Proceedings of the IEEE*, Vol. 66, No. 9, 1978.

[11.20] Scherer, D. "Design Principles and Test Methods for Low Phase Noise RF and Microwave Sources," HP, RF, and Microwave Measurement Symposium, October, 1978.

[11.21] Scherer, D. "The Art of Phase Noise," HP, RF, and Microwave Measurement Symposium, May, 1984.

[11.22] Shoaf, J. H., Halford, D., and Risely, A. S. "Frequency Stability Specification and Measurement: High Frequency and Microwave Signals," *U.S. NBS Technical Note* 632, National Bureau of Standards, Boulder, Colo., 1973.

[11.23] Spilker, J. *Digital Communications by Satellite*, Prentice-Hall, Englewood Cliffs, N.J., 1977, pp. 336–397.

[11.24] Stremler, F. G. *Introduction to Communication Systems*, Addison-Wesley, Reading, Mass., 1977.

[11.25] Tykulsky, A. "Spectral Measurements of Oscillators," *Proceedings of the IEEE*, Vol. 54, No. 2, 1966.

INDEX